STATISTICS:
A GUIDE TO THE UNKNOWN

HOLDEN-DAY SERIES
IN PROBABILITY AND STATISTICS

E. L. Lehmann, Editor

Bickel and Doksum: *Mathematical Statistics*
Hajek: *Nonparametric Statistics*
Hodges and Lehmann: *Elements of Finite Probability, 2d ed.*
Hodges and Lehmann: *Basic Concepts of Probability and Statistics, 2d ed.*
Lehmann: *Nonparametrics: Statistical Methods Based on Ranks*
Nemenyi, Dixon, White: *Statistics From Scratch*
Parzen: *Stochastic Processes*
Rényi: *Foundations of Probability*
*Roberts: *Interactive Data Analysis*
Tanur, Mosteller, Kruskal, et al: *Statistics: A Guide to the Unknown, 2nd ed.*
Tanur, Mosteller, Kruskal, et al: *Statistics: A Guide to Business and Economics*
Tanur, Mosteller, Kruskal, et al: *Statistics: A Guide to Biological and Health Sciences*
Tanur, Mosteller, Kruskal, et al: *Statistics: A Guide to Political and Social Issues*
*Waller: *An Introduction to Numerical Reasoning*

*To be published

STATISTICS:
A GUIDE TO THE UNKNOWN
Second Edition

edited by

JUDITH M. TANUR
State University of New York, Stony Brook

and by

FREDERICK MOSTELLER, Chairman
Harvard University

WILLIAM H. KRUSKAL
University of Chicago

RICHARD F. LINK
Artronic Information Systems, Inc.
and Princeton University

RICHARD S. PIETERS
Phillips Academy, Andover, Mass.

GERALD R. RISING
State University of New York, Buffalo

The Joint Committee on
The Curriculum in Statistics and Probability of
The American Statistical Association and
The National Council of Teachers of Mathematics

Special Editor

ERICH L. LEHMANN
University of California, Berkeley

HOLDEN-DAY
SAN FRANCISCO LONDON DUSSELDORF
JOHANNESBURG PANAMA SINGAPORE SYDNEY

Production Coordination: Michael Bass and Associates
Cover Design: Robert Barringer

PREFACE TO SECOND EDITION

SINCE THE publication of STATISTICS: A Guide to the Unknown (SAGTU) five years ago, three related publications have appeared, each of them consisting of a subset of about a dozen of the SAGTU essays dealing with (i) business and economics; (ii) the biological and health sciences; (iii) political and social issues. Two of these volumes also contained one new essay each (by Gilbert, McPeek and Mosteller on innovations in surgery and anesthetics for (ii) and by Gilbert, Light and Mosteller on political and social innovations for (iii).) In addition, each of the essays in these Mini-SAGTUS was supplemented by a set of problems. The original editors reviewed and approved the new material.

The present, second edition of SAGTU, supervised by Erich Lehmann, differs from the first principally by some minor editorial changes for updating, the addition of the two new essays and the problems mentioned above, and of problems also for the essays not contained in the Mini-SAGTUS. It is the hope of the editors that these problems will help focus the reading of the essays and will make them more useful in the classroom. This supplementary study material is the valuable contribution of David Lane, Donna and Leland Neuberg, Rick Persons, Haiganoush Preisler and Esther Sid.

Judith M. Tanur
Frederick Mosteller
Erich L. Lehmann

PREFACE TO FIRST EDITION

To PREPARE a volume describing important applications of statistics and probability in many fields of endeavor—this was the project that the ASA-NCTM Committee invited me to help with in early 1969. It was the Committee's view that more statistics and its background and probability would be desirable in the school curriculum; thus it would be desirable to show how broadly these tools are applied. The Committee planned this book primarily for readers without special knowledge of statistics, probability, or mathematics. This audience included especially parents of school children, school superintendents, principals, and board members, but also teachers of mathematics and their supervisors, and finally, young people themselves. *Statistics: A Guide to the Unknown* is the result. During the time of the book's preparation several of us who were working on it and teaching simultaneously found much of the material very useful—even inspirational—to undergraduate and graduate students. It would seem that, quite unexpectedly, the book has an additional possible function as an auxiliary textbook.

Instead of teaching technical methods, the essays illustrate past accomplishments and current uses of statistics and probability. In choosing the actual essays to include, the Committee and I aimed at illustrating a wide variety of fields of application, but we did not attempt the impossible task of covering all possible uses. Even in the fields included, attempts at complete coverage have been deliberately avoided. We have discouraged authors from writing essays that could be entitled "All Uses of Statistics in ..." Rather, we asked authors to stress one or a very few important problems within their field of application and to explain how statistics and probability help to solve them and why the solutions are useful to the nation, to science, or to the people who originally posed the problem. In the past, for those who were unable to cope with very technical material, such essays have been hard to find.

To us, this spread of applications gives a renewed appreciation of the unity in diversity that is statistics. On the one hand, we found the same, or similar, statistical techniques being applied in unrelated fields. Authors described the use of correlational analysis in contexts as diverse as a study of the sun, a test of the relative importance of economic variables, an exploration of the components of leadership in the military, and an examination of the effect of registration regulations on voting turnout. Other authors dealt with applications of sampling theory in such disparate fields as accounting, improving the U.S. Census, and opinion polling. And essay after essay discusses experimental design and the necessity, as well as the difficulty, of making inferences from less-than-perfect data. Certainly this is unity in diversity that will help to demonstrate to the general public the wide usefulness of statistical tools.

On the other hand, we found essays grouping themselves into unities of subject matter with differing statistical techniques. For example, two otherwise unlike essays deal with the evaluation of the effectiveness of innovations in traffic control procedures in reducing accidents. At least four essays describe very different methods of studying diseases, their causes and cures—the testing of the value of the Salk vaccine, a mathematical model for disease epidemics, a history of the study of the association between smoking and ill health, and an explanation of the uses of twins in research on illness.

We have tried to emphasize these unities and at the same time avoid unnecessary repetition by a system of cross-referencing. Thus, whenever another essay contains material that will assist the reader to understand the present essay or give him further insight into a particular problem, he is directed to it.

Once our 44 essays had been assembled and edited, we had to decide on their order. Several orderings seemed feasible: we might group the essays by type of statistical tools employed, thus stressing the unity of statistical tools and ignoring the diversity of usual disciplinary lines; we might group essays by the method used for collecting data—sample survey, experiment, Census material, and so on; or we might group them by subject matter of the application.

What we have chosen is the last of these modes of organization. We have classified into four broad areas by field of application, with subdivisions within each. Each subdivision is small enough, cohesive enough, and digestible enough to be read as a single unit and to give an overview of applications within a narrow field. But we were unwilling to forgo the advantages of the other possible methods of classification; following the main table of contents, therefore, are two alternate tables of contents, the first organized by method of collecting the data, the second by statistical tools. In the latter listing, an essay has been listed under a heading whenever the author used that tool, or whenever we felt the reader might learn something about the technique by looking at the essay, or both.

These efforts at classifying emphasized, for us, an aspect of the book we had not deliberately planned or even been aware of earlier. It turned out that we had a large group of essays dealing with public policy, many of them classified under our main grouping entitled "Man and His Social World." We also found that several of this group deal with the evaluation of reforms or changes in policy. On the one hand, we found ourselves with descriptions of two large-scale field experiments: the speed-limit experiments in Scandinavia and the Salk vaccine trials. It seems that in the U.S. until recently, we have done few of these controlled experiments, and it appears to the Committee that one of the jobs that statisticians have been somewhat neglecting is explaining to the public the possibilities and values of experimentation. The public needs such explanations to have a sound basis for deciding whether it wants such experimentation to be carried out. On the other hand, several deal with nonexperimental (or quasi-experimental) evaluations of reforms: Did the Connecticut crackdown on speeding decrease traffic accidents? Did the assignment of more patrolmen to a New York City precinct decrease crime? Is a particular anesthetic dangerous?

We hope that both types of essays will contribute to a greater appreciation of how hard it is to find out whether a program is accomplishing its purposes. Such understanding would give people a little more sympathy for government officials who are trying to do difficult jobs under severe handicaps. It may also, as pointed out above, encourage them to press government to do better-controlled field studies both in advance of and while instituting social reforms.

There is an old saw that a camel is a horse put together by a committee. Our authors supplied exceedingly well-formed and attractive anatomical parts, but to the extent that this book gaits well, credit is due primarily to a most talented and dedicated Committee. In general, the approach to unanimity in the Committee's critical reviews of and suggestions about essays was phenomenal. And, though they may have occasionally been divided about the strong and weak points of a particular essay, they were constantly united in their purpose of producing a useful book, and in their ability to find something more than 24 hours a day to work on it. This dedication, together with my own compulsiveness, has undoubtedly created difficulties for our authors. Nevertheless, our authors persevered and deserve enormous thanks from me, from the Committee, and from the statistical profession at large.

Our thanks go also to the Sloan Foundation whose grant made it possible to put this book together.

There are others to thank as well: for the hard work and advice of George E. P. Box, Leo Breiman, Churchill Eisenhart, Thomas Henson, J. W. Tukey, and the late W. J. Youden; to the office of the American Statistical Association (and, in particular, to Edgar Bisgyer and John Lehman) for invaluable help in all the administrative work necessary to get out a book such as this; and similar thanks to the administration of the National Council of Teachers of Mathematics; to Edward Millman for careful and imaginative editorial assistance; and to other people at Holden-Day, especially Frederick H. Murphy, Walter Sears, and Erich Lehmann, our Series Editor; to Mrs. Holly Grano for acting as a long-distance and long-haul secretary; and to the many friends and colleagues both of the Editor and of the Committee members who so often acted as unsung, but indispensable advisors.

Judith M. Tanur
Great Neck, New York
February 14, 1972

FOREWORD

THE RIGHT Honorable Harold Wilson, Prime Minister of Great Britain, in opening the 37th Session of the International Statistical Institute in London, September 4, 1969, said:

> The list of papers for the Session reflects the ever widening range of application of statistical methods. When I joined the Royal Statistical Society the papers read were still mainly on economic and social statistics. Nowadays the papers read before a society like those for your session of the Institute cover many more topics relating to many disciplines. It means, I am afraid, that as statisticians today you help so many people in so many diverse subject fields that none of your clients can see the overall contribution which you as statisticians make together as a whole. As a result your value is not perhaps sufficiently recognized by any one group of the people with whom you deal nor are your great services fully realized by the general public.

Himself a statistician, the Prime Minister understood well these contributions of statistics. He might be interpreted as calling for statisticians to explain what they do.

Warren Weaver, a great expositor of science, discussed why science is not more widely appreciated and issued a similar call in "The Imperfections of Science" (*American Scientist*, 49:113, March 1961):

> What we must do—scientists and non-scientists alike—is close the gap. We must bring science back into life as a human enterprise, an enterprise that has at its core the uncertainty, the flexibility, the subjectivity, the sweet unreasonableness, the dependence upon creativity and faith which permit it, when properly understood, to take its place as a friendly and understanding companion to all the rest of life.

Dr. Weaver has had a lifelong interest in probability and statistics. He was given the first Archives of Science Award for his contribution to public appreciation of science.

In the U.S., too, government representatives want explanations. For example, Craig Hosmer, a House Member of the House and Senate Committee on Atomic Energy, in discussing the funding of science at a technological conference on March 5, 1968, said: "The scientific community should take greater pains to make clear that its efforts contribute directly and indirectly to the public good."

Thus, these and other important men ask scientists to tell the public about their subject and to explain what contribution science makes to society. Their request is easier made than satisfied. This collection of essays on applications of statistics represents one kind of step toward meeting it.

To find the origins of this work, we might turn back to the great change and advance in mathematics education initiated in 1954 when the Commission on Mathematics of the College Entrance Examination Board brought together, for a sustained study of the curriculum, teachers and administrators of mathematics from several sources: secondary schools, teachers' colleges, and colleges and universities. Prior to that gathering, the several groups of teachers had seldom worked together on the problems of the curriculum. That meeting of minds has developed and continued in many directions; one of its long-run consequences was the establishment of the Joint Committee of the American Statistical Association (ASA) and the National Council of Teachers of Mathematics (NCTM) on the Curriculum in Statistics and Probability. By late 1967, such cooperation between school and college teachers was widespread, and it was easy for Donovan Johnson, then President of NCTM, and me, then President of ASA, to set up the Joint Committee to review matters in the teaching of statistics and probability.

Early in its work the Joint Committee decided that it wanted to encourage the teaching of statistics in schools, for statistics is a part of the mathematical sciences that deals with many practical, as well as esoteric, subjects and is especially organized to treat the uncertainties and complexities of life and society. To explain why more statistics needs to be taught, we need to make clearer to the public what sorts of contributions statisticians make to society. In the field of statistics, we are, indeed, responding to the sort of requests quoted above.

When describing work in the mathematical sciences, one must make a major decision as to what level of mathematics to ask of the reader. Although the Joint Committee serves professional organizations whose subject matter is strongly mathematical, we decided to explain statistical ideas and contributions without dwelling on their mathematical aspects. This was a bold stroke, and our authors were surprised that we largely held firm.

The Joint Committee has been extremely fortunate to find so many distinguished scholars willing to participate in this educational project. The authors' reward is almost entirely in their contribution to the appreciation of statistics. We have been fortunate, too, to have Judith Tanur as editor of the collection and hard-working committee members as her staff.

To teachers, I can report that thus far I have used material from about one-third of the essays in classes and in speeches. Adult students seemed to enjoy discussing the data and reading further. We do not, however, regard the book as a textbook. We have had other favorable reports from adults who were not students and who read the articles voluntarily. Perhaps the most heartening report on readability came from one of our authors, whose secretary told him, after finishing the typing of a revision, that she enjoyed it enormously. When asked what she especially liked, she said that she had finally found out what the work of the office was all about.

In a parallel writing effort, the Joint Committee has also produced a series of pamphlets for classroom teaching entitled *Statistics by Example*. Intended for students whose mathematical preparation is modest, these volumes teach statistics by means of real-life examples. That effort differs from this one in that the student learns specific techniques, tools, and concepts by starting from concrete examples. (The publisher is Addison-Wesley, Sand Hill Road, Menlo Park, California 94025.)

Some readers may wish to know how to become statisticians, and others may have the obligation to advise students about career opportunities. The brochure *Careers in Statistics* (obtainable from the American Statistical Association, Suite 640, 806 Fifteenth Street, Washington, D.C. 20005) provides information about the nature of the work and the training required for various statistical specialties.

The Joint Committee appreciated being able to report on its work at ten meetings of the National Council or its affiliates. We also reported to the American Statistical Association at Chicago, Illinois, and Detroit, Michigan; to a conference called by the National Science Foundation at the University of Minnesota; to the International Statistical Institute workshop on teaching statistics at Oisterwijk, Netherlands; and to the international conference on teaching of probability and statistics of the Comprehensive School Mathematics Program at Carbondale, Illinois. I discussed some of the material in one of my Allen T. Craig lectures at the University of Iowa.

In addition, by its existence at Harvard University, National Science Foundation grant GS-2044X2 has considerably facilitated this project without directly supporting it. Much of the work was done during periods while Frederick Mosteller held a Guggenheim Fellowship and while William Kruskal was a National Science Foundation Senior Postdoctoral Fellow at the Center for Advanced Study in the Behavioral Sciences. We have also benefited from a number of courtesies extended by the Russell Sage Foundation and by the Social Science Research Council. Before resigning to take up the tasks of the presidency of the National Council of Teachers of Mathematics, Julius Hlavaty was a member of the Joint Committee and participated in the decision to create this collection. The national offices of ASA and NCTM have been most helpful, as have representatives of our publisher, Holden-Day, Inc.

Finally, we have no monopoly on the task of explaining statistics to the public. We urge others to provide their views on the purposes, the methods, and the results of statistical science.

Frederick Mosteller, Chairman
Joint Committee of ASA-NCTM
Cambridge, Massachusetts
February 14, 1972

CONTENTS

Part one OUR BIOLOGIC WORLD

STAYING WELL OR GETTING BETTER

Part two OUR POLITICAL WORLD

GOVERNMENT INFLUENCES PEOPLE

PEOPLE INFLUENCE GOVERNMENT

ESSAYS CLASSIFIED BY DATA SOURCES

SAMPLES

SURVEYS AND QUESTIONNAIRES

EXPERIMENTS

QUASI EXPERIMENTS

ESSAYS CLASSIFIED BY STATISTICAL TOOLS

ESTIMATION, HYPOTHESIS TESTING, BAYESIAN AND EMPIRICAL BAYESIAN ANALYSIS, AND DATA ANALYSIS

Estimation

TABLES, GRAPHS, AND MAPS

Tables

Graphs and maps

PERCENTS AND RATES, STANDARDIZATION, AND ADJUSTMENT

SAMPLING AND RANDOMIZATION

CORRELATION AND REGRESSION

Part one

OUR BIOLOGIC WORLD

Staying well or getting better

The biggest public health experiment ever: The 1954
field trial of the Salk poliomyelitis vaccine

Safety of anesthetics

Drug screening: The never-ending search for new
and better drugs

Setting dosage levels

How frequently do innovations succeed in
surgery and anesthesia?

Getting sick and dying

Statistics, scientific method, and smoking

Deathday and birthday: An unexpected connection

Epidemics

People and animals

Does inheritance matter in disease?

The plight of the whales

The importance of being human

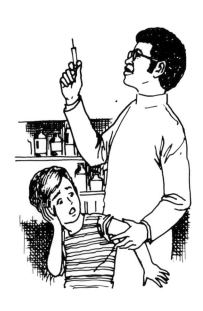

THE BIGGEST PUBLIC HEALTH EXPERIMENT EVER: The 1954 Field Trial of the Salk Poliomyelitis Vaccine

Paul Meier   *University of Chicago*

THE LARGEST and most expensive medical experiment in history was carried out in 1954. Well over a million young children participated, and the immediate direct costs were over 5 million dollars. The experiment was carried out to assess the effectiveness, if any, of the Salk vaccine as a protection against paralysis or death from poliomyelitis. The study was elaborate in many respects, most prominently in the use of placebo controls (children who were inoculated with simple salt solution) assigned at random (that is, by a carefully applied chance process that gave each volunteer an equal probability of getting vaccine or salt solution) and subjected to a double-blind evaluation (that is, an arrangement under which neither the children nor the physicians who evaluated their subsequent state of health knew who had been given vaccine and who got the salt solution).

3

Why was such elaboration necessary? Did it really result in more or better knowledge than could have been obtained from much simpler studies? These are the questions on which this discussion is focused.

BACKGROUND

Polio was never a common disease, but it certainly was one of the most frightening and, in many ways, one of the most inexplicable in its behavior. It struck hardest at young children, and, although it was responsible for only about 6% of the deaths in the age group 5 to 9 in the early fifties, it left many helpless cripples, including some who could survive only in a respirator. It appeared in epidemic waves, leading to summer seasons in which some communities felt compelled to close swimming pools and restrict public gatherings as cases increased markedly from week to week; other communities, escaping an epidemic one year, waited in trepidation for the year in which their turn would come. Rightly or not, this combination of selective attack upon the most helpless age group and the inexplicable vagaries of its epidemic behavior, led to far greater concern about polio as a cause of death than other causes, such as auto accidents, which are more frequent and, in some ways, more amenable to community control.

The determination to mount a major research effort to eradicate polio arose in no small part from the involvement of President Franklin D. Roosevelt, who was struck down by polio when a successful young politician. His determination to overcome his paralytic handicap and the commitment to the fight against polio made by Basil O'Connor, his former law partner, enabled a great deal of attention, effort, and money to be expended on the care and rehabilitation of polio victims and—in the end, more importantly—on research into the causes and prevention of the disease.

During the course of this research, it was discovered that polio is caused by a virus and that three main virus types are involved. Although clinical manifestations of polio are rare, it was discovered that the virus itself was not rare, but common, and that most adult individuals had experienced a polio infection sometime in their lives without ever being aware of it.

This finding helped to explain the otherwise peculiar circumstance that polio epidemics seemed to hit hardest those who were better off hygienically (i.e., those who had the best nutrition, most favorable housing conditions, and were otherwise apparently most favorably situated). Indeed, the disease seemed to be virtually unknown in those countries with the poorest hygiene. The explanation is that because there was plenty of polio virus in the less-favored populations, almost every infant was exposed to the disease early in life while he was still protected by the immunity passed on from his mother. As a result, everyone had had polio, but under protected circumstances, and, thereby, everyone had developed his own immunity.

As with many other virus diseases, an individual who has been infected by polio and recovered is usually immune to another attack (at least by a virus strain of the same type). The reason for this is that the body, in fighting the infection, develops *antibodies*, which are a part of the gamma globulin fraction of the blood, to the *antigen*, which is the protein part of the polio virus. These antibodies remain in the bloodstream for years, and even when their level declines so far as to be scarcely measurable, there are usually enough of them to prevent a serious attack from the same virus.

Smallpox and influenza illustrate two different approaches to the preparation of an effective vaccine. For smallpox, which has long been controlled by a vaccine, we use for the vaccine a closely related virus, cowpox, which is ordinarily incapable of causing serious disease in man, but which gives rise to antibodies that also protect against smallpox. (In a very few individuals this vaccine is capable of causing a severe, and occasionally fatal, reaction. The risk is small enough, however, so that we do not hesitate to expose all our school children to it in order to protect them from smallpox.) In the case of influenza, however, instead of a closely related live virus, the vaccine is a solution of the influenza virus itself, prepared with a virus that has been killed by treatment for a time with formaldehyde. Provided that the treatment is not too prolonged, the dead virus still has enough antigenic activity to produce the required antibodies so that, although it can no longer infect, it is, in this case, sufficiently like the live virus to be a satisfactory vaccine.

In the case of polio, both of these methods were explored. A live-virus vaccine would have the advantage of reproducing in the vaccinated individual and, hopefully, giving rise to a strong reaction which would produce a high level of long-lasting antibodies. With such a vaccine, however, there might be a risk that a vaccine virus so similar to the virulent polio virus could mutate into a virulent form and itself be the cause of paralytic or fatal disease. A killed-virus vaccine should be safe because it presumably could not infect, but it might fail to give rise to an adequate antibody response. These and other problems stood in the way of the rapid development of a successful vaccine. Some unfortunate prior experience also contributed to the cautious approach of researchers. In the thirties, attempts had been made to develop vaccines against polio; two of these were actually in use for a time. Evidence that at least one of these vaccines, in fact, had been responsible for cases of paralytic polio soon caused both to be promptly withdrawn from use. This experience was very much in the minds of polio researchers, and they had no wish to risk a repetition.

Research to develop both live and killed vaccines was stimulated in the late forties by the development of a tissue culture technique for growing polio virus. Those working with live preparations developed harmless strains from virulent ones by growing them for many generations in suitable tissue culture media. There was, of course, considerable worry lest these strains, when used as a vaccine in man, might revert to virulence and cause paralysis or death. (By 1972 it seems clear that the strains developed are indeed safe—a live-virus preparation taken orally is the vaccine presently in widespread use throughout the world.)

Those working with killed preparations, notably Jonas Salk, had the problem of treating the virus (with formaldehyde) sufficiently to eliminate its infectiousness, but not so long as to destroy its antigenic effect. This was more difficult than, at first, had appeared to be the case, and some early lots of the vaccine proved to contain live virus capable of causing paralysis and death. There are statistical issues in the safety story (Meier 1957), but our concern here is with the evaluation of effectiveness.

EVALUATION OF EFFECTIVENESS

In the early fifties the Advisory Committee convened by the National Foundation for Infantile Paralysis (NFIP) decided that the killed-virus vaccine developed by Jonas Salk at the University of Pittsburgh had been shown to be both safe and capable of inducing high levels of the antibody in children on whom it had been tested. This made the vaccine a promising candidate for general use, but it remained to prove that the vaccine actually would prevent polio in exposed individuals. It would be unjustified to release such a vaccine for general use without convincing proof of its effectiveness, so it was determined that a large-scale "field trial" should be undertaken.

That the trial had to be carried out on a very large scale is clear. For suppose we wanted the trial to be convincing if indeed the vaccine were 50% effective (for various reasons, 100% effectiveness could not be expected). Assume that, during the trial, the rate of occurrence of polio would be about 50 per 100,000 (which was about the average incidence in the United States during the fifties). With 40,000 in the control group and 40,000 in the vaccinated group, we would find about 20 control cases and about 10 vaccinated cases, and a difference of this magnitude could fairly easily be attributed to random variation. It would suggest that the vaccine might be effective, but it would not be persuasive. With 100,000 in each group, the expected numbers of polio cases would be 50 and 25, and such a result would be persuasive. In practice, a much larger study was clearly required, because it was important to get definitive results as soon as possible, and if there were relatively few cases of polio in the test area, the expected number of cases might be well under 40. It seemed likely, also, for reasons we shall discuss later, that paralytic polio, rather than all polio, would be a better criterion of disease, and only about half the diagnosed cases are classified "paralytic." Thus the relatively low incidence of the disease, and its great variability from place to place and time to time, required that the trial involve a huge number of subjects—as it turned out, over a million.

THE VITAL STATISTICS APPROACH

Many modern therapies and vaccines, including some of the most effective ones, such as smallpox vaccine, were introduced because preliminary studies suggested their value. Large-scale use subsequently provided clear evidence of efficacy. A natural and simple approach to the evaluation of the Salk vaccine would have been to distribute it as widely as possible, through the schools, to see whether the rate of reported polio was appreciably less than usual during the subsequent season. Alternatively, distribution might be limited to one or a few areas because limitations of supply would preclude effective coverage of the entire country. There is even a fairly good chance that were one to try out an effective vaccine against the common cold or against measles, convincing evidence might be obtained in this way.

In the case of polio—and, indeed, in most cases—so simple an approach would almost surely fail to produce clear cut evidence. First, and foremost, we must consider how much polio incidence varies from season to season, even without any attempts to modify it. From Figure 1, which shows the annual reported incidence from 1930 through 1955, we see that had a trial been conducted in this way in 1931, the drop in incidence from 1931 to 1932 would have been strongly suggestive of a highly effective vaccine because the incidence dropped to less than a third of its previous level. Similar misinterpretations would have been made in 1935, 1937, and other years—most recently in 1952. (On the general problem of drawing inferences from

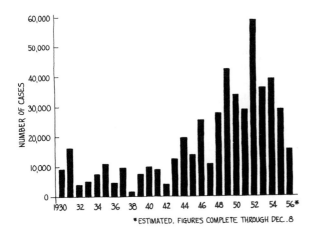

FIGURE 1
Poliomyelitis in the U.S., 1930–
56. Source: Meier (1957)

such time series data see the essay by Campbell.) One might suppose that such mistakes could be avoided by using the vaccine in one area, say, New York State, and comparing the rate of incidence there with that of an unvaccinated area, say, Illinois. Unfortunately, an epidemic of polio might well occur in Chicago—as it did in 1956—during a season in which New York had a very low incidence.

Another problem, more subtle, but equally burdensome, relates to the vagaries of diagnosis and reporting. There is no difficulty, of course, in diagnosing the classic respirator case of polio, but the overwhelming majority of cases are less clearcut. Fever and weakness are common symptoms of many illnesses, including polio, and the distinction between weakness and slight transistory paralysis will be made differently by different observers. Thus the decision to diagnose a case as nonparalytic polio instead of some other disease may well be influenced by the physician's general knowledge or feeling about how widespread polio is in his community at the time.

These difficulties can be mitigated to some extent by setting down very precise criteria for diagnosis, but it is virtually impossible to obviate them completely when, as would be the case after the widespread introduction of a new vaccine, there is a marked shift in what the physician expects to find. This is most especially true when the initial diagnosis must be made by family physicians who cannot easily be indoctrinated in the use of a special set of criteria, as is the case with polio. Later evaluation by specialists cannot, of course, bring into the picture those cases originally diagnosed as something other than polio.

THE OBSERVED CONTROL APPROACH

The difficulties of the vital statistics approach were recognized by all concerned, and the initial study plan, although not judged entirely satisfactory, got around many of the problems by introducing a control group similar in characteristics to the vaccinated group. More specifically, the idea was to offer vaccination to all children in the second grade of participating schools and to follow the polio experience not only in these children, but in the first- and third-grade children as well. Thus the vaccinated second-graders would constitute the *treated* group, and the first- and third-graders would constitute the *control* group. This plan follows what we call the *observed control approach*.

It is clear that this plan avoids many of the difficulties that we listed above. The three grades all would be drawn from the same geographic location so that an epidemic affecting the second grade in a given school would certainly affect the first and third grades as well. Of course, all subjects would be observed concurrently in time. The grades, naturally, would be different ages, and polio incidence does vary with age. Not much variation from grade to grade was expected, however, so it seemed reasonable to assume that the average of first and third grades would provide a good control for the second grade.

Despite the relative attractiveness of this plan and its acceptance by the NFIP advisory committee, serious objections were raised by certain health departments that were expected to participate. In their judgment, the results of such a study were likely to be insufficiently convincing for two important reasons. One is the uncertainty in the diagnostic process mentioned earlier and its liability to influence by the physician's expectations, and the other is the selective effect of using volunteers.

Under the proposed study design, physicians in the study areas would have been aware of the fact that only second-graders were offered vaccine, and in making a diagnosis for any such child, they would naturally and properly have inquired whether he had or had not been vaccinated. Any tendency to decide a difficult diagnosis in favor of nonpolio when the child was known to have been vaccinated would have resulted in a spurious piece of evidence favoring the vaccine. Whether or not such an effect was really operating would have been almost impossible to judge with assurance, and the results, if favorable, would have been forever clouded by uncertainty.

A less conjectural difficulty lies in the difference between those families who volunteer their children for participation in such a trial and those who do not. Not at all surprisingly, it was later found that those who do volunteer tend to be better educated and, generally, more well-to-do than are those who do not participate. There was also evidence that those who agree to participate tend to be absent from school with a noticeably higher frequency than others. The direction of effect of such selection on the incidence of diagnosed polio is by no means clear before the fact, and this important difference between the treated group and the control group also would have clouded the interpretation of the results.

RANDOMIZATION AND THE PLACEBO CONTROL APPROACH

The position of critics of the NFIP plan was that the issue of vaccine effectiveness was far too important to be studied in a manner which would leave uncertainties in the minds of reasonable observers. No doubt, if the vaccine should appear to have fairly high effectiveness, most public health officials and the general public would accept it, despite the reservations. If, however, the observed control scheme were used, a number of qualified public health scientists would have remained unconvinced, and the value of the vaccine would be uncertain. Therefore, the critics proposed that the study be run as a scientific experiment with the use of appropriate randomizing procedures to assign subjects to treatment or to control and with a maximum effort to eliminate observer bias. This plan follows what we call the *placebo control approach*.

The chief objection to this plan was that parents of school children could not reasonably be expected to permit their children to participate in an experiment in which they might be getting only an ineffective salt solution instead of a probably helpful vaccine. It was argued further that the injection of placebo might not be ethically sound, since a placebo injection carries a small risk, especially if the child unknowingly is already infected with polio.

The proponents of the placebo control approach maintained that, if properly approached, parents *would* consent to their children's participation in such an experiment, and they judged that because the injections would not be given during the polio season, the risk associated with the placebo injection itself was vanishingly small. Certain health departments took a firm stand: they would participate in the trial only if it were such a well-designed experiment. The consequence was that in approximately half the areas, the randomized placebo control method was used, and in the remaining areas, the alternating-grade observed control method was used.

A major effort was put forth to eliminate any possibility of the placebo control results being contaminated by subtle observer biases. The only firm way to accomplish this was to insure that neither the subject, nor his parents, nor the diagnostic personnel could know which children had gotten the vaccine until all diagnostic decisions had been made. The method for achieving this result was to prepare placebo material that looked just like the vaccine, but was without any antigenic activity, so that the controls might be inoculated and otherwise treated in just the same fashion as were the vaccinated.

Each vial of injection fluid was identified only by a code number so that no one involved in the vaccination or the diagnostic evaluation process could know which children had gotten the vaccine. Because no one knew, no one could be influenced to diagnose differently for vaccinated cases and for controls. An experiment in which both the subject getting the treatment and the diagnosticians who will evaluate the outcome are kept in ignorance of the treatment given each individual is called a *double-blind* experiment. Experience in clinical research has shown the double-blind experiment to be the only satisfactory way to avoid potentially serious observer bias when the final evaluation is in part a matter of judgment.

For most of us, it is something of a shock to be told that competent and dedicated physicians must be kept in ignorance lest their judgments be colored by knowledge of treatment status. We should keep in mind that it is not deliberate distortion of findings by the physician which concern the medical experimenter. It is rather the extreme difficulty in many cases of making an uncertain decision which, experience has shown, leads the best of investigators to be subtly influenced by information of this kind. For example, in the study of drugs used to relieve postoperative pain, it has been found that it is quite impossible to get an unbiased judgment of the quality of pain relief, even from highly qualified investigators, unless the judge is kept in ignorance of which patients were given which drugs.

The second major feature of the experimental method was the assignment of subjects to treatments by a careful randomization procedure. As we observed earlier, the chance of coming down with a diagnosed case of polio varies with a great many factors including age, socioeconomic status, and the like. If we were to make a deliberate effort to match up the treatment and control groups as closely as possible, we should have to take care to balance these and many other factors, and, even so, we might miss some important ones. Therefore, perhaps surprisingly, we leave the balancing to a carefully applied equivalent of coin tossing: we arrange that each individual has an equal chance of getting vaccine or placebo, but we eliminate our own judgment entirely from the individual decision and leave the matter to chance.

The gain from doing this is twofold. First, a chance mechanism usually will do a good job of evening out all the variables—those we didn't recognize in advance, as well as those we did recognize. Second, if we use a chance mechanism in assigning treatments, we may be confident about the use of the theory of chance, that is to say, probability theory, to judge the results. We can then calculate the probability that so large a difference as that observed could reasonably be due solely to the way in which subjects were assigned to treatments, or whether, on the contrary, it is really an effect due to a true difference in treatments.

To be sure, there are situations in which a skilled experimenter can balance the groups more effectively than a random-selection procedure typically would. When some factors may have a large effect on the outcome of an experiment, it may be desirable, or even necessary, to use a more complex experimental design that takes account of these factors. However, if we intend to use probability theory to guide us in our judgment about the results, we can be confident about the accuracy of our conclusions only if we have used randomization at some appropriate level in the experimental design.

The final determinations of diagnosed polio proceeded along the following lines. First, all cases of poliolike illness reported by local physicians were subjected to special examination, and a report of history, symptoms, and laboratory findings was made. A special diagnostic group then evaluated each case and classified it as nonpolio, doubtful polio, or definite polio. The last group was subdivided into nonparalytic, paralytic, and fatal polio. Only after this process was complete was the code broken and identification made for each case as to whether vaccine or placebo had been administered.

RESULTS OF THE TRIAL

The main results are shown in Table 1, which shows the size of the study populations, the number of cases classified as polio, and the disease rates, that is, the number of cases per 100,000 population. For example, the second line shows that in the placebo control area there were 428 reported cases

TABLE 1. Summary of Study Cases by Diagnostic Class and Vaccination Status (Rates per 100,000)

STUDY GROUP	STUDY POPULATION	ALL REPORTED CASES		POLIOMYELITIS CASES								NOT POLIO	
				Total		Paralytic		Nonparalytic		Fatal polio			
		No.	Rate	No.	Rate	No.	Rate	No.	Rate	No.	Rate	No.	Rate
All areas: Total	1,829,916	1013	55	863	47	685	37	178	10	15	1	150	8
Placebo control areas: Total	749,236	428	57	358	48	270	36	88	12	4	1	70	9
Vaccinated	200,745	82	41	57	28	33	16	24	12	—	—	25	12
Placebo	201,229	162	81	142	71	115	57	27	13	4	2	20	10
Not inoculated*	338,778	182	54	157	46	121	36	36	11	—	—	25	7
Incomplete vaccinations	8,484	2	24	2	24	1	12	1	12	—	—	—	—
Observed control areas: Total	1,080,680	585	54	505	47	415	38	90	8	11	1	80	7
Vaccinated	221,998	76	34	56	25	38	17	18	8	11	2	20	9
Controls**	725,173	439	61	391	54	330	46	61	8	11	—	48	6
Grade 2 not inoculated	123,605	66	53	54	44	43	35	11	9	—	—	12	10
Incomplete vaccinations	9,904	4	40	4	40	4	40	—	—	—	—	—	—

Source: Adapted from Francis (1955), Tables 2 and 3.
* Includes 8,577 children who received one or two injections of placebo.
** First- and third-grade total population.

12

of which 358 were confirmed as polio, and among these, 270 were classified as paralytic (including 4 that were fatal). The third and fourth rows show corresponding entries for those who were vaccinated and those who received placebo, respectively. Beside each of these numbers is the corresponding rate. Using the simplest measure—all reported cases—the rate in the vaccinated group is seen to be half that in the control group (compare the boxed rates in Table 1) for the placebo control areas. This difference is greater than could reasonably be ascribed to chance, according to the appropriate probability calculation. The apparent effectiveness of the vaccine is more marked as we move from reported cases to paralytic cases to fatal cases, but the numbers are small and it would be unwise to make too much of the apparent, very high effectiveness in protecting against fatal cases. The main point is that the vaccine was a success; it demonstrated sufficient effectiveness in preventing serious polio to warrant its introduction as a standard public health procedure.

Not surprisingly, the observed control area provided results that were, in general, consistent with those found in the placebo control area. The volunteer effect discussed earlier, however, is clearly evident (note that the rates for those not innoculated differ from the rates for controls in both areas). Were the observed control information alone available, considerable doubt would have remained about the proper interpretation of the results.

Although there had been wide differences of opinion about the necessity or desirability of the placebo control design before, there was great satisfaction with the method after the event. The difference between the two groups, although substantial and definite, was not so large as to preclude doubts had there been no placebo controls. Indeed, there were many surprises in the more detailed data. It was known, for example, that some lots of vaccine had greater antigenic power than did others, and it might be supposed that they should have shown a greater protective effect. This was not the case; lots judged inferior in antigenic potency did just as well as those judged superior. Another surprise was the rather high frequency with which apparently typical cases of paralytic polio were not confirmed by laboratory test. Nonetheless, there were no surprises of a character to cast serious doubt on the main conclusion. The favorable reaction of those most expert in research on polio was expressed soon after the results were reported. By carrying out this kind of study before introducing the vaccine, it was noted, we now have facts about Salk vaccine that we still lack about typhoid vaccine, after 50 years of use, and about tuberculosis vaccine, after 30 years of use.

EPILOGUE

It would be pleasant to report an unblemished record of success for the Salk vaccine, following so expert and successful an appraisal of its effectiveness, but it is more realistic to recognize that such success is but one step in the continuing development of public health science. The Salk vaccine, although a notable triumph in the battle against disease, was relatively crude and, in many ways, not a wholly satisfactory product that was soon replaced with better ones.

The report of the field trial was followed by widespread release of the vaccine for general use, and it was discovered very quickly that a few of these lots actually had caused serious cases of polio. Distribution of the vaccine was then halted while the process was reevaluated. Distribution was reinitiated a few months later, but the momentum of acceptance had been broken and the prompt disappearance of polio that researchers hoped for did not come about. Meanwhile, research on a more highly purified killed-virus vaccine and on several live-virus vaccines progressed, and within a few years the Salk vaccine was displaced by live-virus vaccines.

The long-range historical test of the Salk vaccine, in consequence, has never been carried out. We do not know with certainty whether or not that vaccine could have accomplished the relatively complete elimination of polio that has now been achieved. Nonetheless, this does not diminish the importance of its role in providing the first heartening success in the attack on this disease, a role to which careful and statistically informed experimental design contributed greatly.

PROBLEMS

1. Using Figure 1 as an example, explain why a control group is needed in experiments where the effectiveness of a drug or vaccine is to be determined.

2. Explain the need for control groups by criticizing the following statement:

"A study on the benefits of vitamin C showed that 90% of the people suffering from a cold who take vitamin C get over their cold within a week."

3. Explain the difference between the observed control approach and the placebo control approach. Which one would you prefer, and why?

4. Why is it important to have a "double-blind" experiment?

5. If "double-blind" experiments provide the only satisfactory way to avoid observer bias, why aren't they used all the time?

6. If only volunteers are used in an experiment, instead of a random sample of individuals, will the results of the experiment be of any value? What can you say about the results?

7. Why did the polio epidemics seem to hit hardest those who were better off hygienically?

8. Why was a *large-scale* field trial needed to get convincing evidence of the Salk vaccine effectiveness?

9. Refer to Figure 1. In which year did the highest polio incidence occur? the lowest? the largest increase? the smallest increase? Give the approximate values of these incidences and increases.

10. Refer to Figure 1. Comment on the use of the *number of cases*. Can you suggest a different indicator of the spread of poliomyelitis in the U.S. during 1930–56. When are the two indicators equivalent? (Hint: refer to Table 1.)

REFERENCES

K. Alexander Brownlee. 1955. "Statistics of the 1954 Polio Vaccine Trials." *Journal of the American Statistical Association* 50:272, pp. 1005–1013.

Thomas Francis, Jr., et al. 1955. "An Evaluation of the 1954 Poliomyelitis Vaccine Trials—Summary Report." *American Journal of Public Health* 45:5, pp. 1–63.

Paul Meier. 1957. "Safety Testing of Poliomyelitis Vaccine." *Science* 125:3257, pp. 1067–1071.

D. D. Rutstein. 1957. "How Good is Polio Vaccine?" *Atlantic Monthly* 199:48.

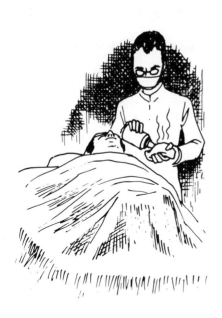

SAFETY OF ANESTHETICS

Lincoln E. Moses *Stanford University*
Frederick Mosteller *Harvard University*

In 1958, AMERICAN hospitals began using a new anesthetic called halothane. It soon became widely accepted for its many desirable properties. Unlike some of the more commonly used anesthetics, it could not catch fire, so fire and explosive hazards did not have to be a concern during surgical operations. Patients found it less disagreeable and recovered from anesthesia more quickly and with less severe aftereffects. Extensive laboratory research and trials on animals and humans in surgery had encouraged belief in its safety. So there were good reasons for halothane to come rapidly into widespread use. By 1962, surgeons used halothane in half of their operations.

After a few years, however, halothane came under suspicion as accounts appeared in the medical literature of some strikingly unusual—but strikingly similar—deaths of patients who had recently had this anesthetic. A few patients recovering from surgery suddenly took a turn for the worse, ran fevers, and died; subsequent autopsies revealed massive fatal changes in their livers. Even though there were only a few such reports, it was natural to ask, "Do these incidents mean that halothane is a poison dangerous to people's livers? Should the use of halothane as an anesthetic in surgery therefore be discontinued?" Some compounds chemically similar to halothane were already known to cause liver damage, so the question arose with all the more force. A national committee was appointed to assemble and examine evidence that might help to answer these questions.

Much information was needed. First, it was possible that the liver changes found with halothane might occur equally often with other anesthetics, but that physicians rarely reported such cases because the old anesthetics would attract less attention. Second, and this is a key point, whether or not halothane had a special adverse effect on the liver, there was the possibility that its other advantages might result in a lower overall postoperative death rate. Frequently things that are easy to use actually work better (this is true of sharp knives, fine violins, and easy-to-read instructions), so it might well be that an anesthetic that is easier for the doctor to use might work better and result in somewhat lower death rates during surgery. Information was needed to confirm or refute this possibility.

These questions could not be answered by doing laboratory experiments with mice, for mice might well behave differently under the anesthetic than human beings; nor could the questions be answered by looking in books, for the information was not even known; nor was it useful to ask experts for their judgments, since different experts had widely different opinions.

THE EVIDENCE

It was necessary to amass a great deal of evidence, so the committee decided to conduct a survey of hospital experience. At that time, halothane had been given to about 10 million people in the U. S. Many of these patients had been in a particular group of 34 hospitals that kept good records and whose staffs keenly wanted to answer the very questions we have asked. They cooperated with the committee by sampling their records of surgical operations performed during the years 1960–64. In addition to recording whether or not the patient died within six weeks of surgery, they gave information on the anesthetic that had been used as well as facts about the surgical procedure and the patient's sex, age, and physical status prior to the operation. Among the 850,000 operations in the study, there were about 17,000 deaths, or a death rate of 2%. The death rates, shown in Table 1, were calculated for each of four major anesthetics and for a fifth group consisting of all other anesthetics. Note that these are death rates from all causes, including the patients' diseases, and are not deaths especially resulting from the anesthetic.

TABLE 1.   Death Rates Associated with Various Anesthetics

HALOTHANE	PENTOTHAL*	CYCLOPROPANE	ETHER	ALL OTHER
1.7%	1.7%	3.4%	1.9%	3.0%

* Nitrous oxide plus barbiturate.

NEED FOR ADJUSTMENT

Table 1 suggests that halothane was as safe as any other anesthetic in wide use, but such a suggestion simply cannot be trusted.   Medical people know that certain anesthetics, cyclopropane, for example, are used more often in severe and risky operations than some other anesthetics (such as pentothal, which is much less often used in difficult cases).   So some or all of the differences between these death rates might be due to a tendency to use one anesthetic in difficult operations and another in easier ones.

Different operations carry very different risks of death.   Indeed, death rates on some operations were found to be as low as 0.25% and on others nearly 14%.   The change in death rates across the categories of patients' physical status was even more dramatic, ranging from 0.25% in the most favorable physical status category to over 30% in the least favorable. Age mattered a great deal: the most favorable 10-year age group (10 to 19) carried a death rate of less than 0.50%, while the least favorable age group (over 90) had a death rate of 26%.   Another factor was sex, with women about two-thirds as likely to die.   Because the factors of age, type of operation, physical status, and sex were so very important in determining the death rate, it was clear that even a relatively small preponderance of unfavorable patients in the group receiving a particular anesthetic could raise its death rate quite substantially.   Thus, the differences in Table 1 *might* be due largely, and possibly even entirely, to discrepancies in the kinds of patients and types of operations associated with the various anesthetics.   Certainly it was necessary to somehow adjust, to equalize for, type of operation, sex, physical status, and age before trying to determine the relative safety of the anesthetics.

If these data had been obtained from a properly planned experiment, the investigators might have arranged to collect the data so that patients receiving the various anesthetics were comparable in age, type of operation, sex, and physical status.   As it was, these data were collected by reviewing old records, so substantial differences in the kinds of patients receiving the various anesthetics were to be expected and, indeed, were found.   For example, cyclopropane was given two or three times more often than halothane to patients with bad physical status and substantially more often than halothane to patients over sixty years of age. There were many such peculiarities, and they were bound to affect the death rates of Table 1.

CARRYING OUT ADJUSTMENTS

The task was to purge the effects of these interfering variables from the death rates corresponding to the five anesthetics. Fortunately, it was possible to do this. We said "fortunately" because it could have been impossible; for example, if each anesthetic had been applied to a special set of operations, different from those in which other anesthetics were used, then differences found in the death rates could perfectly well have come from differences in the operations performed and there would be no way to disentangle operation from anesthetic and settle the question. But in the data of this study, there was much overlapping among the anesthetics in the categories of age, kind of operation, sex, and patient's physical status, so that statistical adjustment, or equalization, was possible. Analysis using such adjustment was undertaken in a variety of ways because the complexity of the problem made several different approaches reasonable and no one approach alone could be relied upon. There was close agreement in the findings regardless of the method used.

A summary of the results is shown in Table 2. Notice that halothane and pentothal after adjustment have higher death rates than their unadjusted rates in Table 1. Also, cyclopropane and "all other," the two with highest rates in Table 1, now have lower death rates. The effects of these adjustments are quite important: halothane, instead of appearing to be twice as safe as cyclopropane—the message in Table 1—now appears to be safer by only about one-fifth.

Figure 1 shows the adjustment effect graphically.

A Special, Simple Case. To understand the nature of the adjustments, it is helpful to look at a very special case with fictitious data.[1] Suppose that halothane and cyclopropane are to be compared, that we are to adjust only for physical status, and that we classify all patients in either good or poor physical

TABLE 2. Adjusted Death Rates for Various Anesthetics
(Adjusted for Age, Type of Operation, Sex, and Physical Status)

HALOTHANE	PENTOTHAL	CYCLOPROPANE	ETHER	ALL OTHER
2.1%	2.0%	2.6%	2.0%	2.5%

[1] This section may be skipped without disturbing continuity.

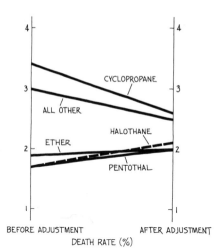

FIGURE 1
Effects of adjusting death rates.
Plotted points are from Tables
1 and 2

status. The fictitious data might look like those shown in Table 3. Thus, halothane was given to patients in good status four out of five times (in this fictitious case), but cyclopropane, just the other way around, was given to patients in good physical status one out of five times. The two anesthetics have exactly the same death rates for each physical status classification separately, but the very different proportions of status for the two make the overall death rate for cyclopropane almost twice that of halothane.

Suppose that we decide to adjust by computing what the overall death rates *would* be if the two anesthetics were given to a population of patients in which 70% were of good status and 30% poor. The death rates for each status group remain the same. The computations would be as displayed in Table 4. For each anesthetic the overall death rate is now 1.6% with these fictitious data.

TABLE 3. Fictitious Data

	NUMBER OF PATIENTS	NUMBER OF DEATHS	DEATH RATE
Halothane			
Poor physical status	200	6	3.0%
Good physical status	800	8	1.0%
All patients	1000	14	1.4%
Cyclopropane			
Poor physical status	800	24	3.0%
Good physical status	200	2	1.0%
All patients	1000	26	2.6%

TABLE 4.   Adjusted Death Rates (Fictitious Data)

	NUMBER OF PATIENTS	NUMBER OF DEATHS	DEATH RATE
Halothane			
Poor physical status	300	9	3.0%
Good physical status	700	7	1.0%
All patients	1000	16	1.6%
Cyclopropane			
Poor physical status	300	9	3.0%
Good physical status	700	7	1.0%
All patients	1000	16	1.6%

In practice, the actual adjustments are far more complicated, partly because fine breakdowns of the data leave many cells with no entries at all and many with small entries.

Consistency over Hospitals.   Even after the adjustments, there was still considerable variability from hospital to hospital.   Our confidence in the validity of the statistically adjusted figures in Table 2 is affected by how consistent the death rates were from hospital to hospital.   After all, if there were many hospitals where cyclopropane had a lower adjusted death rate than did halothane, even though halothane was lower "on the average," we might feel uncertain that halothane was really safer just because of the difference in adjusted death rates.   Some fluctuation from hospital to hospital is to be expected, of course, because of chance factors; absolute consistency is not to be expected.

The question is: "Were the comparisons between the adjusted anesthetic death rates consistent enough over hospitals to warrant taking seriously the apparent differences shown in Table 2?"   Rather complicated statistical techniques were necessary to study this question, but the conclusions were clear: halothane and pentothal both had adjusted death rates definitely lower than the adjusted death rates for cyclopropane and "all other," and those lower adjusted death rates were real in the sense that they could not be explained by chance fluctuations.   The differences were sufficiently consistent to be believed.   On the other hand, ether, with the same adjusted death rate as halothane and pentothal, did not have a consistent pattern of comparison.   Therefore, ether cannot be reliably compared with the others; we cannot tell from the evidence obtained whether it may be somewhat safer than it appears here—or somewhat less safe.   Possibly it is as safe as, or safer than, halothane; possibly it is no safer than "all other."   The findings about ether are indefinite because there were fewer administrations of ether and these were concentrated in a few hospitals; hence there are fewer data to go on.

A SUMMING UP SO FAR

What can we conclude so far from the findings of this study? First, and most important, is the surprising result that halothane, which was suspect at the beginning of the study, emerged as a definitely safe and probably superior anesthetic agent. Second, we see that a careful statistical study of 850,000 operations enabled the medical profession to answer questions more firmly than had previously been possible, despite the much greater "experience" of 10 million administrations of halothane and other 10s of millions of administrations of the other anesthetics.

HOSPITAL DIFFERENCES AGAIN

A brand new question emerged with the observation that the 34 hospitals had very different overall postoperative death rates! These ranged from around 0.25% to around 6.5%. This seemed to mean that the likelihood of dying within six weeks after surgery could be more than 20 times as great in one hospital as in another. Just as before, however, there were strong reasons to approach this startling information skeptically. Some of the hospitals in the study did not undertake difficult operations, such as open-heart surgery, while others had quite large loads in such categories. This kind of difference alone would cause differences in hospital death rates. Further, the age distribution might be different, and perhaps importantly so, from one hospital to another. Indeed, one was a children's hospital, another a veteran's hospital. Some hospitals might more frequently accept surgical patients having poor physical status. So the same interfering variables as before surely affected the differences in hospital death rates. If adjustment was made for them, would the great differences in hospital death rates vanish? Be much reduced? Remain the same? Or, as is conceivable, actually increase?

Adjustment procedures were applied, and the result was that the high-death-rate hospitals, after adjustment, moved down toward a 2% overall death rate, and the low-death-rate hospitals, after adjustment, moved up toward a 2% overall death rate. The adjusted hospital death rates no longer ranged from 0.25% to over 6.5%; instead, the largest of these adjusted death rates was only about three times as great as the smallest. Now, almost any group of 34 such rates will exhibit some variability, and the ratio of the largest to the smallest must be a number larger than 1. Even if a single hospital were measured over several different periods, the rates would fluctuate from chance alone. The fluctuation in rate would be considerable because the death rate itself is basically low and one death more or less makes a difference in the observed rate for the period. The fact that this ratio turned out to be 3 in these data does not, in itself, indicate clearly that there were real, unexplained hospital differences.

Careful statistical study showed that there are probably some real differences from hospital to hospital in postoperative death rates, and that these differences cannot be explained wholly by the hospitals' patient populations in terms of age, sex, physical status, and surgical procedure. Statistical theory showed that the ratio of highest to lowest adjusted rate should be about 1.5 if the hospitals were identical in operative death rate after adjustment (so that the adjusted rates differ only by chance fluctuation).

The position, then, is that we began with the large ratio of about 25 for unadjusted rates, cut the ratio way down to 3 for adjusted rates, and then compared the 3 with 1.5 as a theoretical ratio. Because the adjustments can hardly have been perfect and because there undoubtedly were unadjusted factors that differed among the hospitals, we conclude that the adjusted hospital death rates are indeed close together. Thus, what, on the basis of the unadjusted hospital death rates, looked like a shocking public health problem proved, after statistical investigation, to be quite something else. The apparent problem was mainly, though perhaps not entirely, a dramatic manifestation of differences not in the quality of surgical care, but in the difficulty of the surgical cases handled in the various hospitals.

FOLLOW-UP STUDIES

Concern about hospital differences prompted further extensive studies of hospitals in the United States,* to determine whether the differences in surgical death rate among hospitals are as large as the Halothane Study suggested. These same follow-up studies were designed to gain further information about post-surgical complication rates, and to begin exploring the reasons for hospital differences if such differences were confirmed.

These studies have confirmed the results of the Halothane Study and have added further detailed estimates on the sizes of the differences and on the kinds of surgical procedures which vary most. Furthermore, by carefully adjusting the hospital rates for the characteristics of the patients served by the hospital, and then correlating hospital characteristics with measures of hospital performance, some interesting hypotheses have been developed for further study. For example, the results suggest that a very important determinant of the quality of surgical outcome is the care and power exercised by the hospital physicians in approving surgeons for hospital privileges in their hospital.

* "Comparison of Hospitals with Regard to Outcomes of Surgery" by the staff of the Stanford Center for Health Care Research; *Health Services Research*, pp. 112-127, Summer 1976.

LIVER DAMAGE

Let us return to the question of liver damage. The total deaths from this source were few, and the lack of autopsies for nearly half of the deaths made firm conclusions impossible. Since the study was made, however, an anesthesiologist who had often been exposed to halothane while administering it to patients, has been discovered to be sensitive to halothane; he exhibits symptoms of liver malfunction from breathing it.

Are occasional individuals sensitive to other anesthetics? How many people develop such sensitivity? These are hard questions to answer, and they are especially difficult to study because of the rarity of the occurrences.

CONTRIBUTIONS OF STATISTICS

What were the main contributions of statistics in the program? First was the basic concept of a death-rate study. This needed to be carried out so that the safety of anesthetics could be seen in light of total surgical experience, not just in deaths from a single rare cause. Second, though we have not discussed it, the study used a special statistical technique of sampling records designed to save money and to produce a high quality of information. Third, special statistical adjustments had to be created to appraise the results when so many important variables—age, type of operation, sex, patient's physical status, and so on—were uncontrolled. Fourth, as a result, the original premise of the study was not sustained, but a new result emerged: halothane seemed safer than cyclopropane. The study produced a nonfact as well: the comparative merit of ether is uncertain. We cannot tell if its associated death rate is higher, lower, or nearly the same as halothane, though the indication is that it is nearly the same. Fifth, new evidence on hospital differences showed that the initial, wide variation in death rates, when suitably adjusted for patient populations, left only small unexplained differences in rates. These remaining differences led the medical profession to begin, in 1971, a study of possible causes of hospital-to-hospital differences in postoperative death rates, in the hope of discovering ways to improve postoperative care.

PROBLEMS

1. Explain how it was possible for halothane to be as safe as other anesthetics despite the evidence from autopsies that it caused liver damage.

2. What is meant by equalization or adjustment?

3. Explain the discrepancy between the safety of halothane indicated in Tables 1 and 2.

4. How can you explain the unadjusted differences of the overall postoperative death rates in the 34 hospitals in the study?

5. What were the principal conclusions of the study?

6. What statistical tools were used in the study?

7. Explain why the results about ether were uncertain.

8. Is halothane a cause of liver damage? Explain.

9. Refer to Figure 1. For which anesthetic is the adjustment most drastic? Check your answer by comparing Table 2 with Table 1.

10. In Table 4 explain how the number of deaths in each status group was calculated.

11. Consider a study of the effect of a high cholesterol diet on mortality from coronary heart disease. The mortality rates from CHD in two groups are compared—one group with high cholesterol diet and another with an average diet. What are some of the factors that one needs to adjust for in the two groups before comparing the mortality rates?

12. Why did the statistical study described in this article concentrate on 850,000 operations only although there were over 10 million operations in which halothane was administered and 10's of millions of operations where other anesthetics were used?

REFERENCE

John P. Bunker, William H. Forrest, Jr., Frederick Mosteller, and Leroy D. Vandam. 1969. *The National Halothane Study*, Report of the Subcommittee on the National Halothane Study of the Committee on Anesthesia, Division of Medical Sciences, National Academy of Sciences—National Research Council, published by the U. S. Government Printing Office, Washington, D. C., for the National Institutes of Health, National Institute of General Medical Sciences, Bethesda, Md.

DRUG SCREENING: The Never-Ending Search for New and Better Drugs

Charles W. Dunnett   *Lederle Laboratories Division, American Cyanamid Company*

PROBABLY EVERYONE can recall reading about a small boat lost at sea or an aircraft down in some unpopulated area involving a search for survivors. If the search is successful, both rescuers and rescued receive wide public acclaim. On the other hand, if the search is unsuccessful, the episode is soon forgotten. In neither case do we see much in the press about the planning and organization of the search. We can imagine, however, what a tremendous effort it must be to map out the area to be explored, to marshal the needed resources in aircraft and other vehicles, and to use all the available people and equipment so as to maximize the chance of a successful outcome in the shortest possible time.

Pharmaceutical companies conduct somewhat similar searches for new drugs, but their search is continuous. Just as a rescue team has some idea where to look, perhaps based on a radio message received from the victims, so research chemists often know what types of chemical structures to look for to treat a particular disease, and the chemists can set about synthesizing compounds of the desired type. Sometimes, however, their knowledge may be vague, resulting in such a wide range of possibilities that many, many compounds have to be made and tested. In such a case, the search is very lengthy and requires years of effort by many people to develop a useful new drug.

Researching thousands of compounds for the few that might be effective requires highly organized, efficient testing methods. With an inefficient procedure, progress will be slow and the testing laboratory may never get to the "good" compounds. Of course, the procedure used to test each compound must have a high degree of accuracy and must be capable of detecting a good compound.

Testing a long series of compounds in the search for any that have useful biological properties is known as *drug screening*. Drug screening requires laboratories, technical personnel to operate them, and appropriate apparatus and instruments. Ordinarily, laboratory tests are conducted with animals, which must be housed, cared for, and observed in the laboratory. Space limitations together with the limitations of the staff severely restrict the number of compounds that can be assessed. A single laboratory cannot hope to test more than several hundred, or, perhaps, a few thousand, compounds annually for a specified biological activity. This testing rate cannot even keep up with the rate of synthesis of new compounds. Yet, if all the laboratories throughout the entire pharmaceutical industry are considered, much is accomplished. It is estimated (Arnow 1970), for example, that approximately 175,000 substances are subjected to biologic evaluation each year.

The screening test is only the first hurdle along the path of developing an effective drug; much further testing and investigation needs to be done before the drug can be tried on humans. In fact, only about 20 of the 175,000 substances tested finally become available in the local drug store.

Anything that improves the efficiency of the testing procedure increases the chance of discovering a new cure. Naturally, the biologist designing a new screening procedure attempts to use an animal and test system that reflect the human disease for which a cure is being sought as accurately as possible and yet are easy to use in the laboratory. We will not deal with this aspect of the problem here, but rather with the contribution of statistics to improvement of drug screening procedures.

AN EXAMPLE: ANTICANCER SCREENING

In drug screening, the animals of a sample treated with a particular compound are observed to determine whether the treatment is having a desirable effect. Perhaps all that is noted in each animal is whether or not it is "cured" or shows improvement. The result of the test can be expressed simply as the number of cured or improved animals out of the total number tested. Since not every animal responds to treatment in exactly the same way, the number cured might be 0, 1, 2, or all the way up to the whole sample size.

More often, a measurement of some sort is made upon each animal, the magnitude being indicative of the treatment effect. For example, in anti-cancer screening, after implanting cancer cells in mice, the investigator treats them with the test chemical compound to see whether it retards the growth of the cancer tumors. After treating the mice for a fixed length of time, he removes the tumors and weighs them. For comparison, a similar group of untreated "control" animals is handled in the same way except for the absence of the chemical treatment. If the chemical is effective, the tumor weights of the treated mice should be less than the tumor weights of the control mice. The statistician's job is to help decide, on the basis of the numerical values obtained, whether the tested compound merits further investigation.

The following are actual tumor weights observed in three animals treated with a test compound and in six untreated control animals:

> Treated: 0.96, 1.59, 1.14 grams
> Controls: 1.29, 1.60, 2.27, 1.31, 1.88, 2.21 grams

The reason for having more control animals than treated is that, in one experiment perhaps 30 to 40 different compounds will be tested, each one in a different set of three animals. One control group suffices, however, to compare with the results from all the test compounds; hence it is desirable to have it larger in size in order to obtain a more precise control average.

Note the high variability from one animal to another. This is typical of the biological variation in screening tests that makes it difficult to determine with certainty whether the treated animals have really improved as a result of the treatment. How might a statistician go about deciding whether a compound has any merit?

DRUG SCREENING AS A DECISION PROBLEM

In drug screening, two actions are possible: (1) to "reject" the drug, meaning to conclude that the tested drug has little or no effect, in which case it will be set aside and a new drug selected for screening, and (2) to "accept" the drug provisionally, in which case it will be subjected to further, more refined experimentation.

To abandon a drug when in fact it is a useful one (a *false negative*) is clearly undesirable, yet there is always some risk of that. On the other hand, to go ahead with further, more expensive testing of a drug that is in fact useless (a *false positive*) wastes time and money that could have been spent on testing other compounds.

Thus, we are faced with what is known in statistics as a *decision problem:* how to use available experimental data to choose between alternative courses of action in a rational way.

AN APPROACH TO SOLVING THE PROBLEM

What would the drug screening investigator like to achieve? A virtually unlimited supply of compounds is available for testing, more than he can hope to test over any reasonable period of time. Most of them lack the biological activity he is searching for, but (hopefully) a few of them possess it. His goal is to find as many of these active compounds as possible with the facilities at his disposal.

As a hypothetical, but not entirely unrealistic, example, let us suppose there are 10,000 compounds of which 40 are active and the remaining 9960 are inactive. The investigator is screening the drugs for antitumor activity, and he tests them by treating groups of three tumor-bearing mice with each compound, comparing the resulting tumor weights with the corresponding values for six control animals observed in the same experiment. He wishes to accept or reject each test compound on the basis of the observed tumor weights. Typically, about 50 compounds per week can be tested in this way, so that about four years' work will be required for all 10,000 compounds. Over this period of time, it is inevitable that some of the 9960 inactive compounds submitted to the screening will pass. The follow-up tests required on these false positives will require test facilities that could have been employed to screen some new compounds. Often the next step after a compound passes the screening step is to carry out a "dose-response" study, which consists of treating groups of animals with several dose levels of a drug in order to determine the relationship between dose and response. At this step, the inactive false positive compounds generally will be eliminated. Suppose, for illustration, that 30 animals are required for each compound at this step. This means that to follow up each compound accepted by the screening, it will be necessary to forgo or postpone the testing of ten new compounds.

Consider the experimental data given above. The mean, or average, tumor weight for the treated animals is $(0.96 + 1.59 + 1.14)/3 = 1.23$ grams. Comparing this with the corresponding mean for the six control animals, $(1.29 + 1.60 + 2.27 + 1.31 + 1.88 + 2.21)/6 = 1.76$ grams, we see that a reduction of $1.76 - 1.23 = 0.53$ gram has apparently been obtained. If the drug has no effect, a zero reduction would be "expected," but, of course, the variability of the animals makes it likely that some difference between the two means would occur even if the drug has no effect Thus, the researcher must decide how large a difference he requires the drug to show before he decides to "accept" it.

Let us assume for the moment that a reduction of 0.53 gram in tumor weight is not enough to convince our investigator that the compound is an active one; let's say that he requires a reduction of 0.70 gram before he will permit the compound to pass the screening test. What are the consequences of setting a cut-off value of 0.70 gram in screening the 10,000 compounds?

What he really needs to know is how many of the 9960 inactive compounds and 40 active compounds will pass the test. He could go ahead and screen the compounds using the 0.70-gram criterion, but it would take several years to collect the results: this would be rather late to find out that he made an unwise choice in the cut-off criterion.

Fortunately, there is another way to get at this problem. Statisticians have a unit, or yardstick, called the *standard error*, which they use to determine how often measures like the difference between two means will exceed any specified limit. An estimate of the standard error needed in our case could be calculated from the observed data, but it would be unreliable because of the small number of observations. It is possible, however, to estimate the standard error accurately from other data of the same type, which, in routine screening, are available in large quantities from past records. Suppose that, for tumor weights, the standard error of a difference between a mean of six control tumor weights and a mean of three treated tumor weights is known to be 0.35 gram.

The next step is to divide the difference between the cut-off value and the *expected* value of the test statistic by this standard error. For an inactive compound, our researcher would expect the two mean tumor weights to be the same; hence the test statistic is expected to have the value zero. The cut-off value of 0.70 gram, therefore, is $(0.70 - 0)/0.35 = 2.0$ standard errors from the zero expectation. Consulting a table of the *normal distribution* (available in most statistics texts) he finds that a deviation of 2.0 or more standard errors from the expected value occurs with a probability of .0228. This means that he can expect to observe a misleading reduction in tumor weight exceeding 0.7 gram for approximately 23 out of 1000 inactive drugs submitted to screening. In other words, of the 9960 inactive compounds, .0228 $\times$ 9960 = 227 of them can be expected to pass as false positives.

Consider next the 40 active compounds: how many of them can be expected to pass? This is a more difficult question, because the answer depends on how active they really are. It is not to be expected that even a very active compound will eliminate the tumor completely in the relatively short time the animals are under treatment. Modest decreases in tumor size are the most that researchers can hope for.

Let us assume for now, rather arbitrarily, that the 40 active compounds are each capable of reducing tumor weight by 0.7 gram (the same as that required by our cut-off criterion). As each active compound is tested, some actually will result in a reduction of more than 0.7 gram, while others will give rise to smaller reductions, because of the inevitable variability of the animals. Therefore, only half of the actives can be expected to show a reduction exceeding the stipulated cut-off value and, hence, to pass the test.

This means that on the average, the accepted compounds will consist of 20 actives (true positives) and 227 false positives. The follow-up testing required on each is the equivalent of ten screening tests, so, in addition to the 10,000 screening tests, there will be $247 \times 10 = 2470$ follow-up tests, or 12,470 tests in all. For this effort, the researcher can expect to find 20 active compounds under the assumptions about the composition of the original set of 10,000 compounds and the degree of activity of the actives. It is useful to express this as a "yield" per thousand tests. Using 0.7 gram as the cut-off value, the resulting yield is

$$\frac{20}{12,470} \times 1000 = 1.60 \text{ actives/thousand tests.}$$

Next, consider the consequences of altering the criterion for a compound to pass the screen. Suppose that instead of a 0.7 gram reduction in tumor weight, a cut-off value of 0.6 gram is used. This corresponds to a deviation of $0.6/0.35 = 1.71$ on the normal curve, and the corresponding probability obtained from tables of the normal distribution is .0436. Thus, of the inactive compounds tested, $.0436 \times 9960 = 434$ false positives are expected to occur.

What about the 40 active compounds? The researcher still assumes that each has a degree of activity capable of producing a reduction of 0.7 gram on the average. How many of them will actually produce a reduction of 0.6 gram or more? Expressing the difference of 0.1 gram as a deviation on the normal curve by dividing by the standard error, he obtains $0.1/0.35 = 0.29$; from the normal tables, he finds that the corresponding probability is .6140. Hence, using 0.6 gram as the passing criterion, he can expect $.6140 \times 40 = 25$ active compounds to pass. This results in $434 + 25 = 459$ compounds passing the screen, so the total testing effort is now $10,000 + 459 \times 10 = 14,590$ tests. Thus the yield per thousand tests is

$$\frac{25}{14,590} \times 1000 = 1.71 \text{ actives/thousand tests,}$$

an increase of 0.11 over the yield obtained using 0.7 gram as the cut-off value.

Now it should be clear how to go about "optimizing" the choice of the cut-off value for the given hypothetical structure of the compounds. It is necessary to try various cut-off values to determine the yield for each in the above way. Figure 1 shows a curve of the computed yield plotted against

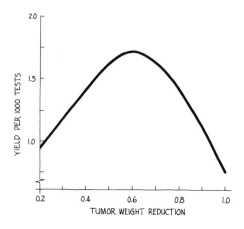

FIGURE 1

Yields for one-stage screening tests

the corresponding cut-off value. It can be seen that the point of maximum yield is easily determined and, in fact, occurs very close to a cut-off value of 0.6 gram reduction. Of course, this applies only to the particular hypothetical mix of active and inactive compounds that we have assumed; we will come back to this point later. For this particular mix, a yield of 1.71 per thousand tests appears to be about the best that can be obtained.

SEQUENTIAL PROCEDURES FOR DRUG SCREENING

It is possible, however, to do still better by using another type of screening. Statistical theory tells us that a kind of test procedure, known as a *sequential test,* is more efficient for reaching decisions of the type we need. In sequential testing, the number of tests is not fixed in advance; rather, the testing proceeds sequentially and the number of tests depends upon the results observed. This gives us the option, after observing a test result on a compound, to forgo making an immediate decision about that compound and to subject the compound to another test. From the results of a second test averaged with those of a first test, we may conclude that further testing is unnecessary, or we may forgo a decision again and wait for a third test result on the compound. This process could go on until a decision is reached to accept or reject the compound. Of course, if too many stages are permitted, the procedure loses efficiency because of the delay and because of the extra trouble involved in keeping a sufficient supply of the compound on hand to repeat the test. In drug screening applications, it has usually been found best to limit the testing to two or three stages.

A sequential procedure is more efficient because it allows us to compile additional data on the compounds about which doubt exists, without much increase in the average amount of testing required, because additional testing needs to be done only on a relatively few compounds.

Suppose that we decide to reject a compound without further testing if a tumor weight reduction of less than 0.2 gram is observed on the first test. If, instead, a reduction of more than 0.2 gram is obtained, we will test the compound again on another group of mice. At the end of this second stage of testing, the tumor weight reduction will be measured as usual, and its value will be averaged with the value obtained for that compound in the first stage. If the average of these two values exceeds 0.5 gram, we will accept the compound; otherwise it will be discarded. For obvious reasons, a procedure of this type is called a *two-stage test* (the procedure discussed in the previous section is called a *one-stage test*).

For illustration, consider our numerical example. For that set of data, the observed tumor weight reduction turned out to be 0.53 gram. Using a one-stage test, we found that the optimum cut-off value was 0.6 gram, which would require this compound to be discarded. With the two-stage test described in the preceding paragraph, however, it would be tested again. Suppose another test (using three treated and six control animals) gave a tumor weight reduction of 0.49 gram. Averaging the two results, 0.53 and 0.49, we obtain 0.51 gram; this is enough for the compound to be accepted.

What are the consequences of using the two-stage test on the hypothetical mix of 10,000 active and inactive compounds? First of all, the amount of testing obviously will be increased because, for some compounds, two tests will have to be run. The amount of increase can be computed. For an inactive compound, the probability of obtaining a tumor weight reduction in excess of 0.2 gram turns out to be .284. Thus, $.284 \times 9960 = 2829$ of the inactive compounds will require two screening tests to permit a decision, the remainder requiring only one test. For an active compound, the corresponding probability is .923; hence, $.923 \times 40 = 37$ of the actives will be tested a second time. Therefore, the total number of screening tests required to screen the 10,000 compounds is $10,000 + 2829 + 37 = 12,866$ (compared to 10,000 in a one-stage procedure).

It is also fairly easy to work out the probability that an inactive compound passes the screening under the sequential procedure; this may be done for an active compound as well. For an active compound, the probability is .77; thus, $.77 \times 40 = 31$ active compounds can be expected to pass. For an inactive compound, the probability is .020; hence, $.020 \times 9960 = 199$ false positives will occur. Therefore, a total of 230 compounds are expected to pass, entailing an additional testing effort equal to 2300 tests. This makes altogether $12,866 + 2300 = 15,166$ tests. The yield per thousand tests for this scheme is

$$\frac{31}{15,166} \times 1000 = 2.04 \text{ actives/thousand tests;}$$

note that this is higher than the best yield that can be obtained with a one-stage test.

By trying various cut-off values for the two stages of the test, an optimum two-stage screening test can be devised. The optimum yield turns out to be 2.15 per thousand tests, achieved with a cut-off value of 0.4 gram reduction at the first stage and 0.5 gram reduction at the second stage. The increase in yield from 1.71 per thousand with the best one-stage test to 2.15 per thousand using the best two-stage test represents an improvement of 26% in the efficiency of the screening test. This is an important advance because it means that useful cures for diseases can be found more quickly.

Further improvements can be obtained by considering more than two stages of testing. We will not go into the details here, but an optimum three-stage test can be devised to yield 2.33 per thousand. More than three stages can be considered, but it is questionable whether the increases in efficiency that theoretically could be obtained are worth the extra complication. (Complicated bookkeeping in a large laboratory has its own high costs in both potential errors and manpower.)

EFFECT OF ASSUMPTION ABOUT THE COMPOSITION
OF THE COMPOUNDS BEING SCREENED

In the forgoing we showed how the statistician can optimize the choice of the cut-off values that determine whether a given compound should be accepted, rejected, or tested again. The optimum, however, was based on a hypothetical mixture of active and inactive compounds being screened. We assumed that 10,000 compounds contained 9960 inactives and 40 actives; moreover, we assumed the active compounds had a degree of activity capable of reducing the animal tumor weights by 0.7 gram.

Of course, the statistician must repeat the whole process for other assumptions about the composition of the compounds being screened to determine the effects on the optimum cut-off values. It turns out that the actual *number* of actives assumed to be in the mix does not affect the screening procedure. In other words, if the 10,000 compounds contained 100 actives, or 4 actives, or only 1, instead of 40, the same screening criteria would produce an optimum yield. (Of course, the actual *magnitude* of the optimum yield would go up or down with the number of actives assumed to be present.) On the other hand, the degree of activity assumed for the active compounds does affect the procedure. All the statistician can do here is to work out the best procedure for various degrees of activity and let the screening investigator decide on the basis of his knowledge what level of activity of interest to him is most likely to occur. Then the statistician can tell him what screening criteria will produce the optimum results for that level of activity. It is also possible to consider a mixture of two, three, or even several different levels of activity among the active compounds.

OTHER VARIABLES IN THE SCREENING PROCEDURE

In the preceding discussion, it was assumed that the test procedure itself was fixed and the only variables were the cut-off criteria. Other factors, however, also can be altered, for example, the number of animals used in each test. Is the choice of three animals for each test compound and six animals for control really best? A reduction in either of these numbers would enable the investigator to test more compounds in each experiment. On the debit side, however, this would increase the standard error of the test statistic, which would have the undesirable effect of increasing the chances for a compound to be misclassified.

The statistician can study the effect of changes in these variables on the theoretical yield, by carrying out calculations similar to those we have already described. The object is to achieve, with the test facilities that are available, the greatest expected yield in terms of active compounds found. Of course, it is impossible to guarantee what results will actually be obtained. No matter how efficient a screening procedure is, in fact, there may be *no* active compounds presented for screening. In the end, we must depend upon the ingenuity of the chemists to produce compounds that have the desired activity.

APPLICATIONS

Statistical studies of the sort described here have improved the efficiency of many of the routine screening procedures in our laboratories. The anticancer screening program is discussed in detail by Vogel and Haynes (1962), who state that an increase in the screening rate from 450 compounds per year to 1300 per year has been achieved. I could provide a happy ending to this tale if I could tell you about the discovery of a new cure for cancer as a result, but although some possible leads have been discovered, it seems that we are still a very long way from this goal. Past successes in other areas, such as the remarkable discovery of the antibiotic Aureomycin after the screening of over 4000 soil samples, are convincing proof, however, that drug screening plays a necessary and important role in the never-ending search for new and better drugs. A recent book by Arnow (1970) contains an interesting account of this and other aspects of drug research.

The statistical aspects of drug screening are similar to screening and selection problems in other fields. For example, in the development of new and improved strains of an agricultural crop, such as wheat, each potential new strain must be planted and grown, and measurements of its yield and other indicators of performance must be taken. On the basis of the results, some strains are chosen to be planted again on a wider scale, and eventually, after many repetitions of the cycle, a new variety may emerge to replace current standard varieties. The plant breeder, like the drug screener, must determine how best to use his facilities for testing various candidates in order to maximize the chances of success. My 1968 article gives a more general discussion of the problems of screening and selection.

PROBLEMS

1. Why are there more control animals than treated animals in the anticancer screening example?

2. What are the two actions possible in the drug screening experiment viewed as a decision problem?

3. Explain what this article means by "false negative" and "false positive."

4. If a cut-off value of .8 gm. is used in the anticancer screening example then from tables of the normal distribution, one finds that .0107 of the inactive compounds are expected to be false positives and .3859 of the active compounds to pass. Using the above information calculate the yield per thousand tests in a one-stage testing procedure. Compare with the value obtained from Figure 1.

5. Suppose that two different laboratories were to screen the same 10,000 compounds using the same screening procedure. Would they necessarily obtain the same number of active compounds? Would they necessarily obtain the same active compounds if they obtained the same number of compounds? Why or why not?

6. In a real experiment the number of active compounds among the total number of compounds tested is unknown. How would a statistician go about deciding the optimal cut-off point?

7. Define a "two-stage test" and a "one-stage test." State the advantages and disadvantages of each type of test.

8. What would be the yield of active compounds per 1000 tests if the investigator requires an average reduction of .80 grams in tumor weight before he will permit the compound to pass the screening test? Can you find a different reduction that gives the same yield? Which of the two would you prefer? (Use Figure 1.)

REFERENCES

L. E. Arnow. 1970. *Health in a Bottle; Searching for the Drugs that Help.* Philadelphia: Lippincott.

C. W. Dunnett. 1968. "Screening and Selection." D. L. Sills, ed., *International Encyclopedia of the Social Sciences* vol. 14. New York: Macmillan and Free Press.

A. W. Vogel and J. D. Haynes. 1962. "Experiences with Sequential Screening for Anticancer Agents." *Cancer Chemotherapy Reports* 22:23–30.

SETTING DOSAGE LEVELS

W. J. Dixon   *University of California, Los Angeles*

MANY OF the problems brought to the statistician for solution require pinpointing a level at which an expected response does or does not occur in order to test the strength and efficacy of drugs, pesticides, hormones, explosives, analgesics, stimulants, fuels, and a wide variety of other materials important to man and his environment. A statistician feels particularly successful if he develops a method for solving such problems that has a wide range of applications. One such method is called the *up-and-down*, or *staircase*, *method*. This is the way it works.

HOW STRONG SHOULD PUNCH BE?

Mr. and Mrs. Smith, aged 22 and 20, respectively, are planning their first cocktail party, inviting all of the people in Mr. Smith's office. They can't afford an elaborate party, but they do want to make a good impression, so they have decided to serve a punch made of gin and cranberry juice. They have little experience with alcohol, and so they don't know the proportion of gin to cranberry juice; they decide to try the punch first on some reliable and courageous friends. Four couples agree to help out.

Mr. Smith thinks all the guests will want at least 8 ounces of punch. Mrs. Smith wants her guests to be happy, but doesn't want anyone to become ill at her party. This is where the courageous friends (guinea pigs) come in. Mr. Smith mixes the first drink with 7 ounces of gin and 1 ounce of cranberry juice, and Mr. Big tosses it down. In about half an hour he is green and staggering. Obviously the punch contained too little cranberry juice. The next drink, 5 ounces of gin to 3 ounces of cranberry juice, goes to Mr. Jones. He is still on his feet half an hour later, but he is behaving strangely. Mrs. Big decides she will be safer if she volunteers for the 3-to-5 mixture. This turns out to be a mistake because she soon feels very warm. Mr. Smith refuses to let Mrs. Smith have a drink, and he believes that he himself must stay out of the testing to keep the record straight. The 1-to-7 drink, therefore, goes to Mr. Average who compliments his hostess on the delicious flavor, but says he really doesn't feel a thing. Mr. Small thinks he should have something stronger and asks for 5 ounces of gin to 3 ounces of cranberry juice. Mr. Small suddenly develops a slight speech defect, so the next drink is again made 3 to 5, and Mrs. Average drinks it. She compliments the hostess, saying the drink is excellent. She is happy and relaxed. Mrs. Small asks for a drink with the same proportions, and the Smiths thank their friends for having solved the problem. They will serve punch made from 3 parts gin and 5 parts cranberry juice.

The basic notion illustrated in this simple example is that of moving some important control level up or down each time, depending on the prior level and the outcome of the prior trial. In the example, when the response of the experimenters suggested too much alcohol, the dose was reduced for the next trial. When the response suggested too little alcohol, the dose was increased. In the end, some judgment was made as to final level. What sort of problems can be easily solved by this method?

APPRAISING STRENGTHS OF OTHER MATERIALS

Some drugs are grown naturally, and they must be tested to determine how strong they are. Penicillin is an example. Most hormones must be similarly assayed for their strength.

(1) Pesticides should be strong enough to destroy insects, but not poison the family cat.

(2) Pain killers should relieve headaches, but not induce palpitations.

(3) Jet propulsion engines must have explosive-type "motors" capable of propelling an airplane, but the explosions must not shake the vehicle to pieces.

How can we design a measurement process that gives us sufficient assurance that we are arriving at a correct dose of a drug, one that will do what is desired? How can we do this in an efficient manner?

In order to test the strength of any given material, we must set up some standard of potency, or performance. With poisons, it is customary to use test animals of similar size and heredity and to inject each with a known concentration of the drug to be studied. A widely used, though arbitrary, choice is that the standard, or threshold, will be that concentration which will kill, on the average, half the animals tested. Obviously, other levels may be used, but if we understand how to handle one level, we are well along toward handling others. Let us examine in detail such a problem and one method for dealing with it.

Curare, a poison that paralyzes the heart and motor-nerve endings in striated muscle, was used for thousands of years by primitive people to destroy their enemies. Doctors now use this lethal substance for the benefit of mankind in certain surgical and medical procedures. Another poison with similar properties is the venom of the scorpion fish. To use this venom properly, we must have a precise measure of the strength of any batch we plan to use. Scorpion fish venom, in fact, has been assayed by the up-and-down method. This is how we went about setting up the trials.

APPRAISING SCORPION FISH VENOM USING FIXED SAMPLE SIZES

An amount of the venom (the stimulus) will be injected into test animals (in this case, mice), and we will record whether a response (death) occurs in a given time period, say, 30 minutes. If we have chosen a dose that is too large, all (or almost all) the animals will die. If the dose is too small, possibly none of the animals will die. How can we find the dose that corresponds to the amount of poison that, if increased, causes more than half of the animals, on the average, to die and, if decreased, causes fewer than half to die? Even if he lives, the same animal cannot be used in a second test because he now may be less able or more able to stand the venom. The task would still not be very difficult if all animals behaved in the same way. Even in carefully selected animals, however, the amount of a drug required to bring about a response differs greatly from animal to animal. The amount of a drug just sufficient to cause a response in a particular animal is called his *threshold level*. We want to estimate some sort of average threshold for the population of animals.

Our measurement for a particular animal given a certain dose will be a response (in this case, death), which we shall record as "X," or a non-response, which will be recorded as "O."

How do we decide on the dose (stimulus) to give each animal? If an individual's threshold is unaffected by the test, we could merely increase a single animal's stimulus gradually until the threshold is reached. Unfortunately, this is not the case; estimating a mean threshold requires careful experimental design.

How can we attack this problem in an efficient manner? One natural design gives an experiment in which the same number of animals are tested at each of a variety of dose levels. To introduce some order into this situation we choose four different dose levels (say, 1, 2, 4, 8 mg) of venom. (It turns out that dosages of a wide variety of chemicals show approximately uniform increments in effectiveness if each dose is a certain percent larger than the preceding dose, i.e., if doses are chosen so that each is a multiple of the preceding one. In our example the doses increase by factors of 2.) We test five mice at each level. If we are so unlucky as to have picked a set of dosages that are all clearly below the threshold of all animals, we learn nothing about the location of the threshold for these animals except that it is greater than the largest dose given, 8 mg. We don't know how much greater. Or, if all animals at all levels respond, we discover only that the threshold is below the smallest dose, 1 mg. Even if the set of dosage levels chosen covers the general threshold level, we may test at a number of doses to which all or none of the animals respond.

Table 1 and Figure 1 show a set of outcomes from such a procedure. Five animals were tested at each of four dosage levels, with the outcomes as shown. Twenty animals were required.

SEQUENTIAL TESTING: THE UP-AND-DOWN METHOD

It is only common sense to seek a testing strategy that leads the experimenter quickly to the proper levels for the tests. We wish to destroy as few mice

TABLE 1. Outcomes for Five Animals at Each Dosage Level in the Order in Which They Are Treated (X Means Death; O Means Survival)

DOSAGE LEVELS	OUTCOMES				
8 mg	X	X	X	X	X
4 mg	X	X	O	X	X
2 mg	O	O	X	X	O
1 mg	O	O	O	O	O

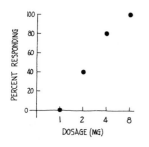

FIGURE 1

Data of Table 1 plotted to show relation between response rate (percent dying) and dosage level

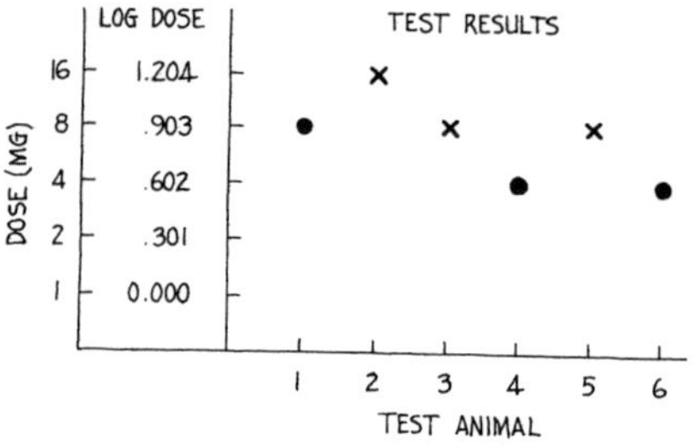

FIGURE 2

Results of a sequence of six tests using the up-and-down method

as possible, and since our supply of scorpion fish venom is limited we must try to get a good estimate from a small number of tests.

One design that has good properties is conducted one test at a time and consists of merely increasing the dose for the next animal if the last one tested did not respond to the dose administered and of decreasing the dose if the last animal did respond.

Let us examine the up-and-down version of our scorpion fish venom experiment in detail, limiting ourselves, for brevity, to animals of the same weight and a fixed venom concentration. For a particular concentration and animal weight, we proceed to test several animals at different dosages following our planned strategy.

We prepare to administer doses of sizes 1, 2, 4, 8, and 16 mg, and we begin by testing the first animal at any one of these levels. We choose, say, 8 mg for the first dose; the animal survives, and we record an O toward the left in Figure 2. The next animal is given the next higher dose of 16 mg. It does not survive; nor does the third animal (tested at 8 mg). In all, six animals are tested following this rule. From the data of Figure 2, because none of the 4 mg and two-thirds of the 8 mg doses produced a response, we might guess that the average threshold is somewhere between 4 mg and 8 mg. We would be uncertain, and we would not have a systematic way of making an estimate. A fairly precise estimate can be obtained, however, by averaging the levels at which the tests were done (in logarithmic units). Greater precision can be obtained by using a special table worked out mathematically to obtain the best estimate possible. By using this strategy, we can obtain about as much information from six animals as we could obtain by testing 20 animals in the design of Table 1 and Figure 1. The sequential character of the new approach tends to concentrate the doses where they are needed to get a good estimate.

The technique is also one of extreme simplicity to use. The sequence of trials in Figure 2 is completely described by indicating the sequence of O's and X's as they are observed and stating the dosage interval and the dosage administered on the last trial. For the example in Figure 2, the series is OXXOXO; the spacing of doses is log 2 = .301, and the final test dose was 4, which, in logarithmic units, is log 4 = .602. The average threshold dosage is estimated to be

.602 + k(.301)

where a value for k may be obtained from a table in Dixon and Massey (1969) for the configuration OXXOXO. The value of k is .831, so the estimate is

.602 + .831(.301) = .852

Because .852 is the logarithm of 7.11 we obtain 7.11 mg for our estimate of the average threshold dose.

SUMMARY

We have illustrated the solution to a measurement or assay problem in which a special design of the sequence to be followed in the collection of the data allows us to get the resulting estimate with high efficiency. Other results of the statistical theory forming the basis of this procedure also show that this is about as good as we can do when we are able to observe only whether we exceed or fall short of the desired level. The theory shows that it requires only twice as many observations to obtain a threshold estimate of the same accuracy as would be obtained if we could measure precisely the exact dosage corresponding to each animal's own threshold. This, of course, is an impossibility for poisons, although we can imagine coming close to it in some problem where repeated measurements of the same animal are possible.

PROBLEMS

1. Give three examples (not mentioned in the article) where appraising strength of material is needed.

2. In the 7th sentence of the example of appraising scorpion fish venom, explain the meaning of the words "on the average."

3. What is the threshold level of an animal?

4. How does this article define the average threshold level for the scorpion fish venom experiment? Would you use the same criterion to

define the threshold level when deciding the dosage level of curare to be used in medical procedures with human beings?

5. What are the advantages of a sequential testing procedure?

6. In Figure 2 we notice that the 1st animal survived when administered a dose of 8 mg. and the 2nd animal died when administered 16 mg. Why don't we conclude at this point that the threshold is between 8 and 16 mg.?

7. In the sequential up-and-down method of finding a dosage level, when does one stop taking samples?

8. Suppose we test 7 animals at the 5 dosage levels of venom described in the text, and obtain the series of responses XXOXOOX, with the final test dose being 8. Draw the results of the sequence of 7 tests similar to Figure 2. Estimate the average threshold, using the value $k = -1.237$ obtained from Dixon and Massey (1969).

9. Draw a graph similar to Figure 2 for the punch example. (Hint: use "parts of gin" as dosage level.) Can you estimate the average desired strength of punch by the up-and-down method described later in the text? Why or why not?

REFERENCES

J. J. Blum, R. Crease, D. J. Jenden, and N. W. Scholes. 1957. "The Mechanism of Action of Ryanodine on Skeletal Muscle." *Journal of Pharmacology and Experimental Therapeutics* 111:477–486.

Lincoln P. Brower, William N. Ryerson, Lorna L. Coppinger, and Susan C. Glazier. 1968. "Ecological Chemistry and the Palatability Spectrum." *Science* 161:1349–1351.

W. J. Dixon and F. J. Massey, Jr. 1969. *Introduction to Statistical Analysis,* Third Edition. New York: McGraw-Hill.

HOW FREQUENTLY DO INNOVATIONS SUCCEED IN SURGERY AND ANESTHESIA?

John P. Gilbert   *Harvard University*
Bucknam McPeek   *Massachusetts General Hospital, Boston*
Frederick Mosteller   *Harvard University*

WHEN THERAPIES are compared for effectiveness, what happens? How often does an innovation appear to be superior to its competitors? When innovations are successful, for example, the Salk vaccine or the development of successful organ transplantation, society gains a major victory. This paper studies the effectiveness of new surgical and anesthetic therapies in their clinical setting.

We reviewed a sample of 107 published papers appraising surgical and anesthetic treatments. Of these therapies sufficiently promising to be tested in human patients, we ask what proportion have proved to be substantial improvements over existing ones? What proportion have been moderately successful? And what proportion have been found to be less effective than had been hoped and expected?

Using these papers, we assess the percentage improvement a new innovation is apt to make, as well as the chance that it will turn out to have been an improvement at all. Thus our aim is to describe the crop of newly tested therapies for effectiveness compared with that of the treatments they are designed to replace.

Except for a major breakthrough like the introduction of antibiotics, we have little reason to suppose that the development of new therapeutic ideas will change drastically. Thus, we assume that a similar distribution of successes and failures will occur in the near future. The results presented here should give realistic expectations at least for the short term.

We drew a sample of papers evaluating different treatments actually given to patients. To get this sample, we turned to the National Library of Medicine's MEDical Literature Analysis and Retrieval System (MEDLARS). Computer-produced bibliographies can be retrieved from this data base which, since January 1964, has provided an exhaustive coverage of the world's medical literature. By searching the system for prospective studies (see the essay by Brown) of specified surgical operations or anesthetic drugs, we were able to gather papers whose authors used human patients to evaluate surgical and anesthetic treatments. The papers appeared between 1964 and 1972.

We considered only papers in English because of our own language disabilities, and papers with ten or more patients in a group because we wanted to study large investigations rather than case studies. Any other bias in the sample selections, then, arose from peculiarities of the MEDLARS indexing system and contents at the time of the search rather than from our prejudices.

The papers included many kinds of studies. To give an idea of the variety, some dealt with ulcers, appendectomy, cirrhosis, cancers, bone operations, colon operations, major vascular operations, stab wounds, antibiotics, clot prevention, drainage, and the impact of anesthetic drugs and techniques.

Our sampled papers reported on three basic types of studies—*randomized controlled trials, non-randomized controlled trials,* and *series.* We use the term randomized controlled trials when the investigator compared two or more treatment groups and assigned patients to the groups by a formal randomization process (such as drawing random numbers to decide which treatment is assigned to each patient). The non-randomized controlled trials did not have such a formal randomization process and varied from comparing groups treated concurrently in the same institution to comparing patients treated previously by one method with patients treated currently with another. The papers reporting on series described sets of patients treated in some specified manner but with no comparison except possibly with other reports in the literature dealing with similar patients. In the rest of this paper we are concerned with only the papers dealing with randomized controlled trials.

If our MEDLARS approach were perfect and produced all the papers, one might think that we have a census rather than a sample of papers. To adopt this attitude would be to misunderstand our purpose. We think of a process producing these research studies through time, and we think of our sample—even if it were a census—as a sample in time from this continuing process. Thus our inference would be to the general process, even if we did have all appropriate papers from a time period.

In appraising the results of comparative investigations, we take several simplifying actions.

First, we classify each therapy as either an *innovation* or as a *standard*. Some diseases have a widely recognized standard therapy against which all others are measured. A good example of this has been (the standard) radical mastectomy for cancer of the breast. In such instances the standard is easy to recognize; all others can be considered as competing innovations regardless of how recent their introduction. We have used the letter "I" to denote the treatment we regarded as an innovation and the letter "S" for the standard.

Ideally one wishes an analysis to produce the maximum amount of information contained in a body of data. It is often impractical to achieve this in practice. Thus in trying to evaluate the difference in performance between standard programs and innovations in our study we were unable to assign a highly accurate and precise value to the observed differences. In many studies we are content with knowing how many differences were positive and how many negative. In our data often the two programs were essentially equal in performance and so it was useful to acknowledge this in the scale. In addition sometimes one program was not only better but was clearly much better, and it was not hard to make a distinction between these two. The five point scale that we have used is a happy solution to this problem because it is relatively simple and easy to apply and it retains most of the relevant information that we need. The five point scale is widely used in both social science and medicine because it allows us to capture much of the information we want in a practical manner.

Second, we speak below of a pair of competitive therapies as having three possible relations: About equal (S=I), the first named preferred to the second named (S>I or I>S), and the first named *highly* preferred to the second named (S>>I or I>>S). We have tried to report on this scale what we think the original investigators would have reported. Usually their words make this clear.

Third, we have divided therapies into two classes: *Primary* therapies intended to cure or ameliorate the patient's primary disease, and *secondary* therapies intended to prevent or treat such complications as infection or thromboembolic disease or to offer improvements in anesthesia or postoperative care. The basic 107 studies included 36 randomized clinical trials. Of these 36 papers, 21 deal with *primary* therapies and 15 deal with *secondary* therapies. For technical reasons* several studies had to be set aside, also

*One study had too many comparisons; another had too small a sample size for its complicated design.

some studies had more than one comparison. In Table I, we deal with comparisons, rather than studies. By coincidence the number of papers equals the number of comparisons in the analysis.

Referring to Table 1 for randomized trials, we see that in five of the 36 comparisons, or about 14%, an innovation was highly preferred to a standard. In 16 comparisons, including the previous five, about 44%, the new therapy was regarded as successful, sometimes because it was no worse than a standard and thus became available as an alternative.

TABLE 1.   Summary for Innovations in Randomized Clinical Trials

		PRIMARY	SECONDARY	TOTAL
I >> S:	Innovation highly preferred	1	4	5
I > S:	Innovation preferred	5	2	7
I = S:	About equal, innovation a success	2	2	4
I = S:	About equal, innovation a disappointment	7	3	10
S > I:	Standard preferred	3	3	6
S >> I:	Standard highly preferred	1	3	4
	Comparisons	19	17	36

In 10, or 28%, the equality of an innovation with a standard could be regarded as a disappointment because, although the innovation was more trouble or more costly or more risky, it did not perform better. In 20 comparisons, about 56%, a standard was preferred (counting innovative disappointments) to an innovation.

Overall, Table 1 shows that innovations highly preferred to standard treatments are hard but not impossible to find, and that almost half of the innovations provided some positive gain. It is worth reflecting on what our attitude might be toward extreme findings in either direction. Suppose that nearly all studies, or even the lion's share, found the innovation highly preferred; one would have to conclude that standard therapies were fairly easy to improve on and indeed that the kind of medicine being appraised was in its infancy or else that a sudden breakthrough had been made on all fronts. This is unlikely with as many different diseases and therapies as occur in the sample. At another extreme, if no substantial gains occurred, the suggestion is that the field has topped out, at least during the period of the study, awaiting some new insights.

Figure 1 summarizes 11 primary studies in which survival was an appropriate measure of outcome and plots the percentage of survivors, often after many years, for the standard therapy against that for the innovation. Two papers had two comparisons of a standard against an innovation, making 13 comparisons in all. The seven points below the 45° diagonal line show the

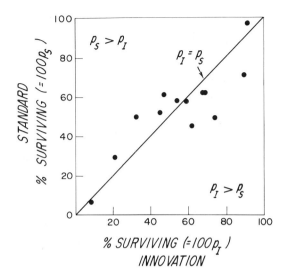

FIGURE 1

Primary Therapies—Survival Percentages

Points falling below the 45° line indicate higher survival rates for the innovation than for the standard. Primary therapies are intended to cure the patient's disease.

innovation performing better; the six points above show the standard performing better. The greatest observed gain (shown by the point farthest below the line with coordinates (74%, 49%)) comes from a study of therapeutic portacaval shunt. Curiously, the point farthest above the line (showing the greatest apparent loss) corresponds to a study of the same operation performed prophylactically, in advance of urgent need (32%, 50%). The overall impression given by the figure is one of points rather closely hugging the diagonal line. The degree of scatter from the line depends in part upon

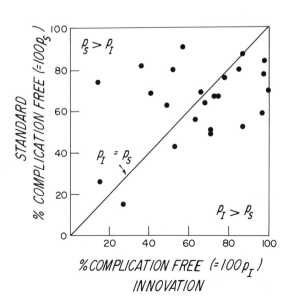

FIGURE 2

Secondary Therapies—Percentage Free of Complication

Percentages avoiding specific postoperative complications. Secondary therapies are intended to reduce the frequency of postoperative complications. Points below the 45° line indicate fewer complications of a specific sort accompanying the innovation than the standard. The innovations have 15 below, 8 above, and 1 on the line.

the size of samples (number of patients used in the studies), and we explore this idea further later.

Figure 2 shows 24 comparisons based on 11 secondary studies (five had 1 comparison, four had 2, one had 3, one had 8). The 15 points below the line indicate the innovation as an improvement over the standard treatment; the 8 above indicate the reverse, and the 1 on the line gives a tie.

The overall scatter about the 45° line in Figure 2 is large, encouraging us to believe that larger percentage differences have been found here than in the studies of Figure 1. By and large, the changes in rate of complications are larger than the changes in survival rate. We make this more quantitative below.

In the work reported so far, some innovations performed better than a standard, others worse. We next regard these outcomes as a sample from the population of all those surgical innovations developed by our medical system and tested by randomized clinical trials. Every study has its uncertainties associated with sampling variability and other sources of unreliability. We want to allow for sampling variability in our description of the gains and losses. The general idea is that if we focus on a particular sort of performance, we may be able to gather strength from several studies even though they deal with disparate operations. For example, among the primary studies we focus on those where the main hope from the operation is the extension of life. Then we might ask about the distribution (variety) of improvements actually achieved by this type of innovation.

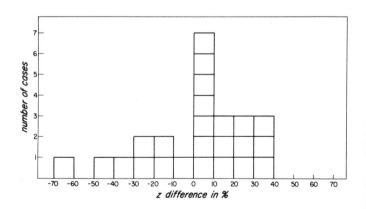

FIGURE 3

Histogram of 24 observed differences in percentages avoiding a complication for several operations. This illustrates one way of representing a frequency distribution function.

We use the idea of distributions, and so we want to illustrate what these are. In our study of post-operative complications we had 24 differences of the form: the percentage of patients who did not develop complication under the innovation MINUS the percentage who did not develop the complication under the standard. We use these 24 differences for illustration. We can ask of the data how many differences were in intervals of length 10 such as between 0 and 9%, 10% and 19%, or between –20% and –29%. This information is presented graphically in Figure 3.

Seven of the differences fell in the interval 0 to 9%, three in each of the intervals 10 to 19%, 20 to 29%, and 30 to 39%, while one was so low that it fell in the interval –60% to –70%. Thus Figure 3 gives us an idea of how these differences are distributed with regard to their values. If we had had many values we could have made much smaller intervals and we could think of it being very like an idealized smooth curve that might look like:

This curve is called the density function of the distribution.

Often it is more relevant or convenient to ask how many data points were larger than a particular value on the scale, rather than asking how many data points were within a particular interval as we did above. If we represent this way of looking at the data graphically we obtain Figure 4.

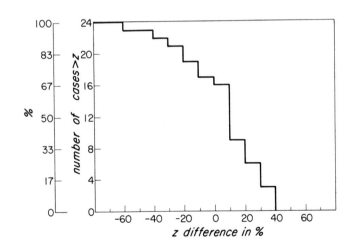

FIGURE 4

Frequency distribution cumulated from the right to show the frequency of getting a given percent difference at least as large as the one on the horizontal axis.

Figure 4 was obtained by adding up the number of squares, i.e. observations, to the right of each point on the horizontal scale of Figure 1. For this reason it is called a cumulative distribution function. If again we think of having many more points and using much smaller intervals we might find the figure to approach a continuous curve that is the cumulative distribution function that corresponds to the continuous density pictured above. This curve looks like:

When the vertical axis is a percent or a proportion, we can read off the estimated probability of a difference at least as large as the one on the horizontal axis.

If every study were based on an enormous sample of patients, so that sampling errors would be very small, the reports of gains and losses would give us the distribution of differences in true performance between innovations and standards in our sample of papers. In turn that sample distribution would estimate the distribution of gains in the population—the process generating these studies and comparisons. But studies are of necessity limited in size, and, in reports of small studies, differences vary more due to sampling error than in large ones. We need to have a way to pool the results of such studies, large and small, that will give an idea of the distribution of *true* gains and losses in the trials.

One such method is to allow for the sampling variability associated with specific randomized trials and come up with a pooled figure. The observed difference may be thought of as having two additive components—the true difference plus the sampling error. A special statistical technique called analysis of variance produces estimates of the average and standard deviation of the sample of true differences. (The standard deviation measures how spread out the distribution is.) We can estimate how often various sizes of gains can be expected to occur by making an assumption about the true differences, namely that the differences approximately follow a normal distribution. (The word normal here applies to a particular shape of distribution—it is not being used in the sense of normal versus abnormal. Heights of adults and scores on achievement tests are examples of distributions that are approximately normal in shape. A normal distribution looks something like

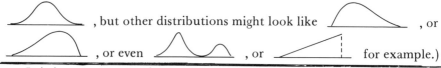

It is important to understand that this method develops summary statistics for *true* gains and losses *across* studies. The statistics reported are (a) the

estimated average true gain, averaged across comparisons, and (b) the estimated standard deviation of the true gain, averaged across comparisons. If, for example, the average gain were 0% and the standard deviation of the gain 6%, our assumption of a normal distribution allows us to calculate that gains of 10% or more could occur in about one-twentieth of the opportunities. It would still be true that the gain would be positive for about half the innovations and negative (that is, a loss) in half, in agreement with Table 1. It then becomes the goal of clinical research to identify the favorable and unfavorable innovations so that we may use the former and avoid the latter.

The statistics in Table 2 summarize the results.

TABLE 2   Analysis of Variance Estimates of Average
and Standard Deviation of True Gains

	ESTIMATED AVERAGE GAIN	ESTIMATED STANDARD DEVIATION OF GAINS
Primary Therapies	1.5%	8%
Secondary Therapies	0.4%	21%

The average gain for the primary therapies is not far from zero, a result that agrees with our more qualitative analysis of Table 1. A zero average gain is consistent with some innovations having substantial improvements balanced by others having substantial losses or with other mixes such as many small gains and a few large losses. The size of the estimated standard deviation of effects of innovations lends added support to such interpretations. (A zero standard deviation would imply that all innovations give essentially the same amount of improvement.) And we know that some of these innovations do produce substantial improvements even when sample size is taken into account.

These figures also yield a rough guess about the proportion of comparisons having true differences favoring the innovation as great as, say, 10%. For the primary therapies, the probability that a new therapy has a positive gain of at least 10%, if the sample represents the future well, is about 0.13, or about 13 chances in 100

For the secondary therapies, a gain of at least 10% (a 10% reduction in a specific complication) has a probability of 0.32.

The above procedure is rough and ready and leans hard upon an assumption of a normal distribution in its calculation, but the real distribution may not be normal. A new approach called "empirical Bayes" (Efron and Morris 1973) offers an alternative.

If each comparison were based on an infinitely large experiment, we would know the true gain exactly for that comparison. Then to estimate the proportion of gains of more than 10% we would count the number of comparisons with gains larger than 10% and divide by the total number of comparisons. And so if we had 25 comparisons and 5 had gains greater than 10%, we would estimate the probability of a gain of more than 10% as 5/25 or 0.20. This approach does not lean on any assumption about the shape of the distribution of true gains. But we can't use it because we do not have infinitely large experiments.

The new method takes note of the uncertainty associated with each observed gain, primarily using the sample sizes. Instead of regarding an observed gain as greater than 10%, or not greater, it estimates the probability that the true gain is greater than 10%. And so each comparison yields a probability of being greater than 10%, and we average these probabilities from all the comparisons to get our estimate of the overall probability of a gain of more than 10%.

If the observed gain is very large, say 30%, then its probability is nearly 1 (0.99, for example, or 99 chances out of 100) of having a true gain of more than 10% because the variability of the experimental observation is very much less than the 20% difference between 10% and 30%. This 0.99 corresponds to the 1 we would have counted toward the numerator (our 5 of the 5/25) had we known the true gain exactly. If the observed gain is negative, the probability that the true value exceeds 10% will be small, perhaps 0.01. This 0.01 is like the 0 we would have counted for this comparison had we known it to be exactly the true gain. When the observed gain is exactly 10%, the probability is 0.5 that the true gain is larger than 10% and 0.5 that it is smaller.

The same technique applies to finding the chance of gains greater than 5% or 0% or –7%, and so on. The resulting set of probabilities are conveniently graphed and thereby summarized by a cumulative distribution as shown in Figure 3.

Figure 5 shows the estimated cumulative distributions of the true gains in percentages for the primary and for the secondary therapies. By picking a gain in percentage, z, and reading the corresponding vertical axis on the appropriate curve, one can estimate the probability of a new therapy producing a gain as large as or larger than the chosen value of z.

For examples we have:

(a) For primary therapies the chances (i) of a 10% gain or more in survival are about 4 in 100 (we regard this as a better estimate than that given earlier [13 in 100] because we prefer the method rather than because we prefer the answer), (ii) of a 0% gain or more are about 48 in 100, (iii) of a loss of no more than 10% are about 98 in 100 which means that the chances of a loss in excess of 10% are about 2 in 100.

(b) For secondary therapies, the chances (i) of a gain of 10% or more in a specially chosen complication are estimated as 38 in 100 which is close to the earlier 32 in 100, (ii) of a 0% gain or more 57 in 100, (iii) of a loss of no more than 10% as 72 in 100, which means that a loss of 10% or more has chances of about 28 in 100.

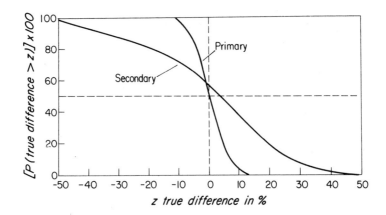

FIGURE 5

The Probability of a New Treatment Producing a Gain as Large as or Larger Than a Chosen Value, z

To find the probability of a difference in percentages greater than a given number z, say z = 10%, erect a perpendicular from 10% on the horizontal axis to the appropriate curve, and read its ordinate off the vertical axis, about 0.04 for the primary and about 0.38 for the secondary.

Is there some reason that secondary therapies are more likely to succeed than primary therapies? Is there something special about a treatment aimed at the disease process itself? We think the difference arises in large measure because in our quantitative analysis we chose to analyze primary therapies in which survival was an appropriate measure of outcome, while for secondary therapies the measure was avoidance of a specific complication. In a way the incidence of a specified complication is a much more discrete measure. One can envision a treatment having a large effect on a specific complication, whereas the difference between life and death may be the sum of the effects of a variety of factors—the primary treatment, the primary disease process, secondary treatments, and a variety of other disease processes and factors like old age and inter-current disease. Over the last generation the expected length of life has increased only slightly, but great changes have occurred in the variety and extent of postoperative complications as a result of changes in therapy such as the introduction of antibiotics, and of newer anesthetic agents, and techniques.

The sample of published papers is objectively chosen, and we think rather a good one for reflecting the sorts of differences analyzed here. What is less clear is how good a sample it is of therapeutic surgical research on patients generally during this period. First it is likely, and those we have talked with agree, that published papers are reports on better work on the average than that in unpublished research. Second, research that turns out well, our discussants agree, is more likely to be published. This reasoning suggests that the mass of unpublished research, insofar as it might produce measures comparable to those described in this paper, would have a lower average performance for innovations compared with standards than those in our sample. We suppose then that the innovations assessed by randomized clinical trials and reported in the surgical literature, and here, are biased upwards—that is, they present a more promising picture for innovation than if all innovations were subjected to randomized clinical trials. No doubt some innovations are so unsatisfactory that they are quickly abandoned, along with whatever trials were initiated on them. These conjectures suggest that if one were to consider adjusting the distributions shown in Figure 3 to report on all surgical innovations versus standards, the mean of the distribution would be lower and the standard deviation would probably be larger to allow for more frequent large negative differences. We have no grounds but speculation for the amount of such changes.

In a recent review of randomized trials used in evaluating social programs (Gilbert, Light, and Mosteller, 1976), the authors concluded that many new programs do not work and the effects of those that do are usually small. In contrast to these findings in the area of social innovations, this review provides strong evidence for a more optimistic view of the rate of progress in surgery and anesthesia. Almost half of the innovations reported in this series of controlled trials were at least as good as the standard, and a fair number were substantially better. Thus the analyses suggest that four out of ten innovations in secondary therapy produce a reduction in complication rates of 10% or more while two or three out of ten innovations in primary therapy produce a 5% or greater increase in survival. These estimates are for the distribution of the underlying true effects of the innovations. In a sense these results describe the clinical judgment that chooses those innovations as promising enough to test. If innovations were successful in a high proportion of trials, it would suggest that new therapies were being delayed until we were absolutely sure of their success, while if almost none were successful it would suggest a scarcity of new ideas in the field. Thus these distributions also describe research productivity and its effects on the development of better clinical care.

Another view is that this research process tries to reject all the innovations that produce losses and to keep the ones with gains. If this were done, then the median gain (the middle gain) retained for the primary therapies would be about 4%, and that for the secondary therapies would be about 15%. Of course, this would be an idealized state, for we cannot hope to weed out *all* the losses and detect *all* the gains. But it gives an estimated upper limit to what could be accomplished.

We further emphasize that to say that a proportion of innovations are substantial improvements does not serve to identify which they are.

Our findings give us an idea of the sorts of gains that can be made from selecting the better of pairs of therapies that are tested by randomized trials (we do not discuss the others here). The left sides of the curves warn us also that innovations may lose rather than gain, and so evaluation is needed. For example, in secondary therapies losses of as much as 20% could occur about one-fifth of the time. These curves emphasize the size and frequency of the losses as well as the gains. Thus as physicians well know, one cannot assume in advance that a new treatment is an improvement over an old, even when it looks promising enough to warrant a clinical trial. Our distributions show that some innovations provide important gains for the clinical care of patients, such as reducing a death rate by 5%.

To give an idea of the risks represented by a five percent change in death rate, we note that among all the people who died in recent years, 5% were in the ten year age range 40–49. Thus we can think of this rate as corresponding to the natural losses over a ten year period at middle age. Another way to think of 5% is that it is about four times the average surgical death rate from all operations over the country as a whole. Thus its importance is not small.

Reducing a death rate from 35% to 30% may be an important improvement in patient care, but this does not mean that it will be easily identified in the everyday setting of clinical practice. Indeed, statistical theory shows that a well-run randomized controlled trial would need 1,105 patients in each group to be 80% confident of detecting such a difference. Without a large formal trial, the uncontrolled effects of patient selection, of concurrent treatments, and of other factors make the detection of such differences even more difficult.

Since relatively small, even though important, numerical gains or losses are to be expected from most innovations, clinical trials must regularly be designed to detect these small differences accurately and reliably. Our sampled papers, taken as a group, provide an optimistic picture of progress in surgery and anesthesia. This progress depends on a judicious combination of continued development of new therapeutic ideas and their evaluation in good-sized unbiased clinical trials.

PROBLEMS

1. What are the three basic types of studies reported in the sampled papers reviewed by the authors?

2. Describe the scale used in the evaluation of the innovations.

3. What is meant by primary and secondary therapies?

4. Why is it not appropriate to consider a complete enumeration of papers during a certain period of time (say, 1964–1972) as a census rather than as a sample?

5. Using Figure 3, calculate the estimated probability of an innovation being an improvement over the standard. (Hint: In what fraction of the 24 observed differences was I an improvement over S?)

6. Find the estimated probability of an innovation being an improvement over the standard using Figure 4.

7. Does a 0% average gain imply that all the innovations had exactly the same effect as the standards? Why, or why not?

8. Why do the authors prefer the empirical Bayes method to the normal distribution method?

9. Refer to Figure 5. In the primary therapies and in the secondary therapies what are the chances of
    a. a gain of 30% or more?
    b. a loss of no more than 20%?
    c. a gain of more than 0%? Compare with the result for Problems 5 and 6.

10. Why do the authors suppose that the innovations assessed by randomized clinical trials and reported in the surgical literature are "biased upwards"? What is meant by "biased upwards"?

REFERENCES

J. P. Gilbert, R. J. Light, and F. Mosteller, 1976. "Assessing Social Innovations: An Empirical Base for Policy." C. A. Bennett and A. R. Lumsdaine, eds., *Evaluation and Experiment: Some Critical Issues in Assessing Social Programs.* New York: Academic Press.

B. Efron and C. Morris, 1973. "Stein's Estimation Rule and its Competitors: An Empirical Bayes Approach." *Journal of the American Statistical Association,* 68: 117–130.

With permission of the Oxford University Press, this chapter is based on the longer article: J. P. Gilbert, B. McPeek, and F. Mosteller, "Progress in Surgery and Anesthesia: Costs, Risks, and Benefits of Innovative Therapy." J. P. Bunker, B. A. Barnes, and F. Mosteller, eds., *Costs, Risks, and Benefits of Surgery,* Oxford University Press, 1977.

WARNING: The Surgeon General Has Determined That Cigarette Smoking Is Dangerous to Your Health.

STATISTICS, SCIENTIFIC METHOD, AND SMOKING

B. W. Brown, Jr.   *Stanford University School of Medicine*

AFTER HUNDREDS of years of tobacco use, smoking has been condemned on a scientific basis as a serious hazard to the health of the smoker. Future generations may regard the scientific indictment of smoking as a major contribution to preventive medicine and the health of the western world. Statisticians, statistical principles of scientific thought, and statistical methods of scientific study played essential parts in evaluating the effects of smoking.

Today a majority of health experts believe (though there are dissenters) that smoking is bad for the health—that it causes cancer and heart attacks and has other deleterious effects on the body. Governments have taken action to modify or suppress advertising by the tobacco industry, and to educate people about the hazards of smoking. Private health organizations, such as the American Cancer Society, actively propagandize against smoking. The effects of this new knowledge concerning smoking can readily be seen in surveys, which indicate that a growing number of people have quit smoking, that there is a decreasing proportion of smokers among young people, and that a decreasing number of cigarettes per capita are being sold. Over the last decade, the decrease in smoking at large parties, at meetings, and in public places especially among people in the health sciences, has been strikingly apparent to even casual observers.

How was it established scientifically that smoking is hazardous to the health? Why has this hazard been established only recently although smoking has been a widespread custom of the Western World for 300 years? When specific data on the question of hazard were published more than 40 years ago, why did the question spark such controversy, even among scientists most knowledgeable on the issues involved? And what role did statistics play in resolving the controversy?

EARLY VIEWS OF SMOKING AND HEALTH

Tobacco smoking is an ancient habit of man. Crude cigarettes have been found among the artifacts left by cave dwellers in Arizona; Columbus carried tobacco from the New World to the Old, with an endorsement by the Indians for its medicinal effects; Sir Walter Raleigh was a strong advocate of the use of tobacco; and many others have lauded its merits as cure and comfort for most of the diseases and distresses afflicting mankind. Cigarette smoking became so prevalent in the Western World that, according to a survey of the U.S., in the mid-sixties only 30% of males 17 years of age and older reported that they were not and had never been regular smokers. Fifty-one percent of the men and 34% of the women in the U.S. reported that they were currently regular smokers. Together, these smokers consumed more than a billion cigarettes each day in the year 1969.

Though tobacco has had its advocates from the beginning and has enjoyed an increasing, currently overwhelming popularity in the Western World, it has had opposition also, from the beginning. Swinburne said "James the First was a knave, a tyrant, a fool, a liar, a coward; but I love him, because he slit the throat of that blackguard Raleigh, who invented this filthy smoking." Some objected to the filth, while others felt the habit was sinful; still others objected that the habit was not good for the health. We can trace speculations concerning the ill effects of tobacco in records and writings as far back as three centuries, but these comments, at best, offered the authority of the experienced physician giving his impressions. They were not based on systematically gathered and evaluated scientific evidence. For example, a century ago Dr. Oliver Wendell Holmes, distinguished professor of the Harvard Medical School and father of Supreme Court Justice Holmes, said "I think tobacco often does a great deal of harm to the health. I myself gave it up many years ago." But Dr. Holmes gave no evidence to justify his conclusion.

Papers on the effects of smoking that have appeared in medical journals over the last 100 years show the general tendency of medical scientists to collect their information more systematically and to evaluate it more carefully as scientific evidence. In 1927, Dr. F. E. Tylecote, an English physician, wrote that almost every lung-cancer patient he had known about had been a regular smoker, usually of cigarettes. In 1936, Drs. Arkin and Wagner reported more specifically that 90% of 135 men afflicted with lung cancer were "chronic smokers."

DIFFICULTIES OF RESEARCH

With the accumulation of such anecdotal reports by serious medical men it became clear that the question of the harmful effects of tobacco should be subjected to scientific study. But how could the question be examined scientifically? Although scientific method in the physical sciences had been thoughtfully discussed and developed over the past several centuries, the principles that had been evolved for the physics laboratory could not be adapted easily to scientific study in the biological and medical sciences. Specifically, the biological scientist fell far short of the physicist in his attempt to hold fixed all factors except those under investigation. Instead, he was faced with investigations involving many uncontrolled, or partially controlled, factors that caused unwanted variation in the data. Conclusions had to be made in the face of this biological variation in the study material.

The problem of drawing valid scientific conclusions in the life sciences attracted an assortment of extremely able English men of science; among them were Francis Galton, a genius and creative psychologist; Karl Pearson, trained as an engineer and ultimately an outstanding philosopher of science; and R. A. Fisher, a mathematician who laid down new principles of great importance to scientific study. It was Fisher, building on the work of Galton and Pearson, who suggested, in the twenties, an alternative to the impossible requirement that all factors be held constant except the factors under investigation.

Fisher suggested that if two fertilizers, for example, are to be compared, each should be allocated to a number of different plots of ground, and that the allocation be random, according to some sort of lottery or coin-tossing system. Fisher pointed out that such random allocation would tend to balance all differences between the plots receiving the two different fertilizers and thus would yield an unbiased experiment—that is, one fair to both fertilizers. He further pointed out that any differences in the results obtained with the two fertilizers could be evaluated, using the probability theory that had been developed for gambling games, to decide how likely it was that so large a difference might have come about simply through a chance allocation of the more fertile plots to one of the treatments.

This new scientific methodology, employing randomization and probability theory, was a tremendous contribution to the young science of statistics. It spread quickly through the agricultural sciences and then to the medical sciences. Today a new drug will not be approved by the U.S. Food and Drug Administration for marketing until it has been compared with other treatments in what is called a *randomized clinical trial*. In the thirties, however, these rigorous standards for scientific study were new and scientists were just becoming familiar with their application in the medical sciences. How could the new methodology be applied to the smoking question?

RETROSPECTIVE STUDIES

Clearly it would be difficult, though perhaps not impossible, to conduct a randomized experiment on human beings to determine whether smoking is harmful. The first attempt at some sort of satisfactory substitute approach was to identify cases of lung cancer and, at the same time, to select some other persons comparable to the lung-cancer patients in age, sex, and other characteristics, and then to determine whether the persons with lung cancer smoked more heavily than the artificially constructed "control group." The first study of this kind was reported by Müller in 1939, who found much more smoking among the lung-cancer cases. Of course, it was clear to statisticians and others that this kind of data must be regarded with caution. The "control group" in such a study can only be selected somewhat arbitrarily and it is quite easy to imagine that the selection might tend to omit, or fail to recognize, smokers because of conscious or unconscious bias on the part of either the investigator or the persons responding to questions regarding their own smoking habits or those of their deceased relatives. Similarly, biases might exaggerate smoking experience among the lung-cancer cases.

Because the retrospective approach did not measure up to the standards of the randomized clinical experiment, other approaches were tried. Raymond Pearl (1938), an eminent medical statistician at Johns Hopkins University, had been keeping close records of health experiences of hundreds of families in the Baltimore area for some years. Pearl's work was going on at the time that serious study of the tobacco and health question began, and he decided to compile information about the smoking habits and longevity of all males in his files. Using the statistical methods common in computing life tables for life insurance purposes, Professor Pearl constructed a life table for nonsmokers, another for moderate smokers and a third for heavy smokers (see Figure 1). His data showed that 65% of the nonsmokers survive to age 60, whereas only 45% of the heavy smokers live that long. Indeed, Figure 1 shows that at every age between 30 and 90, proportionately more nonsmokers survive

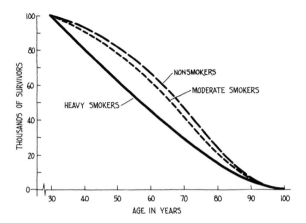

FIGURE 1

The survivorship lines of life tables for white males falling into three categories relative to their tobacco usage: (a) non-smokers, (b) moderate smokers, smokers, and (c) heavy smokers.
Source: Pearl (1938)

than moderate smokers, who, in turn, survive longer than heavy smokers. The impressive study hinted at smoking effects far beyond an increase in the relatively rare disease of lung cancer.

When Pearl's results were published, they added substantial weight to the evidence against smoking.   But his claims were based on a special kind of study, so workers who did not have this sort of longitudinal data continued using the retrospective method of identifying lung-cancer patients and demonstrating that these patients were tobacco smokers to a much larger extent than any comparable control group that might be chosen.   In the next decade many such retrospective studies were published, but the mounting number of scientific reports did not add conviction in proportion to their number. They all used the same method and they all might be drawing the same erroneous conclusion.   Meanwhile, the consumption of tobacco per capita continued to increase, smoking among teenagers continued to rise, and the medical profession as a whole did not advise the public against the habit.

Medical scientists recognized the tremendous importance of the question. Tobacco was an intimate and welcome part of the lives of a substantial portion of the people.   The tobacco industry was a vital part of the economy of a huge region of the U. S. and furnished the livelihood of additional thousands through related industries such as advertising, transportation, and sales.   If tobacco were as harmful as Pearl's 1938 paper indicated, it would be the obligation of the medical profession and government health agencies to help the public protect itself against this hazard. On the other hand, if the studies were erroneous, it would be extremely unfortunate if the medical profession or the government were to interfere in private lives and disrupt large segments of the economy on the basis of unsound scientific work. Thus medical scientists were deeply concerned about assessing the strength of the evidence against tobacco. Statisticians, especially in the biological and medical sciences, found themselves at the center of the dilemma facing the question. "Is the retrospective study a sound scientific approach?"

PROSPECTIVE STUDIES

It was felt that a new kind of study—more closely resembling the randomized clinical study—must be done. Pearl's choice of subjects and study were unsatisfactory in that his methods for choosing families and for choosing subjects within the families were never clearly set down. The chance for substantial biases in selection of the subjects and in the interviews for smoking experience seemed unacceptably large, and the report was not taken seriously. His prospective approach, however, in which smokers and nonsmokers are identified and then studied until death is an appealing one. It is a much closer approximation of a real experiment, in which a population would be split randomly into two groups; one group would be given one treatment (in this case, smoking) and the other group would receive an alternative treatment (no smoking). Both groups would be studied over time to determine the effects of the treatments. Even in the prospective study, the lack of random assignment of persons to be smokers and nonsmokers means that the conclusions are subject to the reservation that the smokers may be different from the nonsmokers in some systematic way that is entirely unrelated to their smoking habits, but that is causally related to some other factor that is deleterious to their health. For example, suppose that people who *like* smoking tend to get cancer and people who *don't like* it tend not to. Despite this inherent weakness in the prospective study approach, statisticians and other scientists called for this kind of study, and several were conducted in the fifties.

The first two prospective studies of the tobacco question were done by eminent statisticians in England and the U.S. In England, Dr. Richard Doll and Sir A. Bradford Hill, whose careful retrospective studies of the question had led them to conclude that smoking does cause lung cancer, were the first to report on a prospective study. They sent questionnaires to all members of the medical profession in the United Kingdom, roughly 60,000 men and women, and received about 40,000 replies. They asked about present smoking habits and a few other characteristics, such as age and sex. Then, through the Registrars-General in the United Kingdom, they determined the survivorship of these men and women over a period of several years. Their study results were as anticipated: there were remarkably more lung-cancer deaths among the smokers than among the nonsmokers. Of course, this report, based on a different study approach in which the information on smoking habits was obtained first, rather than after the fact of death or disease, was regarded as extremely important additional evidence against smoking.

The prospective study offered new information that could not be obtained from the retrospective study. The retrospective study offered counts of smokers among lung-cancer cases. The prospective study offered counts of lung-cancer deaths among smokers and, thus, a direct measure of the death rate for lung cancer among smokers. The Doll and Hill data showed the rate for heavy smokers to be 1.66 per 1000 men per year, compared to 0.07 for nonsmokers; thus, heavy smokers had a lung-cancer death rate 24 times higher than nonsmokers. But even more important, the prospective study recorded all deaths and the recorded causes of these deaths, whereas the retrospective studies focused on the cause of death that was suspect. With information on all deaths that occurred in the group it was found that there was also a surplus of deaths due to heart attack among the smokers, and that this surplus was not readily explainable as chance variation in the data. Medical investigators had reported the effects of tobacco on the body (for example, the immediate cooling of the skin through contraction of the blood vessels when a person begins to smoke a cigarette), but the Doll and Hill study provided rather direct evidence that such effects on the heart and blood system could increase the risk of early death. The death rate for heart attacks among smokers was 5.99 per 1000 men per year for heavy smokers and 4.22 for nonsmokers—not a large difference, but extremely important, if real, because of the large number of deaths from this cause.

The second prospective study was reported by two American statisticians, Dr. E. Cuyler Hammond and Dr. Daniel Horn. Through the American Cancer Society, these men enlisted the aid of about 22,000 woman volunteers, each of whom was asked to select 10 healthy men between the ages of 50 and 69 and have each of them fill out a smoking questionnaire. Then these women reported on the health status of each of these men each year. Death certificates were obtained for each death reported. About 200,000 men were followed through a period of almost four years during which some 12,000 of them died. The Hammond and Horn report confirmed the findings of the Doll and Hill prospective study. Hammond and Horn reported a lung-cancer rate 23.4 times higher among heavy smokers (more than one pack a day) compared to nonsmokers and a death rate for heart and circulatory diseases 1.57 times as high among heavy smokers as among nonsmokers. Hammond and Horn emphasized the fact that although lung-cancer rates were strikingly high among smokers, the greater share of the excess deaths among smokers occurred in the heart and circulatory disease category, simply because this is a much more common cause of death.

NEW OBJECTIONS

Further prospective studies, involving different investigators and large groups of subjects chosen from various sources and followed for periods of up to a decade, were reported in the late fifties and early sixties. All of these studies substantiated the results of the first two studies. The evidence, which had been carefully gathered and evaluated by eminent men of medicine and statistics, seemed to condemn tobacco. But two factors were yet to be reckoned with: the impressive economic importance of tobacco as a prime industry and as contributor to other industries and the overwhelming strength of the smoking habit among the population. Again the statisticians played a major role in the drama, but this time, surprisingly, their influence was in defense of the "vile weed." One of the defenders was Sir Ronald Fisher, the man whose ideas were behind the randomized clinical trial and perhaps the greatest statistician who ever lived (Fisher died in the mid-sixties). Another statistician who came to the defense of tobacco was Dr. Joseph Berkson, Chief of Medical Statistics for the famed Mayo Clinic until he retired in the mid-sixties and one of the most creative people in medical statistics, having been trained in both medicine and medical statistics. Fisher and Berkson, both colorful and persuasive scientists, in the past had often found themselves on opposite sides of arguments concerning scientific methodology. This time they were on the same side, however, and Berkson was heard to say in jest that this was the only point that caused him serious doubt about his position on the tobacco-health question.

Although the positions taken by Fisher and Berkson gave aid and comfort to the habitual smoker, to the tobacco industry, and to the people who made their living through the tobacco industry, it must be stressed that these men had a larger and more important issue in mind, namely, the methodology of science itself. And, though the short-term effect of their writings was a delay in the scientific indictment of tobacco, the long-term effects will more than counterbalance these by tightening up scientific standards for medical studies in which appropriate randomized clinical experiments cannot be carried out because of possible danger to participants. In the late fifties, Fisher and Berkson both repeatedly pointed out that the evidence against tobacco as a hazard to the health was only circumstantial because clinical experiments could not be done. They pointed out, in detail, weaknesses in the evidence, even in the evidence from the prospective studies. They proposed other explanations that would account for the data if tobacco had no deleterious effects on the health.

Berkson's main point was that the apparent ill effects of smoking were too pervasive. The death rates seemed to be higher for too many different causes of death. Such a universally bad effect from tobacco was never anticipated, and there were no theories to explain how smoking could cause death through each of the multitude of diseases implicated by the reports from the prospective studies. Berkson suggested that a more likely explanation was that there were biases in sample selection, or in the data collected, or both, and that these biases affected all of the prospective studies because they were inherent in the methodology of the prospective studies and they affected all causes of death. Otherwise, Berkson suggested, we must hypothesize some sort of general constitutional effect of smoking on the body, or some sort of accelerated aging caused by smoking; he was dissatisfied also by the absence of known mechanisms causing the disorders.

Fisher, who had been knighted for his work in genetics and who had always been concerned with the problems of interpreting nonexperimental (i.e., nonrandomized) studies, put these two interests together and argued that if persons with a hereditary tendency toward smoking also had a hereditary predilection for disease, the result would be exactly the kind that had been repeatedly demonstrated, both retrospectively and prospectively. Then, Fisher argued, if the explanation lay in such hereditary predilections, a smoker who quit could not change his genetic makeup or his peculiar higher risk of disease thereby, and smoking could not be regarded as harmful to health. Fisher proposed some (nonrandomized) twin studies that might shed light on his hypotheses, but little could be done because large groups of people were needed for definitive results. (See the essay by Reid, especially Table 2 and its discussion.)

Neither Fisher nor Berkson has been completely answered. In fact, the questions they raised cannot be answered definitively without randomized experiments. Of course, Fisher and Berkson knew this as well as anyone, but such conspicuous critiques, made by widely known men of high prestige, in the context of a scientific question of intense public interest, served to underline the dangers in drawing conclusions from nonrandomized studies.

FURTHER RESEARCH

The immediate effect of the Fisher and Berkson critiques was to spur on statisticians and medical scientists to an even more careful look at the tobacco-health question. Although the questions raised by Berkson and Fisher could never be answered directly, additional indirect evidence has been marshaled on many points. There is space here only to mention some of the approaches, without detailing their results.

Because experimental verification of smoking as a cause of ill effects in man did not seem practicable, scientists had attempted to investigate the effects of tobacco smoke on animals, where random assignment to smoking and nonsmoking groups was feasible. Although much experimental work had been done, until the early sixties the actual smoking experience of the human had not been adequately simulated in the laboratory and no direct positive evidence had been obtained. In the sixties, however, studies demonstrated that tobacco smoke and some of its constituents do cause lesions in animal tissue that are similar to those seen in human lung cancer. When, finally, an experimental setup was attained that satisfactorily simulated the smoking habit in dogs, a few cases of lung cancer were induced in what seemed to be well-controlled, randomized experiments.

At the same time, in partial answer to Berkson's point that there was no theory or information that would lead one to suspect that smoking would affect the diversity of organs and systems cited in the prospective reports, much work in animals and humans has been carried out to determine what physiological effects tobacco has on the organ systems. One of the most important results has been the discovery that tobacco smoke does have deleterious effects on the blood vessels of animals and the verification of this same damage in the blood vessels of heavy smokers.

The evidence against tobacco has been strengthened by more detailed information and closer statistical evaluation in ongoing prospective studies. It has been found that smoking is not associated with all causes of death, but that for the causes implicated, the death rates increase regularly and convincingly with the amount smoked. When the details of smoking habits and smoking experience were analyzed, including especially the results for persons who had smoked but quit the habit for varying lengths of time, the death rates have been found to be closely consistent with what might be expected if smoking were indeed a causal factor in death.

THE SURGEON GENERAL'S REPORT

By the mid-sixties, scientific opinion had swung far toward the conclusion that smoking is harmful. Despite the realization that the evidence against smoking can only be indirect, scientists were convinced that some conclusion must be reached and acted on for the good of the public health. A special committee, appointed by the Surgeon General of the U.S. Public Health Service, made a long and thorough study of all aspects of the tobacco-health question. The committee of ten men included experts on chemistry, pharmacology, internal medicine, surgery, pathology, epidemiology, and statistics. The statistical representative was William G. Cochran, Professor of Statistics at Harvard University and a world-renowned expert in the application of statistical methodology to problems of the life sciences. In 1963 this Advisory Committee submitted to the Surgeon General a report that cited smoking as a cause of lung cancer and several other cancers and stated that the evidence pointing to smoking as a cause of death due to heart attack was strong enough to justify acting on this presumption.

LATER DEVELOPMENTS

In the sixties, independent health organizations and governmental health agencies in this country and throughout the world publicly announced their conclusions that tobacco is a health hazard. Along with these indictments came a call for further action. Many health officials felt that it wasn't enough to inform the public through the usual scientific publications, carried along by brief news reports and by word of mouth. Rather, they felt that a public habituated to tobacco and under constant bombardment by advertisements that associated cigarettes with youth, health, attractiveness, and elegance must be protected. The issue became one of how far the responsibility and legal authority of the government can extend to protect the citizen against himself and against the influences of deleterious propaganda, especially when the evidence did not measure up to the standards of true experimental science.

By 1970, there were still scientists (a few of them statisticians) who found the evidence against smoking less than convincing and who were willing to so testify at congressional hearings. But, for the most part, statisticians will play a new part in the tobacco-health arena as the issue ceases to be a question of science and becomes one of public policy. In 1969, Bernard Greenberg, Professor of Biostatistics at the University of North Carolina and a noted public health statistician, published a paper discussing the strength of evidence against smoking and the extent to which such evidence might justify various levels of government action in protection of the individual. These possible actions ranged from rather modest propaganda campaigns to legal judgments against tobacco companies for causing untimely death and illness through the vending of their product. Professor Greenberg's paper was not a philosophical discussion, concerned with ethical, political, and legal principles; it was an attempt to use modern statistical theory—this time a branch of mathematical statistics called *decision theory*—to arrive at reasoned advice, based on the costs to the public of actions that might be taken and the gains and losses that might accrue to the public as the result of such actions. This new kind of thinking in public health requires a different kind of statistical theory than that for experiments.

SUMMARY

This essay has traced the role of some eminent statisticians and the statistical methods they used to determine that smoking is harmful to the public health. The work that has been done on this problem and the debates on scientific method that it has stimulated will have untold benefit both through the eventual elimination of smoking as a general habit of the people and in the development of better techniques and higher standards for the scientific study of other hazards to the public health in the future.

PROBLEMS

1. What did R. A. Fisher suggest to use as an alternative to holding all factors constant except the ones under investigation?

2. Explain what is meant by an "unbiased experiment."

3. Explain the terms "retrospective study" and "prospective study." Give the pros and cons.

4. Refer to Figure 1. Approximately what percentage of males survived up to age 50 in the three different categories?

5. Refer to Figure 1. What was the approximate median lifetime (i.e., the age up to which 50% of the people survived) in each of the three categories?

6. What were the objections of Berkson and Fisher to the results of both the prospective and retrospective studies?

7. If a prospective study shows that there is a higher incidence of lung cancer among smokers than among nonsmokers, can we conclude that smoking causes lung cancer? Why or why not?

REFERENCES

Joseph Berkson. 1958. "Smoking and Lung Cancer: Some Observations on Two Recent Reports." *Journal of the American Statistical Association* 53:28–38.

Harold Diehl. 1970. *Tobacco and Your Health: The Smoking Controversy.* New York: McGraw-Hill.

Alfred Dunhill. 1954. *The Gentle Art of Smoking.* New York: Putnam.

R. A. Fisher. 1959. *Smoking: The Cancer Controversy.* Edinburgh: Oliver & Boyd.

B. G. Greenberg. 1969. "Problems of Statistical Inference in Health with Special Reference to the Cigarette Smoking and Lung Cancer Controversy." *Journal of the American Statistical Association* 64:739–758.

A. E. Hamilton. 1927. *This Smoking World.* New York: Appleton–Century.

Raymond Pearl. 1938. "Tobacco Smoking and Longevity." *Science* 87:216–217.

U. S. Public Health Service. 1963. *Smoking and Health.* Report of the Advisory Committee to the Surgeon General.

DEATHDAY AND BIRTHDAY:
An Unexpected Connection

David P. Phillips   *State University of New York, Stony Brook*

IN THE movies and in certain kinds of romantic literature, we sometimes come across a deathbed scene in which a dying person holds onto life until some special event has occurred. For example, a mother might stave off death until her long-absent son returns from the wars. Do such feats of will occur in real life as well as in fiction? If some people really do postpone death, how much can the timing of death be influenced by psychological, social, or other identifiable factors? Can deaths from certain diseases be postponed longer than deaths from other diseases?

In this essay we shall see how dying people react to one special event: their birthdays. We want to learn whether some people postpone their deaths until after their birthdays. If we compare the date of death with the date of birth for a large number of people, will we find fewer deaths than expected just before the birthday?   If we do find a dip in deaths, we may conclude that some of these people are postponing death until after their birthdays.

We shall use elementary statistical methods in approaching the problem. For example, the comparison of an actual number of events with the number that might be expected is one of these methods; others will be noted later.

NATURE OF THE DATA TO BE INVESTIGATED

We shall examine only the deaths of famous people.   There are two reasons for this.   First, it seems likely that ordinary people look forward to their birthdays less eagerly than do famous people because a very famous person's birthday, generally, is celebrated publicly, and he may receive a substantial amount of attention, gifts, and so on.   In contrast, much less attention is paid to the birthday of an ordinary person, and he may have relatively little reason to look forward to it.   Hence famous people may be more likely to postpone deaths than less famous ones.   Second, it is easier to examine the deaths of the famous than of other people.   To discover whether there is a dip in deaths before the birthday, we need information on the birth and death dates of individuals.   This type of information is not available from conventional tables of vital statistics; therefore, we cannot easily determine the birth and death dates of large numbers of ordinary people.   On the other hand, we can easily determine the birth and death dates of famous people because there is much biographical information available about them.

In all, we shall examine the birth and death dates of more than 1200 people.   It is tedious to classify these dates by *day,* so we shall examine the *month* of birth and *month* of death.   Thus, for the purpose of this analysis, we shall be concerned, not with the relationship between the birthday and the day of death, but rather with the relationship between the birth month and the death month.   For our purposes, a person is said to have died in his birth month if the month of his death has the same name as the month of his birth.   For example, if a person was born on March 1, 1897, and died on March 31, 1950, he died in his birth month.   On the other hand, if he was born on March 1 and died on February 28, he did not die in his birth month; rather, he died in the month just before his birth month.   Although we gain convenience by examining events by month rather than by day, we lose precision; if we find a dip in deaths in the month before the birth month, we cannot tell whether a dying person is hanging on for a few days or for a few weeks.

IS THERE A DIP IN DEATHS BEFORE THE BIRTH MONTH?

Table 1 shows the month of birth and month of death of people listed in *Four Hundred Notable Americans.*   For example, we can see from the first column that one person who was born in January died in January, two people who were born in January died in February, and so on.   The column labeled "Row Total" gives the total number of people who died in each month and the row labeled "Column Total" gives the total number of people born in each month.

Table 1 enables us to compare two hypotheses.   The first hypothesis states that the death month is related to (is dependent on) the birth month in that some people postpone death in order to witness their birthdays.   This will be called the *death-dip hypothesis*.   The second hypothesis states that no deaths are being postponed and that the month of death is not related to (is independent of) the month of birth.   This will be called the *independence hypothesis*.   (When we formulate our problem in terms of two hypotheses, one that we wish to disprove in order to lend credence to the other, and try to decide which hypothesis seems more consistent with the data, we are using a standard statistical testing procedure.   The concept of "independence" is also an important part of many statistical hypotheses.)

Our general plan is to see whether the month immediately preceding the birth month has fewer deaths than the independence hypothesis suggests.   As we explain below, it turns out that if the independence hypothesis were true, about $\frac{1}{12}$ of the deaths would occur in each of the six months preceding the birth month, $\frac{1}{12}$ in the birth month, and $\frac{1}{12}$ in each of the five following months.   Although this may seem obvious, it actually depends upon detailed calculations because we must take into account that some calendar months produce more deaths than others and some produce more births than others. It is intuitively satisfying, nevertheless, that for the independence hypothesis, the calculations give nearly equal expected numbers of deaths for each of the twelve months preceding, during, and following the birth month.

We now compare the actual number of deaths before the birth month with the number of deaths that are expected, on the average, if the independence hypothesis is true.   If the observed number of deaths is noticeably less than the expected number, there is a dip in deaths before the birth month.

First, we count the actual number of deaths in the month just before the birth month.   If we sum the numbers in the starred cells in Table 1, we will have the total observed number of deaths in this period.   This number is 16.

Now we calculate the total expected number of deaths in the month just before the birth month.   If the independence hypothesis is true, then the death month is independent of the birth month.   This means that the deaths of those born in any given month should be distributed throughout the year in the same way as the deaths of those born in any other month. Thus, because 6.32% [this is $(22/348) \times 100$] of all the deaths in Table 1 fall in December, 6.32% of those born in January should die in December, 6.32% of those born in February should die in December, and so on.   In Table 1, we see that 11.2% [$39/348) \times 100$] of all deaths fall in April.   Then,

TABLE 1. Number of Deaths by Month of Birth and Month of Death (Sample 1)

MONTH OF DEATH	MONTH OF BIRTH												ROW TOTAL
	Jan.	Feb.	Mar.	Apr.	May	June	July	Aug.	Sept.	Oct.	Nov.	Dec.	
Jan.	1	1*	2	1	2	2	4	3	1	4	2	4	27
Feb.	2	3	1*	3	1	0	2	1	2	2	6	4	27
Mar.	5	6	5	3*	1	0	5	1	2	5	3	1	37
Apr.	7	6	3	2	1*	3	3	1	3	2	4	4	39
May	4	4	2	2	1	2*	4	1	3	2	1	5	31
June	4	0	4	5	1	1	1*	2	1	2	4	0	25
July	4	0	3	4	3	3	4	1*	6	4	2	5	39
Aug.	4	4	4	4	2	2	3	3	1*	1	2	0	30
Sept.	2	2	1	0	2	0	2	4	2	0*	5	2	22
Oct.	4	2	2	3	2	2	2	3	3	1	4*	5	33
Nov.	0	2	0	2	1	1	0	3	3	3	1	0*	16
Dec.	1*	2	2	1	2	1	4	1	4	0	2	2	22
COLUMN TOTAL	38	32	29	30	19	17	34	24	31	26	36	32	
TOTAL													348†

Source: R. B. Morris, ed., *Four Hundred Notable Americans* (New York: Harper & Row, 1965).
* Deaths corresponding to month preceding birth month.
† The total number of deaths is less than 400 because (1) some of those in the source volume have not yet died; (2) for some of those in the volume, the month of birth and/or death is not known.

if independence holds, we would expect 11.2% of those born in any month to die in April. For example, there are 19 people born in May; we expect 11.2% of these people, that is, $(11.2/100) \times 19 = 2.1$, to die in April. In a similar fashion, we can work out the expected number of deaths in each of the 12 starred cells in Table 1. If we sum these 12 numbers, we will have the total number of deaths that we expect to occur in the month before the birth month *if* the independence hypothesis is true. This expected number is 28.3.

The more intuitive method mentioned earlier estimates the total expected number of deaths in the starred cells to be simply $348/12 = 29.0$. In other words, we expect about $\frac{1}{12}$ of all deaths to occur one month before the birth month. In fact, we expect about $\frac{1}{12}$ of all deaths to occur in the birth month or in any month before or after it. In general, this rough-and-ready method of estimating the total expected number for any month gives results very close to those provided by more precise methods.

We can now compare the observed number of deaths before the birth month with the expected number in this period. No matter which "expected number" we use, it is considerably higher than the number of deaths observed just before the birth month: we expect about 28 or 29 deaths, but we observe only 16—about 12 fewer than expected. In other words, we observe a dip in deaths in the month before the birth month as predicted by the death-dip hypothesis. As we shall soon see, the discrepancy is much more than might reasonably be explained by chance.

IS THERE A DEATH RISE AFTER THE BIRTH MONTH?

If the death-dip hypothesis is true, what will become of the 12 or so people who presumably have postponed death until their birthdays? When are they expected to die? There is no way of answering this question a priori because, even if the death-dip hypothesis is true, the death dip might have come about in a number of different ways; some of these different ways imply differing periods of survival for those who have postponed death. For example, the death dip could result entirely because some people who were hovering between life and death unexpectedly recovered; in this case, it might be years before these people die, and we could expect no rise in deaths immediately after the birth month. On the other hand, the death dip could appear solely because those who do not die just before their birthdays live a few days or weeks longer than expected; in this case, there should be a peak in deaths soon after the birth month. Depending on the way in which the death dip came about, we would or would not expect a rise in deaths after the birth month: we cannot tell what to expect on the basis of the death-dip hypothesis.

Although the death-dip hypothesis is not helpful here, practical experience with another sample (of famous Englishmen) suggests that we look for a rise in deaths in the four-month period consisting of the birth month and the three months thereafter. Thus, although we searched for a *death dip* in a *one-month* period, we shall search for a *death rise* in a *four-month* period, because past experience, not theory, makes this approach seem promising.

Table 2 gives the observed number of deaths six months before the birth month, five months before, and so on, down to zero months before, one month after, and so on up to five months after the birth month. Because $n = 348$ is the total number of people in sample 1, $n/12 = 29.0$ is the number expected to die six months before the birth month, five months before, and so on. From this table it is evident that not only is there a dip in deaths before the birth month, but there is also a rise in deaths during the birth month and during the three months thereafter. We expect about $\frac{4}{12}$ of all deaths in the first sample [$348 \times (\frac{4}{12}) = 116$] to fall in this four-month period, but we observe 140 deaths during this time.

COULD THE DEATH DIP AND DEATH RISE BE DUE TO CHANCE?

We know that surprising phenomena sometimes occur just by chance and for no other reason. For example, a person might deal himself a straight in poker; ordinarily, we attribute this happy event to the vagaries of providence, not to the dishonesty of the dealer. In much the same way, we might wonder whether the death dip and death rise have arisen by chance and for no other reason.

Now suppose our poker player were to deal himself not just one straight, but four straights in a row in the four times he deals while playing with us. This *could* have happened by chance, but it is so unlikely that we would prefer some other explanation. The less likely an event is to occur by chance, the more we prefer some other explanation. Similarly, if we find a death dip and death rise in, say, four samples and not just one, there is a small possibility that these phenomena could have occurred by chance, but another explanation might be more plausible.

In sample 1 we observed that there are fewer deaths than expected in the month before the birth month, and more deaths than expected in the four-month period consisting of the birth month and the three months thereafter. Can we find a similar death dip and death rise in other samples of people?

MORTALITY IN THREE MORE SAMPLES

Three new samples were taken, consisting of people who are famous for two reasons. First, they achieved high status in their lifetimes: they were listed in *Who Was Who in America*. Second, they came from well-known families (e.g., Adams, Vanderbilt, Rockefeller) which are listed in "The Foremost Families of the U.S.A.," an appendix to *Royalty, Peerage and Aristocracy*

TABLE 2. Number of Deaths, Before, During, and After the Birth Month (Sample 1)

	6 MONTHS BEFORE	5 MONTHS BEFORE	4 MONTHS BEFORE	3 MONTHS BEFORE	2 MONTHS BEFORE	1 MONTH BEFORE	THE BIRTH MONTH	1 MONTH AFTER	2 MONTHS AFTER	3 MONTHS AFTER	4 MONTHS AFTER	5 MONTHS AFTER
NUMBER OF DEATHS	24	31	20	23	34	16	26	36	37	41	26	34

$n = 348$
$n/12 = 29.0$

Source: Table 1.

of the World (vol. 90, 1967). We have ensured that the new samples do not overlap each other or sample 1.

Three volumes of *Who Was Who* were examined, those for the years 1951–60, 1943–50, and 1897–1942. Sample 2 contains all who are listed in the first of these volumes and have their surnames in "Foremost Families." Sample 3 contains all who are listed in the second *Who Was Who* volume and have their surnames in "Foremost Families." Sample 4 contains *every other* person who is in the third volume of *Who Was Who* and has his surname in "Foremost Families." We chose every other person rather than every person because the third volume is so much larger than the other volumes that it would be tedious to examine every person who meets our selection criteria.

Table 3 gives the observed number of deaths before, during, and after the birth month for samples 2, 3, and 4. The last number in each row of this table is the expected number of deaths six months before the birth month, five months before, and so on. We can see that the death dip and death rise evident in sample 1 also appear in samples 2, 3, and 4. In each of these samples there are fewer deaths than expected in the month before the birth month and more deaths than expected in the four-month period consisting of the birth month and the three months thereafter.

If we now combine the data in all four samples, we get the results seen in Table 4. A graph of these results appears in Figure 1, where we can see a death dip just before the birth month and a death rise during and after it.

In summary, just before the birth month, in each of the four samples presented, we found fewer deaths than would be expected under the hypothesis that the month of death is independent of the month of birth. In each of the four samples presented, more deaths occur during and immediately after the birth month than would be expected if independence held. The similarity of results in each of the four samples helps to convince us that the death dip and death rise are real phenomena and are not merely chance fluctuations in the data.

THE SIZE OF THE DEATH DIP AND DEATH RISE

Let us estimate the size of the death dip before the birth month and the size of the death rise thereafter for the aggregate sample of 1251 people. Out of 1251 people, 86 died in the month before the birth month. Given independence between the birth month and the death month, we would expect about $\frac{1}{12}$ of all 1251 deaths, or approximately 104 deaths, to fall in this period. Thus, just before the birth month only about 83% (86/104) of the deaths expected actually occurred. To put this another way, in the month before the birth month there were about 17% fewer deaths than we would expect under independence.

TABLE 3. Number of Deaths Before, During, and After the Birth Month Samples 2, 3, and 4

NUMBER OF DEATHS	6 MONTHS BEFORE	5 MONTHS BEFORE	4 MONTHS BEFORE	3 MONTHS BEFORE	2 MONTHS BEFORE	1 MONTH BEFORE	THE BIRTH MONTH	1 MONTH AFTER	2 MONTHS AFTER	3 MONTHS AFTER	4 MONTHS AFTER	5 MONTHS AFTER	TOTAL	TOTAL / 12
Sample 2*	17	23	26	27	28	28	42	32	31	34	36	30	354	29.5
Sample 3†	10	14	12	11	8	12	15	15	15	13	20	13	158	13.2
Sample 4‡	39	32	29	35	31	30	36	35	38	26	31	29	391	32.6

* Sample 2 excludes those listed in *Who Was Who in America 1951–1960* who died outside of that period and those listed in *Four Hundred Notable Americans*.
† Sample 3 excludes those listed in *Who Was Who in America 1943–1950* who died outside of that period or during World War II and those listed in *Four Hundred Notable Americans*.
‡ Sample 4 excludes those listed in *Who Was Who in America 1897–1942* who died outside of that period or during both World Wars and those listed in *Four Hundred Notable Americans*.

TABLE 4. Number of Deaths Before, During, and After the Birth Month, (All Samples Combined)

	6 MONTHS BEFORE	5 MONTHS BEFORE	4 MONTHS BEFORE	3 MONTHS BEFORE	2 MONTHS BEFORE	1 MONTH BEFORE	THE BIRTH MONTH	1 MONTH AFTER	2 MONTHS AFTER	3 MONTHS AFTER	4 MONTHS AFTER	5 MONTHS AFTER
NUMBER OF DEATHS	90	100	87	96	101	86	119	118	121	114	113	106

$n = 1251$
$n/12 = 104.3$

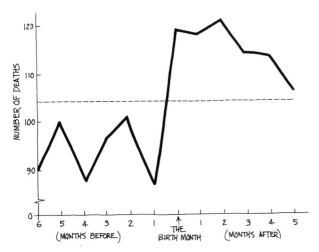

FIGURE 1

Number of deaths before, during, and after birth month (all samples combined)

Similarly, we can estimate the size of the death rise during the birth month and the three months thereafter. There were 472 deaths in the period including the birth month and one, two, and three months thereafter. Given independence between birth and death months, the expected number of deaths in this four-month period is estimated to be $\frac{4}{12}$ of all deaths, or 417 deaths [($\frac{4}{12}$) $\times$ 1251 = 417]. Thus the actual number of deaths in and just after the birth month is $\frac{472}{417}$ = 11% more than the number expected.

We can see that the death dip and death rise for sample 1 are larger than the death dips and death rises for the other three samples examined. In the month before the birth month, sample 1 has 45% fewer deaths than independence leads us to expect. In the remaining three samples combined, we observe 70 deaths in the month before the birth month; given independence, we expect about 75.25 [($\frac{1}{12}$) $\times$ 903] in this period. Thus, just before the birth month, the observed number of deaths in samples 2, 3, and 4 combined is approximately 7% less than the number expected under independence.

Similarly, it is evident that the death rise in sample 1 is larger than the death rises in the remaining samples. We estimate that in sample 1, in the birth month and in the three months thereafter, there are 20% more deaths than expected. The equivalent figure for samples 2, 3, and 4 is 10%.

The death dip and death rise in the sample 1 may be larger than in the other samples because the members of sample 1 are considerably more famous than the members of the other samples. The 348 people in sample

1 are supposed to be the most famous people in American history. The larger number of people in the remaining samples were less stringently selected.

RELATION BETWEEN FAME AND THE SIZE OF THE DEATH DIP AND DEATH RISE

We have referred several times to the notion that a group of famous people is expected to produce a larger death dip before the birth month than a group of ordinary people. Now we shall assess this idea more carefully. We classify the members of sample 1 into three groups according to how famous they are and examine the sizes of the death dip and death rise produced by each of these groups. If we are right, we should find that the more famous a group is, the larger its death dip and death rise.

There are obviously many ways to classify groups by fame. The method used here is convenient and seems plausible. The best-known members of "the four hundred" are those whose names have become household words in the "common culture." The "common culture" may be said to consist of the knowledge shared by almost all the members of a society—in other words, some sort of lowest common denominator of knowledge. To find which of the four hundred is "in" the common culture, we must find a set of people who know only what is in that culture. The members of the four hundred who are known to this group of people have names that are part of the common culture.

Of all the people in a society, children come closest to having no more knowledge than is in the common culture. If a child has heard of someone in the four hundred, he is very famous indeed. Thus the members of the four hundred who appear in children's biographies may be judged to be better known than members who do not appear in such biographies.

Two series of children's biographies were examined: Dodd, Mead's (1966) and Bobbs-Merrill's (1966). The criterion of coverage or noncoverage in these series can be used to classify members of the four hundred into groups of differing fame. Three different subgroups were formed from the original four hundred.

> Group 1 consists of those of the four hundred whose names are found in both of the children's biography series. For example, George Washington, Thomas Jefferson, Benjamin Franklin, Mark Twain, and Thomas Edison are in group 1.

> Group 2 consists of those of the four hundred whose names are in only one of the series. For example, John Quincy Adams, John Hancock, Jefferson Davis, Edgar Allen Poe, and Alexander Graham Bell are in group 2.

Group 3 consists of those of the four hundred whose names are in neither series. For example, Samuel Adams, Millard Fillmore, Rutherford B. Hayes, H. L. Mencken, and Nikola Tesla are in group 3.

For our purposes, the members of group 1 are judged to be more famous, on the average, than the members of group 2, and the members of group 2 are judged to be more famous, on the average, than the members of group 3. We say "on the average" because single individuals could be moved readily from one group to another if we used a third or fourth biography series to judge fame. But we think that the groups as a whole are ordered with respect to fame in the way we want them to be.

We can now measure the death dip and death rise produced by each of these groups. Table 5 gives the relevant information.

As predicted, the more famous a group is, the larger is its death dip. Group 1 produces a larger death dip than group 2, and group 2 produces a larger death dip than group 3. The death dip for the most famous group is quite large. About 78% of the deaths that are expected to fall in the month before the birth month do not do so. It should be stressed, however, that the number of people in group 1 is not large.

From Table 5 we can see also that the more famous the group is, the larger is the death rise that it produces. In group 1 (the most famous group) the observed number of deaths during and just after the birth month is about 58% greater than the number expected. Note that there is no death rise at all for group 3, the least famous group.

SUMMARY

We have noted two sets of findings that are consistent with the notion that some people postpone death to witness a birthday because it is important to them. There is a death dip before the birth month and a death rise thereafter in four separate samples. We have noted also a consistent relation

TABLE 5. The Size of the Death Dip and Death Rise for Groups of Differing Fame

GROUP	SIZE OF DEATH DIP (%)	SIZE OF DEATH RISE (%)	TOTAL NO. IN GROUP	NO. OF DEATHS IN THE BIRTH MONTH, AND 1, 2, 3 MONTHS THEREAFTER	NO. OF DEATHS IN THE MONTH BEFORE THE BIRTH MONTH
1	−78	58	55	29	1
2	−63	23	129	53	4
3	−20	−3	164	53	11

between the fame of a group and the size of its death dip and death rise: the more famous the group, the larger the death dip and death rise it produces. These results might be due to chance, but this possibility is sufficiently small that we would prefer some other explanation of these phenomena.

There are indications that some people postpone dying in order to witness events other than their birthdays. There are fewer deaths than expected before the Jewish Day of Atonement in New York, a city with a large Jewish population. In addition, there is a dip in U.S. deaths, in general, before U.S. Presidential elections.

We have anecdotal evidence that the timing of death might be related to other important social events. For example, many people have noted that both Jefferson and Adams died on July 4th, 50 years after the Declaration of Independence was signed. We may find it easier to believe that this is not coincidental if we read Jefferson's last words, quoted by his physician.[1]

> About seven o'clock of the evening of that day, he [Jefferson] awoke, and seeing my staying at his bedside exclaimed, "Oh Doctor, are you still there?" in a voice however, that was husky and indistinct. He then asked, "Is it the Fourth?" to which I replied, "It soon will be." These were the last words I heard him utter.

PROBLEMS

1. Why does the article examine the deaths of famous people only?

2. State and describe the two hypotheses being tested.

3. For the data in Table 1 is the assumption that "1/12 of the deaths occur each month" reasonable? Give reasons.

4. Calculate the expected number of deaths in each of the twelve starred cells in Table 1 under the assumption of independence, using the first method described in the text. Check that the expected total number of deaths in the month before the birth month is 28.3.

5. When would the two methods of estimating the total expected number of deaths in the month just before the birth month give exactly the same answer?

6. Refer to Figure 1.
    a. What is the meaning of the dotted line?
    b. Read from the graph the approximate number of deaths 1.5 months before the birth month. Is this figure meaningful? Why or why not?

---

[1] Merrill Peterson, *Thomas Jefferson and the New Nation* (New York: Oxford University Press, 1970), p. 1008.

c. Suppose that the death month of famous people is really independent of the birth month. Would you then expect the graph to be a horizontal line? Why or why not?

7. Consider the following alternative explanation (which might be called the birthday bash theory) for the death rise during and after the birth month:

"The reason for a death rise during these specific months is that famous people tend to exhaust themselves during the heavy celebrations on their birthdays thereby increasing the chance of death shortly after that date."

From the data presented in this article can one distinguish between this explanation and the one proposed in the article?

8. Do you think sample 4 is an appropriate sample of the people whose names appear in the 1897–1942 volume of "Who Was Who" and whose surnames appear in "Foremost Families"? Comment.

EPIDEMICS

Maurice S. Bartlett   *Oxford University*

[*Editorial comment.*   Many advisors have suggested that this volume include an application in which a statistical or stochastic model is especially constructed to fit a real world process.   These advisors want the volume to show the reader the model builder at work, to show how the model is tested, and to explore the new things it demonstrates.   To appreciate the whole process requires some mathematics.   We also might have to see the worker's wastebasket to appreciate how much effort may go into attempts he found unsatisfying.

Even the reader who has little mathematical equipment can gain considerable insight into the mathematical study of the birth, death, and maintenance of epidemics by skipping the harder mathematics in the piece Professor Bartlett has so kindly provided.   For those who wish to skip along, the more mathematical parts have been set off and indented, and a few words, sometimes redundant, of transition have been inserted.—J.T.]

A CLASSIC book by the late Professor Greenwood, a medical statistician who was an authority on epidemics, has the title *Epidemics and Crowd Diseases,* which emphasizes the relevance of the population, or community, in determining how epidemics arise and recur.   One of the first to realize the need for more than purely empirical studies of epidemic phenomena was Ronald Ross, better known for his discovery of the role of the mosquito in transmitting malaria to human beings.   Since those early years, the mechanism and behavior of epidemics arising from infection that spreads from individual to individual, either directly or by an intermediate carrier, have been extensively studied,[1] but it is perhaps fair to say that only in recent years have some quantitative features become understood and even now much remains to be unraveled.

MATHEMATICAL MODELS AND DATA

The statistical study of epidemics, then, has two aspects—on one hand, the medical statistics on some infectious disease of interest and, on the other hand, an appraisal of the theoretical consequences of the mathematical model believed to be representative, possibly in a very simplified and idealized fashion, of the actual epidemic situation.   If the consequences of the model seem to agree broadly with the observed characteristics, there is some justification for thinking that the model is on the right lines, especially if it predicts some features that had been unknown, or at least had not been used, when the model was formulated.

How do we build a mathematical model of a population under attack by an infectious agent?   There is no golden rule for success.   Some feel that everything that is known about the true epidemic situation should be set down and incorporated into the model.   Unfortunately, this procedure is liable to provide a very indigestible hotchpotch with which no mathematician can cope, and while in these days of large-scale computers there is much more scope for studying the properties of these possibly realistic, but certainly complicated, models, there is still much to be said for keeping our model as simple as is feasible without too gross a departure from the real state of affairs.   Let us begin then at the other extreme and put in our ingredients one at a time.

We start with a community of individuals susceptible to the infection, let us say a number S of them.   We must also have some infection, and we will confine our attention to the situation in which this can be represented by a number of individuals already infected and liable to pass on the disease. The astute reader will notice that even if we are concentrating on epidemic situations with person-to-person infection, we perhaps ought not to amalgamate *infected* and *infective* persons.   Some people may be already infected, but

[1] Other pioneering workers in this field include W. Hamer, A. G. McKendrick, H. E. Soper, and E. B. Wilson.

not yet infective; some might not be infected, at least visibly, and yet be infective—so-called *carriers*.   As we are considering the simplest case, however, we merely suppose there is a number I, say, of infective persons.   When, as is to be hoped in real life, these persons recover, they may become resusceptible sooner or later (as for the ·common cold) or permanently immune, as is observed with a very high proportion of people in the case of measles.

While these ingredients for our epidemic recipe are very basic and common to many situations, there is a better check on our model if we are more definite and have one disease in mind, so let us consider only measles from now on.   This is largely a children's complaint, mainly because most adults in contact with the virus responsible have already become immune and so do not concern us.   Measles is no longer as serious an illness as it used to be (even less now that there is a preventive vaccine), but is a convenient one to discuss because many of its characteristics are fairly definite: the permanence of subsequent immunity, the incubation period of about a fortnight, and the requirement of notifiability in several countries, including the U.S., England, and Wales.   The last requirement ensures the existence of official statistics, though it is known that notifications, unfortunately, are far from complete.   Provided we bear this last point in mind, however, and where necessary make allowance for incomplete notification, it should not mislead us.

A DETERMINISTIC MATHEMATICAL MODEL

To return to our mock epidemic, we next suppose that the infective persons begin to infect the susceptibles.   If the infectives remain infective, all the susceptibles come down with the infection eventually, and that is more or less all there is to be said.   A more interesting situation arises when the infective persons may recover (or die, or be removed from contact with susceptibles) because then a competition between the number of new infections of susceptibles and the number of recoveries of infectives is set up.   At the beginning of the epidemic, when there may be a large number of susceptibles to be infected, a kind of chain reaction can occur, and the number of notifications of new infected persons may begin to rise rapidly; later on, when there are fewer susceptibles, the rate of new notifications will begin to go down, and the epidemic will subside.

If we drew a graph of the number of infectives I against the number of susceptibles S at each moment, it would look broadly like the curve in Figure 1.   The precise path, of course, will depend on the exact assumptions made on the overall rate of infection, on whether this is strictly proportional to the number I, and on whether also proportional to S, so that the rate at any moment is, say, calculated from the formula aIS where a is a constant. The path will depend also on the rate of recovery of the infected population,

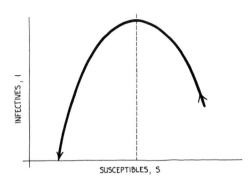

FIGURE 1

General path of an epidemic beginning with many suscepti- bles S, increasing at first the number of infectives I, then decreasing

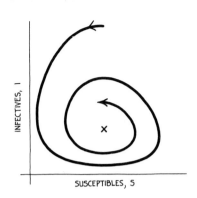

FIGURE 2

Deterministic model. Approach to equilibrium point (at cross) of I, S curve

likely to be proportional to I and given by bI, say, where b is a constant. Without going into too much detail yet, we can note one or two distinct features in the figure: (1) the susceptible population may not be reduced to zero at the time that the sources of infection have been eliminated; (2) because the path has no "memory," we could start at any point on it and proceed along the same curve—if our starting point were to the right of the maximum point, our infectives I would rise, if to the left, they would fall.

Actually, our more detailed assumptions are equivalent to the pair of recur- rence relations for calculating I_{t+1}, S_{t+1} at any time $t + 1$ in terms of the values I_t, S_t at the previous time t (the unit of time should be small, say, a day or week):

$$S_{t+1} = S_t - aI_tS_t, \qquad I_{t+1} = I_t + aI_tS_t - bI_t.$$

We notice that $I_{t+1} - I_t$ is positive or negative according to whether $aS_t - b$ is positive or negative. The value b/a for S (on our assumptions it is independent of I) is called the *critical threshold* for reasons that will become clear.

Thus far we have a model for a "closed epidemic," which terminates when I is zero. Can we turn it into a model for measles, which has been claimed to come in epidemics about every two years? One ingredient still missing is an influx of susceptibles, due, in the case of childhood illnesses, to births within the community.

Let us, therefore, add a term c, say, to the right-hand side of the first equation above. If we follow the course of events in Figure 1, the path will start turning right before it reaches the axis $I = 0$ and can be shown to proceed in an ever-decreasing spiral (Figure 2) till it finally arrives at an equilibrium point, which is determined by the equations

$$c - aIS = 0, \qquad aIS - bI = 0.$$

The second of these gives $S = b/a$ (the critical threshold value), and the first then yields $I = c/b$. These results are partly encouraging, but partly erroneous. The encouraging feature is the tendency to recurrent epidemics; we can even find the period of the cycle, which is found to be approximately

$$\frac{2\pi}{\sqrt{ac - \frac{1}{4}a^2c^2/b}}.$$

Sir William Hamer, who first put forward the above model for measles in 1906, took $b = \frac{1}{2}$ when t is reckoned in weeks, corresponding to an average incubation period of a fortnight, and c for London at that time as 2200. The value of a is more uncertain, but one method of arriving at it is to note that the *average* number of susceptibles, which was put at 150,000, should be around the theoretical equilibrium value b/a giving $a = 1/300,000$. Notice that c will tend to be proportional to the size of the community, so that if ac is to remain constant, a must be inversely proportional to the population, but this is not an unreasonable assumption; it is consistent, for example, with effective infectivity over a constant urban area, the entire town being regarded as an assemblage of such units.

The introduction of an influx of susceptibles showed that instead of following the simple curved path of Figure 1, an epidemic might follow a spiral until it finally settled down with a particular number of susceptibles. The time to go around the spiral once, called the *period,* is estimated for London data at 74 weeks, in reasonable agreement with the average period of somewhat less than two years that has been observed for large towns in England and Wales (see Table 2), the U.S. and comparable countries in the present century. We would hardly expect the epidemic pattern to remain precisely the same under very different social conditions, though the annual measles mortality figures for London quoted from John Graunt for the seventeenth century (Table 1) suggest a similar pattern even then (with perhaps a slightly longer average period of 2 to 3 years).

TABLE 1.   Deaths from Measles in London in the Seventeenth Century

1629	1630	1631	1632	1633	1634	1635	1636	1637–46	1647	1648
41	2	3	80	21	33	27	12	Not recorded	5	92

1649	1650	1651	1652	1653	1654	1655	1656	1657	1658	1659	1660
3	33	33	62	8	52	11	153	15	80	6	74

UNSATISFACTORY FEATURES OF THE MODEL

The erroneous feature of the improved model is that actual measles in London or other large towns recurs in epidemics without settling down to a steady endemic state represented by the theoretical equilibrium point. What aspects of our model must we correct? There are some obvious points to look at:

(1) Our assumptions about the rate of recovery correspond to a gradual and steadily decreasing fraction of any group infected at the same time, whereas the incubation period is fairly precise at about two weeks, before the rash appears and the sick child is likely to be isolated (this being equivalent to recovery).

(2) We have ignored the way the children are distributed over the town, coming to school if they are old enough or staying at home during a vacation period.

(3) Measles is partly seasonal in its appearance, with a swing in average notifications from about 60% below average in summer to 60% above average in winter.

INTRODUCING CHANCE INTO THE MODEL

We will consider these points in turn. The effect of point (1) is to lessen the "damping down" to the equilibrium level, but not, when correctly formulated, to eliminate it. Point (2), on the movement over the town, raises interesting questions about the rate of spread of infection across different districts, but is less relevant to the epidemic pattern in time, except for its possible effect on (3). If we postulate a $\pm 10\%$ variation in the "coefficient of infectivity" a over the year, it is found to account for the observed $\pm 60\%$ or so in notifications. There seems to be little evidence of an intrinsic change in a due, say, to weather conditions, and it may well be an artifact arising from dispersal for the long summer vacation and crowding together of children after the holidays. Whatever its cause, it does not explain the persistence of a natural period; only the seasonal variation would remain and give a strict annual period, still at variance with observation.

To proceed further, let us retrace our steps to our closed epidemic model of Figure 1. To fix our ideas, suppose we initially had only one infective individual in the community. Then the course of events is not certain to be as depicted; it may happen that this individual recovers (or is isolated) before passing on the infection, even if the size of susceptible population is above the critical threshold. This emphasizes the chance element in epidemics, especially at the beginning of the outbreak, and this element is specifically introduced by means of probability theory. To examine the difference it makes, let us suppose the *chance* P of a new infection is now proportional to aIS and the chance Q of a recovery proportional to bI. Denote the chance of the outbreak ultimately fading out without causing a major epidemic by p. We shall suppose also that the initial number S_0 of susceptibles is large enough for us not to worry about the proportionate change in S if the (small) number of infective persons changes. Under these conditions two infective persons can be thought of as acting independently in spreading infection, so that the chance of the outbreak fading out with *two* initial infective persons must be p^2.

Now consider the situation after the first "happening." Either this is a new infection or a recovery, and the relative odds are $P/Q = aS_0/b$. If it is a new infection, I changes from 1 to 2, and the chance of fade-out is p^2 from now on, or I drops to zero, and fade-out has already occurred. This gives the relation

$$p = \frac{P}{P+Q} p^2 + \frac{Q}{P+Q},$$

whence either $p = 1$ or $p = Q/P = b/(aS_0)$. If $b \geq aS_0$ (that is, if we are below the critical threshold) the only possible solution is $p = 1$, implying as expected that the outbreak certainly fades out. However, the ultimate probability of fade-out can be envisaged as the final value reached by the probability of fade-out up to some definite time t, this more general probability steadily increasing from zero at $t = 0$ to its limiting value, which therefore will be the *smaller* of the roots of the above quadratic equation. This is $b/(aS_0)$ if this value is less than one, providing us with a quantitative (nonzero) value of the chance of fade-out *even if the critical threshold is exceeded* and stressing the new and rather remarkable complications that arise when probabilistic concepts are brought in.

The mathematics shows that if the initial number of susceptibles is smaller than a value determined by the ratio of some rates used in the model, then the epidemic will certainly fade out. If it is larger than this critical value, then there is still a positive probability that the epidemic will fade out.

When new susceptibles are continually introduced, represented by c, the complications are even greater. For small communities, however, the qualitative features can be guessed. Once below the threshold, the number of infectives will tend to drop to zero, and though the susceptibles S can increase because of c, it seems unlikely that the number will pass the threshold before I has dropped to zero. The epidemic is now finished, and cannot re-start unless we introduce some *new* infection from outside the community. This is exactly what is observed with measles in a small isolated community, whether it is a boarding school, a rural village, or an island community. For such communities, the period between epidemics depends partly on the rate of immigration of new infection into the area and not just on the natural epidemic cycle. Moreover, when new infection enters, it cannot take proper hold if the susceptible population is still below the threshold, and even if

TABLE 2.   Measles Epidemics for Towns in England and Wales (1940–56)

TOWN	POPULATION (THOUSANDS)	MEAN PERIOD BETWEEN EPIDEMICS (WEEKS)	TOWN	POPULATION (THOUSANDS)	MEAN PERIOD BETWEEN EPIDEMICS (WEEKS)
Birmingham	1046	73	Newbury	18	92
Manchester	658	106	Carmarthen	12	79
Bristol	415	92	Penrith	11	98
Hull	269	93	Ffestiniog	7.1	199
Plymouth	180	94	Brecon	5.6	149
Norwich	113	80	Okehampton	4.0	105
Barrow-in-Furness	66	74	Cardigan	3.5	>284
Carlisle	65	75	South Molton	3.1	191
Bridgwater	22	86	Llanrwst	2.6	>284
			Appleby	1.7	175

Source: Bartlett (1957), Tables 1 and 2.

above, new infection may have to enter a few times before a major outbreak occurs.   The average period tends to be above the natural period for such communities.   If we assume that the rate of immigration of new infection is likely to be proportional to the population of the community, the average period between epidemics will tend to be larger the smaller the community, and this again is what is observed (Table 2).

Consider now a larger community.   We expect random effects to be proportionately less; there is still, however, the possibility of extinction when the critical threshold is not exceeded.   Nevertheless, before all the infectives have disappeared, the influx c of new susceptibles may have swung S above the threshold, and the stage is set for a new epidemic.   Under these conditions (and provided ac remains constant, as already assumed), the natural period will change little with the size of community.

CRITICAL SIZE OF COMMUNITY

How large does the community have to be if it is to begin to be independent of outside infection and if its epidemic cycle is to be semipermanent?   Exact mathematical results are difficult to obtain, but approximate solutions have been supplemented by simulation studies of the epidemic model, using computers. An example of one such series plotted to extinction of infection after four epidemics (an interval representing nearly seven years) is shown in Figure 3.   This particular series has no built-in seasonal incidence, but some internal

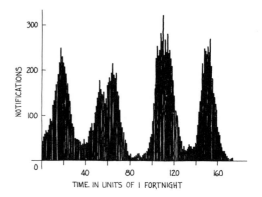

FIGURE 3

Results of simulation of four epidemics of measles over a seven-year period for a town whose average susceptible population is 3700

migration within its population boundaries; its average susceptible population is 3700.   It appears from all such results that the critical size of the susceptible population is, for the measles model, of the order of 7000, or if the factor of 40 estimated for Manchester, England, between total and susceptible population is used, over a quarter of a million people in the community.

Now we do not need to use this theoretical figure as more than an indication of what to look for.   By direct examination of measles notifications for any town, we can see whether notifications have been absent for more than two or three weeks.   In view of the rather well-defined incubation period, we would infer from this lack of notifications that the infection had disappeared if we knew that notifications were complete.   Incomplete notification is a complication, but not one that is likely to affect these quantitative conclusions very much, for fade-out of infection is found to increase rather rapidly as the community size decreases and soon becomes quite recognizable from the detailed statistics.   In this way, it was ascertained that in England and Wales, during the period 1940–56, cities of critical size were Bristol (population about 415,000) and Hull (269,000).   This investigation was supplemented by an examination of U.S. statistics for the period 1921–40, from which it was found that some comparable North American cities were Akron (245,000), Providence (254,000), and Rochester (325,000).   Therefore, there is an observed critical community size of around 300,000, in reasonable agreement with what we were expecting.

Of course, towns of such size are not completely isolated from other communities as assumed in our model; this could tend to lessen the observed critical size, especially if the isolation is comparatively slight.   In Table 3 the fade-out effect is shown for aggregates of individual "wards" in Manchester to demonstrate how it decreases with the population aggregate.   The critical size (defined precisely in terms of 50% probability of fade-out after an epidemic) is, again as expected, smaller than for complete towns due to the

TABLE 3.   Observed (Aggregate) Fade-Out Effect in Manchester Wards

WARDS	CUMULATIVE POPULATION (THOUSANDS)	NUMBER OF EPIDEMICS FOLLOWED BY FADE-OUT	PROBABILITY OF FADE-OUT (%)
Ardwick	18.4	12	100
St. Mark's	38.2	12	100
St. Luke's and New Cross	71.8	9	75
All Saints, Beswick, and Miles Platting	140.8	4	33
Openshaw, Longsight, N. and S. Gorton, Bradford, and St. Michaels	254.1	1	8
Medlock St., W. and E. Moss Side, Rusholme, Newton Heath, Colly-hurst, Harpurhey, and Cheetham	419.3	0	0

Source: Bartlett (1957), Table 3.

extensive migration across the ward boundaries; it is estimated to be 120,000 total population living within the area.

CONCLUSIONS

If we review these results, we may justifiably claim that our theoretical model for measles, idealized though it inevitably is, has achieved some fair degree of agreement with the observed epidemic pattern.   In particular:

(1) It predicts a "natural" period between epidemics of rather less than two years.
(2) A small ($\pm 10\%$) seasonal variation in infectivity (whether or not an artifact of seasonal pattern in school-children's movements) accounts for the larger ($\pm 60\%$) observed seasonal variation in notifications.
(3) It predicts extinction of infection for small communities, with consequent extension (and greater variability) of periods between epidemics.
(4) It predicts a critical community size of over a quarter of a million necessary for the infection to remain in the community from one epidemic to the next.

Epidemic patterns, of course, will be very sensitive to changing customs and knowledge; and the introduction of a vaccine for measles will inevitably change its epidemic pattern, and perhaps in time eliminate the virus completely.   However, the greater understanding of epidemics that follows from appropriate models may be applied to other epidemic infections, and should assist in predicting and assessing the consequences of any changed medical practice or social customs even for measles.

PROBLEMS

1. Explain the difference between a deterministic mathematical model and a chance model.

The following problems refer to the material in small print.

2. Explain what is meant by "critical community size."

3. The deterministic model in Figure 2 predicts that there is an equilibrium point of I,S; i.e., the epidemic will never fade out. What is the concept that had to be introduced into the model to alter this prediction and thus better explain the data in Table 3 where we notice that in some communities nearly all the epidemics eventually fade-out?

4. Stated in words the formula $S_{t+1} = S_t - aI_t S_t$ says that "the number of susceptibles at the time $t+1$ = number of susceptibles at time t− _____ at time t.

5. What is the equilibrium point (S_e, I_e) for the deterministic model shown in Figure 2?

6. Check by substitution that $p = 1$ and $p = Q/P$ are the solutions of the equation $p = p^2 P/(P+Q) + Q/(P+Q)$.

REFERENCES

M. S. Bartlett. 1957. "Measles Periodicity and Community Size." *Journal of the Royal Statistical Society*, A120: 48–70.

J. Graunt. 1662. Reprint, 1939. *Natural and Political Observations upon Bills of Mortality.* Baltimore: Johns Hopkins University Press.

M. Greenwood. 1935. *Epidemics and Crowd Diseases.* London: William and Norgate.

R. Ross. 1910. *The Prevention of Malaria,* Second Edition. London: J. Murray.

DOES INHERITANCE MATTER IN DISEASE?
THE USE OF TWIN STUDIES IN
MEDICAL RESEARCH

D. D. Reid   *London School of Hygiene and Tropical Medicine*

THE CONTROVERSY about the relative importance of nature and nurture that goes on in fields such as psychology also goes on in medicine. The crucial question is "Do we go mad or develop heart disease because we inherit a special susceptibility to a disease from our parents, or are these diseases the result of a stressful environment throughout our lives?" With most diseases, we cannot decide because patients almost always differ in both their nature, or genetic endowment, and their nurture, or environment, during the formative years of childhood, and of course, both can be important in particular diseases. Yet it is important to know whether in coronary heart disease, for example, it is worth persuading a patient to alter his environment by stopping smoking. If he both smokes heavily and has heart disease because he has inherited an anxious, worrying temperament, then altering his smoking habit alone might have disappointingly little effect. For these reasons, medical research in this area has concentrated on the unique and unusual opportunities presented by the occurrence of disease in twins.

The idea for this approach came from the English biometrician Francis Galton in 1875. He pointed out that there are two kinds of twins, identical and nonidentical. Identical twins come from the same fertilized egg and therefore have the same set of genes, which determine their physical and mental characteristics. Nonidentical twins, like ordinary siblings born at different times, come from separate eggs and their gene patterns are no more (and no less) alike than those of ordinary brothers and sisters (see Figure 1). Both types of twins, however, usually share the same family environment during their childhood and are usually treated alike by their parents. If, then, one of the pair develops a disease that is transmitted through the genes, it is more likely to appear in his identical twin than in a twin who is genetically different. On the other hand, in diseases due to diet or some other aspect of family life that is not genetically determined in the strict sense, the identical and nonidentical twin siblings of affected children are equally likely to develop the disease. Thus, we can get some indication of the relative importance of heredity and environment in causing a specific disease by comparing the

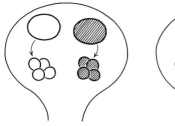

FIGURE 1

Genes determining light hair are indicated (O) and dark hair (●). Nonidentical twins come from separate eggs which carry different genetic elements that determine color of hair in the two children. It may or may not be different in the case of nonidentical twins. Part (a) illustrates an instance in which hair color is different. Identical twins result from the division of the same egg and so retain the same genetic element. Both children thus have the same hair color

TABLE 1.   Concordance Example
(O = Unaffected; ☉ = Affected)

PAIR NUMBER	FIRST TWIN	SECOND TWIN
1	☉	☉
2	☉	☉
3	☉	☉
4	☉	☉
5	☉	O
6	☉	O

relative risk of the twin being affected when they are identical twins to that risk when they are nonidentical.   The difference in risk will be high in diseases where inheritance is important and low where it is not.

MEASURES OF DISEASE "CONCORDANCE"

These risks of the coincidence of disease appearing in both twins are measured in two ways.   Both methods aim to assess the degree of *concordance*, or agreement, between the disease experience of identical and nonidentical twins. Table 1 is a model population of six sets of identical twins.   One measure of the risk of both being affected is the "pair-wise concordance rate."   In this example, in four out of the six pairs, both twins are affected, so the "pair-wise concordance rate" is $\frac{4}{6} \times 100\%$ or $66\frac{2}{3}\%$.   Pairs in which neither twin has the disease do not enter this diagram.   The second measure depends on the relative frequency or risk of the twin of an affected person also suffering from the disease.   This "proband concordance rate," or "affected concordance rate," in the example is given by the ratio of the number of twins affected as pairs (8) to the total number of affected people (10), that is, $\frac{8}{10}$ or 80%.   The "affected concordance rate" is perhaps the more widely used in twin studies.   (Knowing one of these measures, we can readily obtain the other, so no important choice is being made here; it is more a question of custom.)

SOME RESULTS FROM THE DANISH TWIN REGISTRY

Table 2 compares the "affected concordance rates" for identical and nonidentical pairs of Danish twins of whom at least one is affected by one of the diseases listed.   The high rates in identical twins versus the lower ones for nonidentical twins for tuberculosis (54 versus 27), rheumatoid arthritis (50 versus 5), bronchial asthma (63 versus 38), and epilepsy (54 versus 24) suggest a strong genetic element in these diseases.   On the other hand, death

TABLE 2. Occurrence of Selected Somatic Diseases in the Danish Twin Register Based on a Survey of 4368 Same-Sexed Pairs

DISEASE	AFFECTED CONCORDANCE RATES (%)			
	Identical Rate	No.*	Nonidentical Rate	No.*
Cerebral apoplexy	36	120	19	164
Coronary occlusion	33	122	27	179
Tuberculosis	54	185	27	309
Rheumatic fever	33	178	10	238
Rheumatoid arthritis	50	63	5	73
Death from acute infection	14	137	11	235
Bronchial asthma	63	94	38	125
Epilepsy	54	37	24	49

* Numbers refer to number of affected individuals, not pairs.

from acute infections other than tuberculosis and rheumatic fever (14 versus 11) shows no such disparity between the identical and nonidentical pairs of twins. Chance exposure to infection, perhaps outside the home, seems, therefore, to be much more important than any inherited susceptibility.

THE UNUSUALNESS OF TWINS

These examples have shown the potential of this method of distinguishing between genetic and environmental factors in different diseases. A word of caution is needed, however, about the risks of generalizing from the twins to the population as a whole, for twins are unusual in more senses than one. They are unusual in that multiple births occur relatively infrequently. A recent survey in the U.S. has shown that the proportion of twins has been falling in recent years. In 1964, 1 in 96 deliveries were of twins. (Triplets and quadruplets occurred once in 9977 and once in 663,470 deliveries— compared with the ratios of 1 in 9216 and 1 in 884,730 expected on the basis of the Helin–Zeleny hypothesis, that is, if twins occur once in 96 deliveries, triplets and quadruplets will occur once in 96×96 deliveries and $96 \times 96 \times 96$ deliveries respectively. It does not concern us here.)

Most important from the medical point of view is the fact that twins are more likely to be born to black than to white Americans and especially to older mothers who have already had several children. In a condition like mongolism, which is particularly common in children born to older women who have had large families, twins are thus likely to be more often affected than are singletons. Because twins have to depend on nutrients designed for a single child in the womb, they start at a disadvantage and are more likely to be born prematurely and to have difficult births. It is hardly surprising that their death rate in infancy and early childhood is higher than average. As they grow older, their disadvantage lessens, and in respect to the diseases of adult life, their experience is probably close to that of singletons. Thus, assessments based on observing disease in middle-aged twins are likely to be reasonably applicable to people in general.

PRACTICAL ASPECTS OF TWIN STUDIES

Sampling Populations of Twins. Early studies of disease in twins were often based on the patient with an unusual disease who came to the hospital and was found to have a twin suffering from the same condition. As in clinical medicine in general, the apparently unusual tends to be noted and published. Series of such coincidences in twins have thus been given prominence in the medical literature. But, because of the haphazard method of collection, such series are unlikely to give a true picture of the incidence of disease in the population of twins as a whole. Volunteer series, recruited perhaps by appeals in press or radio, are also likely to be biased. The ideal is to collect information either on all twins born in a generation or at least on a randomly selected, and thus truly representative, sample of them. Particularly in Scandinavia where vital records have long been accurate and complete, national twin registers have been established on this basis to serve medical and social research.

Some results from the Danish Twin Register have already been given. This Register comprised all twins born in Denmark during a defined period (1870–1910). Of the 37,914 twin births that occurred during that time, over half of the pairs had been broken by the death of one or both twins before their sixth birthday; these were not followed up. About 40% of the pairs were of different sex and not so useful for investigating diseases occurring in one sex more than another. Of the remainder, some 60% were nonidentical twins of the same sex and 40% identical and so like-sexed pairs. (In other words, the elimination of mixed-sex pairs from the surviving sets of twins changed the identical-nonidentical ratio of twin births from the usual 20:80 to 40:60.)

Establishing Type of Twins. Having identified and traced the twins through the population registers and other means, a problem arises in establishing their type by methods that can be widely and simply applied. In studies of small numbers, refined techniques can be used to compare inherited characteristics such as blood groups or blood-protein patterns to see whether in all such respects twins are truly identical. Fingerprints, voice sounds, and other physical traits can also be used. For large scale surveys in which subjects cannot be examined, but only interrogated by postal questionnaires, simpler methods are needed.   It is fortunate, therefore, that the reply to one question is surprisingly effective in distinguishing identical from nonidentical pairs of twins.   That question may take the form of "Were you as like as two peas in a pod?" and it is encouraging to note that when this question indeed was asked, over 95% of pairs in which both answered "yes" proved, on blood and other examination, to be identical, and when both answered "no," over 95% were fraternal; finally, 2% disagreed.

Once the pairs of twins have been classified, their disease experience must be ascertained.   This can be done by collecting hospital records or death certificates over a period of years or by asking questions directly about either past illnesses or the presence of symptoms of chronic diseases such as coronary heart disease or rheumatism.

Other Applications of Twin Studies.   Twin studies also may help to detect the effects of environmental factors in disease.   Cigarette smoking, for example, is believed to cause other lung diseases as well as lung cancer.   As in the case of lung cancer, it could be argued that an individual inherits both a liability to take up smoking and a specific susceptibility to lung diseases such as chronic bronchitis.   If this were true, the apparent association between cigarette smoking and the presence of bronchitis could be dismissed as an effect arising at least in part from other causes rather than as proof that smoking caused bronchitis.

Surveys of smoking habits in identical and nonidentical twins have shown that there is indeed some evidence of a genetic element that affects smoking habit.   Regarding smoking as a disease, we can, as before, compare the affected-concordance rate" in identical sets of twins with that in nonidentical sets.   This shows that, if one of a set of twins smokes, his twin is more likely also to be a smoker if he is an identical twin than if he is a fraternal twin and, thus, no more closely related than an ordinary brother.

Because of this genetic element in smoking, the independent effect of smoking on bronchitis has to be assessed by comparing the frequency of bronchitis in identical twins who share the same genetic endowment but smoke different amounts.   A survey based on the Swedish National Twin Study has done just this and the results for both types of twin are given in Table 3.   These are set out simply in the form of prevalence rates which give the percentage of people in each twin and smoking group who have chronic coughs.   Clearly, within each group of identical twins, smokers have higher prevalence rates for chronic cough than nonsmokers.   In other words, even when the genetic background is identical, smoking appears to be associated with more bronchitis and thus is likely to be its cause.

Related Ideas.   The use of twins is not confined to studies of medicine or biology; they have been used in studies of reading, for example.   The idea of using identical twins as a control in the smoking investigation is a special

Table 3.   Prevalence of Cough Among Smokers and
Nonsmokers in Smoking Discordant Twin Pairs

	COUGH PREVALENCE (%)		TOTAL NUMBER OF CASES
	Smokers	Nonsmokers	
Identical twins			
Men	14.6	7.7	274
Women	13.6	7.6	264
Nonidentical twins			
Men	12.3	5.5	733
Women	14.5	5.7	653

case of the important statistical idea of using homogeneous "blocks" to test different effects.   First, keeping close to the twin idea, in agricultural experiments, littermates are sometimes assigned to different treatments such as feeding regimes to improve the precision of the resulting average weight gains under the different diets.   Of course, many litters may need to be used, but fewer individuals will be needed because of the "matching" provided by the litters.

Future International Collaboration.   The results from the Scandinavian Twin Registries show how useful such data can be.   Unfortunately, for less common diseases even large national registries may not uncover enough cases of a specific disease (as in different forms of heart disorder) to make detailed statistical analysis possible.   The World Health Organization therefore has set up a "registry of registries" to collect data in a uniform fashion in many countries and assemble and analyze them centrally.   In this way, we hope to reap the full benefit of the unique research opportunities that studies of twins in sickness and health can provide.

PROBLEMS

1.   Why are volunteer series in twin studies likely to be biased? (Hint: refer to the essay by Meier.)

2.   By comparing the risk of an identical twin being affected to that of a non-identical twin, an experimenter would eliminate any _____ factors, and thus any observed differences between the two kinds of twins could be attributed to _____ factors.

3. In the data of Table 3, only those pairs of twins are considered for which:
    a. Either both are smokers or nonsmokers.
    b. One twin is a smoker and the other a nonsmoker.
    c. Either both or one or none are smokers.
Which answer is correct?

4. Refer to Table 3. Can you convert the data of cough prevalence from percentage into incidence (number of cases)? Carry out this conversion for the "Identical twins" if you can; explain what other information you need, if you can't.

5. Explain how one should use twins in studies to detect:
    a. The effect of genetic factors in disease.
    b. The effect of environmental factors in disease.

REFERENCES

M. G. Bulmer. 1970. *The Biology of Twinning in Man.* New York: Oxford University Press.

A fuller account of "The Use of Twins in Epidemiological Studies" is given in a WHO report in *Acta Genetica et Gemellologia,* 15(1966):2, pp. 109–128. The Danish Registry and its work are described in the same journal, 18(1968):2, pp. 315–330. The report "Multiple Births USA 1964" from the National Center for Health Statistics is published by the U.S. Public Health Service.

THE PLIGHT OF THE WHALES

D. G. Chapman   *University of Washington*

BETWEEN THE end of World War I and 1960, several species of whales in the ocean around the Antarctic continent were the basis of an important industry. These giant mammals, the largest that have ever existed on the earth, were sought for animal oil and, to a lesser extent, meal and meat extract (the latter for human consumption) as well as a myriad of byproducts. In antiquity, whalers went out in small boats and endured great risks to capture such large sources of meat. Men continued to hunt whales in small boats with primitive weapons, as portrayed in *Moby Dick,* until late in the nineteenth century. In the twentieth century, whaling has been highly modernized with explosive harpoons, large ships, and powerful radar-equipped catcher boats, which enable the whaling industry to operate in the stormy and inhospitable oceans next to the Antarctic ice cap.

This area of the world, while unfriendly to man, is very inviting to whales, for during the southern summer the waters bloom with small plants which, in turn, feed myriads of minute animals, known generally as *krill*. Certain species of whales catch these by straining large volumes of water in their huge mouths through sievelike filters called *baleen plates* (hence this group of whales is referred to as baleen whales). These whales have no teeth and do not eat fish or other marine mammals. The largest of the baleen whales, and indeed of all whales, are the blue whales, which may reach a length of 100 feet, though 70 to 80 feet is a more usual size.

BLUE WHALES

Immediately following World War II, Europe and Japan were in desperate need of many things, including animal oil. It was not surprising, therefore, that the number of Antarctic whaling catcher boats increased; furthermore, technologies developed during the War made whaling more efficient. As a result, some conservationists feared that Antarctic whales, particularly the blue whales, would be completely eliminated. Figure 1 shows the annual catch of blue whales in the southern oceans in the decade before the War and in the postwar period to 1960. The basis for concern for the blue whales was easy to document, but the catch of other species was stable or increasing. Some of those associated with the industry suggested reasons other than a decline in population for the decline of the blue whale catch and were re-

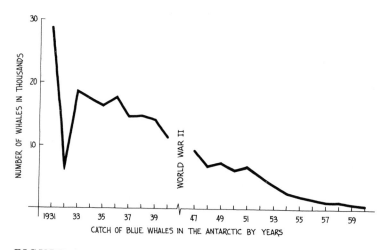

FIGURE 1

Catch of blue whales in the Antarctic, seasons 1930–31 to 1959–60 excluding World War II period

luctant to accept restrictions on catches. Thus the International Whaling Commission, set up in 1946 to manage the resource, found itself the center of controversy. The Commission has representatives from all interested countries; it establishes size regulations and quotas, and also has the authority to ban hunting of species that appear to be endangered.

The Commission set up a study group to bring together all the data and develop statistical methods for attacking such questions as: How many whales are in the stock that feed in the Antarctic? How many young are born each year? How many whales die from natural causes each year? How are these birth and death rates affected by factors over which man has some control?

COUNTING METHODS

Let us consider the first, and perhaps most basic, question: How many whales are there of a particular species? Whales unfortunately don't stay still to be counted. They roam over large areas, spending most of their time under water, though they do surface at regular intervals to breathe. Furthermore, the southern oceans cover a vast part of the world; the whaling area exceeds 10 million square miles, an area larger than all of North America.

Marking Method. There are several standard ways to estimate wild animal populations, all of which involve some statistical techniques. We shall describe three of them. The first involves marking a number of whales; a foot-long metal cylinder is fired into the thick blubber that lies just under the skin. If and when marked whales are later caught, some information is available on their movement, on their rate of capture, and on the proportion of marked members in the whole herd. The usefulness of the latter information is easily seen by simulating such an experiment with a can of marbles. Assume that like the whale population, the number of marbles in the can is unknown. Now pick a few marbles (say, ten) out of the can, mark them, and return them to the can. Next, stir the whole can thoroughly and draw another sample. Count separately the marked and unmarked marbles. If the unmarked ones are four times as numerous in the sample as the marked ones, we reason that the same is true of the whole canful; but because there is a total of ten marked marbles, we infer there are 40 unmarked marbles, or 50 marbles in total.

This simple scheme has been used with many animal populations, though there are many obvious complications in practice, and for whales this is especially true. How do we know, for example, that the metal mark fired into the blubber actually penetrated and did not ricochet off? Did the crew who cut up the captured whale carefully look for the mark—even a foot-long metal cylinder is easy to overlook in cold, stormy working conditions when the volume being cut up is approximately the size of a house. Also, unlike the marbles, whales are born and die over a period of years. All of these complications require refinements and extensions of the simple experiment outlined here. It is necessary to have a series of experiments extending over many years and to use comparative procedures. For example, if a group of whales is marked in year 1 and a group of the same size is marked in year 2, then, after year 2, the ratio of recoveries of whales marked in year 1 to the recoveries of whales marked in year 2 reflects the proportion of marked whales of group 1 that died in the intervening year. These deaths may have been natural or caused by hunters. Moreover, the *ratio* is a valid measure of this mortality because its numerator and denominator are equally affected by the possible errors listed above. Such a comparative study is only one of the several statistical procedures used to analyze whale-marking data.

Catch-per-Day Method. The second estimation method is based on changes in the rate of catching whales. The rate of catching depends mainly on the frequency with which whales are seen, and other things being equal, this depends on their density. Thus the catch per day reflects the density. How can this be translated into absolute numbers? If the change in catch per day is entirely a result of the removal by man, then it is easy to make this translation; if catching 25,000 whales in one season lowers the catch rate for the next season by 10% then at the outset of the first season there must have been 25,000/0.10, or 250,000 whales.

Again, the situation is more complex than this simple example. Whaling ships hunt over a vast area in difficult conditions, so that the catches fluctuate violently. Whaling companies introduce new technology to improve their efficiency. Moreover, we reemphasize that there are other causes of whale mortality, and that there are new births as well; both of these factors must be taken into account in adjusting the population estimate. One way to overcome some of these difficulties is to adjust for changes in efficiency and also to follow the change in catch per day (adjusted) over a period of several seasons. Figure 2 shows the catch per day of blue whales plotted against the cumulative catch by the whaling-factory ships over the seasons 1953–54 to 1962–63 when natural deaths and births were numerically quite small. As more whales were caught, the catch per day went steadily down. This graph suggests that there were only 10,000–12,000 blue whales in 1953 and this number declined to about 1000 in 1963. As pointed out, there are statistical refinements, and the result obtained in this way must be combined with estimates obtained in other ways.

Age Analysis. The catch-per-day method works well with rapidly declining populations, but in other situations, the complications and corrections make it less useful. Still a third method is available, however, that uses the ages of whales. Just as trees have annual rings in their trunks and fish have

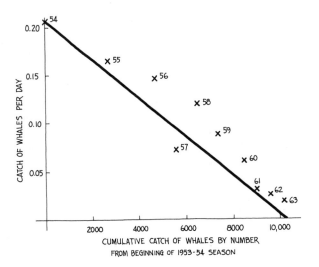

FIGURE 2

Blue whale catch per day (ad-
justed for efficiency improve-
ments) versus cumulative catch,
1953–54 to 1962–63

annual rings in their scales, whales have annual rings in a waxy secretion in the ear (earplugs). The ages of a sample of the whales killed each year were determined by the rings of their earplugs. In addition, information on the length of every whale killed commercially made it possible to relate age to length and to calculate an estimated age for every captured whale.

It was thus possible to make a statistical estimation of the number of four-year-old whales in any season and the number of five-year-olds in the following season. Because one year's five-year-olds are the survivors of the previous year's four-year-olds, a survival rate or, conversely, a mortality rate can be determined. Because all ages are estimated and because some adjustments have to be made, the estimated mortality rates fluctuate wildly. However, by averaging over several year's classes, over areas and seasons, useful results can be obtained. Furthermore, with additional statistical analysis, it is even possible to assess the magnitudes of possible errors in such estimations. These mortality rates help us to predict the future of the whale population. (See the essay by Keyfitz for an explanation of this method as applied to human populations.)

Results. Thus we have three methods of estimating population sizes and mortality rates: the marking method, the catch-per-day method, and age analysis. The results of different methods were checked against one another, and fortunately the different estimates were in good agreement. Sources of error were carefully checked and ruled out, so that the study group finally concluded that the blue whales numbered at most a few thousand and might total even less than 1000. Thus there was and is a real danger of extinction of this species in the Antarctic (there are also small numbers of blue whales in the northern oceans). Fortunately, the International Whaling Commission banned the taking of blue whales as soon as the study was finished—first, in a large part of the southern oceans and eventually in all waters south of the equator. It is too soon to predict the long-term survival of the species; blue whales are occasionally seen, but these are probably the survivors noted above (whales can live, in the absence of hunting, to more than 40 years of age). We can ask whether the population has been reduced to such low levels that reproduction is reduced below the level necessary for species continuation, but it will be a number of years before this can be answered.

FIN WHALES

The second-largest whale species in the world, also part of the baleen family, is the fin whale. It averages about ten feet less than the blue whale in length. During the fifties this stock annually yielded in excess of one million barrels of oil per year. With the decline of the blue whales, fin whales bore the brunt of the exploitation. The same methods of analysis used for the blue whales were applied to the fin whales; in fact, the analysis was more critically needed because the condition of the fin whale stock was not obvious as was that of the blue whales. Moreover, fin whale catches were still very high: in the 1961–62 season over 27,000 fin whales were killed. The study group recommended that the fin whale catch should be reduced to 7000, or less, if the fin whale stock was not to be further depleted. The proposal for such a drastic reduction came as a shock to the Commission; the study group forecast that the next season's catch, regardless of quotas, would drop to 14,000. When actual figures proved the forecast right, most countries wanted to move toward the drastic reductions required, but some of the whaling nations were able to block action. Another disastrous season caused a revision in the thinking of the commissioners, and in 1965, a substantial schedule of reductions in the quota was agreed upon. Nevertheless, the delay in reaching this agreement and the delay in reducing the quota subsequently, meant that the presently permitted catches have had to be lowered even further. Subsequent analyses using additional data show that, as of 1971, catches in excess of 3000 per year are likely to mean further reduction of the stock. This drastic reduction in the permitted catch has had a severe impact on the industry. Of the five countries that hunted whales with factory ships in the Antarctic in 1960, only Japan and the Soviet Union are still actively engaged in the industry. Further, these have diverted some of their effort to catching smaller whales of little interest in other times and have transferred some factory ships to other oceans or to other types of fishing.

THE FUTURE OF THE WHALE STOCK

The study group was asked to advise the Commission not only on the state of the whale stocks, but also on what the optimum stock size might be. The optimum size is that which will yield the maximum number of whales each year on a continuing basis. It is now recognized that whales should be managed like our fisheries and forestry resources on a sustained-yield basis. So far, no one has found a way of harvesting the plankton production of the southern oceans except via the whales. In 1969, the Soviet Union sent a ship to the Antarctic waters to harvest the krill directly. They succeeded in catching a reasonable quantity, which they converted to a krill paste, reported to be quite tasty. Unfortunately, to meet the costs involved, they were forced to sell it on the Moscow market at about the same price as beef. Faced with a choice between beef and krill paste, the Moscow housewife did exactly what her Western counterpart would do, so the krill harvest was a failure.

The whale stocks should be allowed to increase because the statistical analysis shows that the optimum levels are much higher than the present depleted stock sizes. The steps to permit stocks to increase are difficult, but at least the Commission has now fully accepted the methods of analysis that were first applied to whales in this study. The present quota on catches, not only in southern oceans but also in the North Pacific, is based on the best scientific evidence as reviewed and analyzed by the Scientific Committee of the Whaling Commission, which includes scientists (particularly experts in population statistics) from Canada, Great Britain, Japan, Norway, the Soviet Union, and the U.S., as well as the Food and Agriculture Organization of the United Nations. One loophole remains: each whaling country has the responsibility for enforcing the quota and other restrictions without any supervision by international observers. Steps are being taken to change this. It is to be deplored that scientific methods were not introduced sooner and that even now stricter enforcement is necessary, but recent restrictions by the Commission represent a major accomplishment in management of a world resource, one that will survive only if man and nations cooperate to save it.

PROBLEMS

1. Refer to Figure 1. What was the approximate catch of blue whales in the Antarctic in the season 1931–32? 1937–38? What is the percentage reduction between the two seasons? When was the catch the highest? The lowest?

2. What are the three counting methods discussed in the text? Describe briefly.

3. Describe an experiment whose purpose is to estimate the size of a human population in an isolated island, using one of the methods mentioned in the article. What assumptions are you making?

4. Refer to Figure 2. What was the approximate catch of blue whales per day (adjusted) in the season 1955–56 (plotted as 1956)?

5. Refer to Figure 2. Can you explain the relative "high" for the 1958 figure?

6. From Figure 2 it was concluded that there were only 10,000–12,000 blue whales in 1953. How was this number obtained?

7. Suppose it was determined by the age analysis method that the mortality rate of blue whales is .25 per year. Furthermore by the marking method it was estimated that there were about 7000 blue whales at the start of 1960. How many whales would you expect to be still living at the end of 1961 assuming that the mortality rate does not change over the years?

8. What is meant by "optimum stock size" of whales? Why should there be an optimum size at all?

REFERENCES

D. G. Chapman, K. R. Allen, and S. J. Holt. 1964. "Reports of the Committee of Three Scientists on the Special Antarctic Investigations of the Antarctic Whale Stocks." *Fourteenth Report of the International Whaling Commission*. London. Pp. 32–106.

J. A. Gulland. 1966. "The Effect of Regulation on Antarctic Whale Catches." *Journal du Conseil*, 30:308–315.

N. A. Mackintosh. 1965. *The Stocks of Whales*. London: Fishing News.

Scott McVay. 1966. "The Last of the Great Whales." *Scientific American*, 215:2, pp. 13–21.

THE IMPORTANCE OF BEING HUMAN

W. W. Howells   *Harvard University*

IN THE summer of 1965, my paleontological colleague, Bryan Patterson, was in charge of a Harvard expedition working near the shore of Lake Rudolf in northern Kenya. At a locality called Kanapoi, searching in deposits believed to be of the early Pleistocene (the last geological epoch before the present), he picked up an important fossil. The broken lower end of a left humerus (the upper arm), it was easily recognized as *hominoid;* that is to say, it came from a creature of the group formed by man and his closest living relatives, the apes, but not from a monkey.

What was the special importance of the fossil? From shape and size it could be seen at once not to belong to a gorilla, an orangutan, or a gibbon (and the last two have never been present in Africa anyhow). It was extraordinarily similar to the same piece in modern man, in fact, it was indistinguishable. But the date of the deposit was certainly before the existence of anything like modern man, and after the field season was over, volcanic basalt from a bed lying above the deposit gave an age estimate, by radioisotope dating, of about 2½ million years.   The oldest human stage that had been established so far was that of the erect-walking but small-brained and large-jawed australopithecines found by Leakey at Olduvai Gorge, which had been dated at about 1¾ million years.   If this small piece of arm bone were "human," or *hominid,* in the sense of belonging to such a creature, it would extend the continuous record of human evolution backward three-quarters of a million years at a single bound.

But there was one problem.   This piece of elbow joint in man can easily be told from that in orang, gorilla, and gibbon, but not from that in the chimpanzee.   Although the rest of a chimpanzee's bone is shorter and stouter, this region is so similar in the two that many, if not most, specimens defy classification as one or the other on examination.   In spite of different uses of the arm, this particular part shows such slight, subtle, and inconstant distinctions in size and shape as to baffle ordinary methods of study even by experts. The problem, therefore, was this: either the bone was that of the earliest australopithecine yet found in our direct ancestral history or it was simply that of an ancestral chimpanzee, in which case we could breathe normally. What about testing something old with something new?   Could an electronic computer tell us anything useful?

A computer, of course, does not really "tell" anything.   It merely makes possible answers to mathematical questions that we would not live long enough to answer if we tried to work them out with simple calculating machines. With its enormous capacities and speed, a computer transfers the effort from getting the right answer to getting the right question.   Biological material—bones or skulls are good examples—lends itself to particular kinds of questions.   Because the genes they inherit are capable of a virtually infinite number of different combinations, no two individuals of a population or species are exactly alike (with the spectacular exception of identical twins).   So, quite apart from different habits of use, diet, or other accidents of growth, human elbow joints vary normally in size and details of shape, though they vary within a limit of form that is basic to the actions of human elbow joints.

Quite different species of animals, of course, have quite different forms in various body parts.   Any beginner can distinguish between a cheek tooth of a mammalian carnivore, with its narrow, knifelike shearing crown, and that of a herbivore, which has a broad surface for grinding vegetable matter. These are marked evolutionary divergences.   Within herbivores the differences are smaller, and within groups of herbivores such as pigs (for example, domestic pigs, wild boars, warthogs, etc.) or elephants, species distinctions are matters for experts, who can obtain a wealth of information from fossils as to the history of pigs and elephants or as to the exact species of animals present at a given time in the past at a fossil locality such as Kanapoi.   Finally, for particular parts, such as the elbow joint in chimpanzee and man, the species distinctions may be so slight as to be eclipsed by the variation *within* each species, already described.   That is the situation we are faced with here.

This is not just a matter of impression: it may be viewed quantitatively. Some time ago, Professor William L. Straus of Johns Hopkins University, a man with much experience in such studies, tried to deal with the same problem, when the same piece of the humerus of a species of australopithecine was found at the site of Kromdraai, near Pretoria, South Africa.   In this case, it was plain that the bone belonged to *Paranthropus,* the species in question, because other skeletal and cranial parts of the same species had been found at the site, and the bone could hardly be assigned to anything else.   Here the problem was whether the bone was more manlike or more apelike, since the hominid ("human") position of the australopithecines at that time was less clear.   Professor Straus made a number of typical measurements of human and chimpanzee bones in an attempt to find differences between them.   He found statistically significant differences[1] in the averages of certain of the measurements, but the absolute differences were slight, and the overlap in each measurement between man and chimp was so great that the *Paranthropus* fragment could not be allocated to either.   In no case did its measurements lie outside the range of either man or chimpanzee, though the figures were more often closer to the mean, or average, figures for the latter.

This was no solution and led to no decision as to the relationships of *Paranthropus,* insofar as the arm could shed light on them.   In such a case, we need a method that is not limited to comparisons of single measurements, but that somehow takes account of the whole shape of the bone, or part, as the eye tries to do, and also has some way of emphasizing the really telling differences in shape between two species, if they exist.   Now here is an important point: in the end, any such problem comes down to a mathematical question because the eye itself (though very seldom consciously) attempts to assess the *average* differences in proportions and complex aspects of shape, to rate the varying importance of these, and finally, to judge the probability that a given total shape, in a single case, falls nearer to the essential basic form within the variation of one population than to that of another.   These are questions of quantity and probability, whether measured or not, and are thus statistical in nature.   After all, educated opinion is always the weighing of probabilities.   And here is another important point: biologists and anthropologists—and members of many other sciences—are not often strong in mathematics of a higher order, though they may see only too acutely the limits of their own ways of solving problems.   At the same time, mathematicians, although they have hearts of gold, are not usually sufficiently conversant

[1] The reader may recall that, when a statistician can detect a difference likely to be a real effect and not one stemming from chance variation, he calls it "statistically significant."   By saying "statistically" he warns that the absolute size of the difference may be small and seemingly unimportant because it depends on the objects being studied.

with the niceties of biological problems to understand just what the biologist is trying to gain by using a mathematical analysis.   When the two really get together, however, the rewards in the way of new solutions may be great.   And I must say that mathematical training among biologists who see better what such training can offer has increased notably in recent years.

Fortunately, the particular problem of the Kanapoi fossil is not exceedingly complex, and the solution was provided some years ago by the great English statistician and geneticist R. A. Fisher in the form of the *discriminant function*.   The discriminant function eliminates the futile business of looking at measurements one at a time, of finding that the overlap prevents discrimination of two sets of specimens such as human and chimp elbow joints even though they are known to be from quite different animals, and of being unable to place something such as the *Paranthropus* specimen logically nearer one group than the other.   It has a set of weights with which to multiply a number of different measurements of a specimen, the sum of the products being a single *discriminant score* that makes the best attainable use of all the information in the several measurements.   Given two groups, such as men and chimpanzees, the computation develops the optimum set of weights possible from the measurements used: the effect is to sift out important differences—often quite invisible to the eye or in average figures—so as to emphasize precisely the aspects of shape and size that will best discriminate between the two groups.   That is to say, compared to just that variation *within* a set of human elbow joints, or chimpanzee elbow joints, the distinctions *between* the sets are searched out mathematically so that the discriminant scores of the two groups are segregated one from the other to the maximum degree possible, limited only by the information contained in the measurements.   Thus the overlap, acting as a mask to hide any real group differences, is reduced or removed.

The basic idea of the discriminant function may be appreciated graphically in the case of *two* measurements, represented by the two axes of Figure 1.   (The measurements might be heights and girths of two different groups of men.)   The oval areas correspond to groups of individuals from two populations A and B.   If we just look at measurement 1 by dropping projections on the horizontal axis, we find considerable overlap between the two populations.   The same holds for measurement 2.   On the other hand, the slanted line perfectly separates the two populations.   This is not the place for mathematical detail, but to write the previous sentence is to say that looking at something like

$$(\text{Measurement 1}) + 2 \times (\text{Measurement 2})$$

gives us a new score, the discriminant score, which permits much better separation of the populations than either of measurements 1 or 2 alone.   If there are more than two measurements, as in the present case, there are great potential gains in combining measurements.

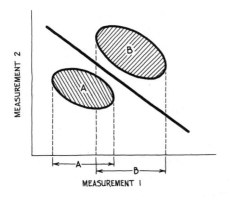

FIGURE 1

Two measurements together separate groups better than either separately

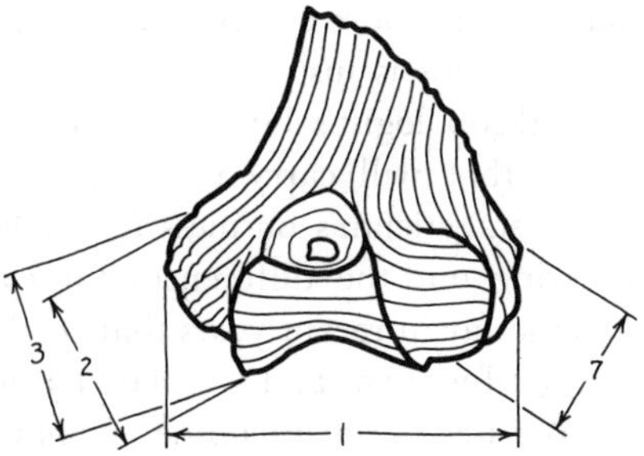

FIGURE 2

Kanapoi humeral fragment and measurements taken

Professor Patterson and I felt fairly strongly that the Kanapoi fragment was hominid—on the human, not the ape side, of the hominoid group as a whole. But we wanted to demonstrate this statistically, not merely to voice an opinion to which opposing opinions could be raised by others. As strategy, we examined human and chimpanzee humeri to see what measurements would most likely reflect such differences as we thought appeared, whether frequently or not. Figure 2 shows the fragment itself and some of the measurements. To begin with, we took the total breadth across the whole lower end, as a matter of general size (measurement 1). Second, the more projecting inner, or medial, epicondyle (at the left in the figure) has a snub-nosed, or slightly turned-up, aspect in some chimpanzees, and we hoped to register this effect by measuring from the lowest point on the trochlear ridge both to the "beak" of the epicondyle and to the nearest point on the shoulder just above it (measurements 2 and 3). The idea was that a slightly greater difference between these two would reflect a deeper curve and more upturned epicondyle. We also measured the backward protrusion of the central, or trochlear, ridge of the joint, the length and breadth of the oval inner face of the medial epicondyle (none of these is shown in Figure 2), and an oblique height of the opposite, or lateral, epicondyle. We thought these measurements showed some tendency to vary one way in man, the other in chimpanzees, though not being the rule in either (if there *were* regular distinctions, obviously the problem of discrimination would be much less). We were not certain of the functional meaning of the possible differences, but they logically could be related mostly to muscle attachments connected with simpler and more powerful use of the flexor and extensor muscles of the hand in the chimp, in hanging by the arms or supporting the body in ground-walking by the characteristic resting on the middle knuckles, all as contrasted with the more general, but more complex and varied, use of the hands in man.

Now this is just where the cooperation comes in. It is the paleontologist's or anthropologist's business, from his background knowledge, to find measurements that will carry important and real information as to differences. It is the statistician's business to say how the measurements can be put together to bring out the differences for evaluation. Here, cooperation had already gone so far in recent years that the biologists knew in advance what statisticians could offer them and we planned our work accordingly.

We measured 40 human bones in the Peabody Museum at Harvard and those of 40 chimpanzees in the Harvard Museum of Comparative Zoology and the American Museum of Natural History in New York. As in Straus's measurements, the overlap of man and chimp was great, but the mean differences, resulting from the special selection of measurements, were in most cases better defined. The means for the two groups and the figures for both the Kanapoi and Kromdraai fragments (the latter taken on two casts) are given to $\frac{1}{10}$ mm in Table 1.

The chimpanzee specimens, as a sample, may be accidentally a little large on the average. The *Paranthropus* fragment is obviously small in all dimensions and so appears "human" when we glance at this list; however, this does not necessarily mean that the shape relations conform to those of man. The correspondences of the Kanapoi measurements to the human means (of this particular sample) are very close throughout—closer than we might expect any random human bone to be in all its measurements.

To assure ourselves of this apparent closeness, we computed a discriminant function from the human and chimp figures. For only seven measurements and such small samples, the calculations could be done by hand, though at the cost of no little labor. In technical language, matrices have to be formed of the sums of all the cross multiplications of all the measurements of all the individuals both within each group and of the total lot; other

TABLE 1.   Measurements

MEASUREMENT	CHIMP MEAN	HUMAN MEAN	KANA-POI	PARANTH-ROPUS	SCALED VECTOR
1. Bi-epicondylar width	64.1	58.0	60.2	53.6	−.09
2. Trochlea-med. epi. dist.	44.8	40.7	41.7	33.6	+.40
3. Trochlea-supracond. dist.	41.3	38.8	39.4	32.1	−.62
4. Posterior trochlear edge	26.4	22.1	22.2	19.9	+.11
5. Med. epi. length	24.7	20.3	20.8	15.5	+.19
6. Med. epi. breadth	12.8	12.6	13.9	10.4	−.32
7. Lat. epi. height	31.5	26.7	27.6	24.9	+.56

steps require the inversion of one matrix and the determination of the latent roots of another. Inversion by hand of a matrix of even the modest size of 7 $\times$ 7 is a tedious business and one open to error. This all leads to finding the discriminant function, which takes the seven measurements from a specimen, multiplies each measurement by a weight specific to that measurement, and then adds these products to give the discriminant score. This is a great deal of arithmetic and we can only say that to have a computer handle such a job from punched cards in a matter of minutes is very welcome. Waiting for paint to dry or for a film to be developed now seems long and drawn out by comparison, and such easy computation has obviously greatly encouraged undertakings such as the one described here.

The last column in Table 1 gives, not the actual weights in the discriminant function as used, but rather a rescaled form of the weights with their relative importance in proper perspective (because, for example, a small measurement, such as thumb length, might require a much larger weight in the function than a large measure, such as stature, to make it effective). These figures show how a number of measurements combine to form a single pattern of greatest difference between the two groups. As might have been expected, the two measurements to register the snub-nosed effect of the medial epicondyle, or its opposite, are useful, as shown by the large size of the scaled vector values. The plus value of measurement 2 and the minus value of measurement 3 combine to make the total discriminant score higher when the epicondyle is most turned up; that is, when measurement 2 is high relative to measurement 3 (see Figure 1), the function creates a greater positive value to add and a smaller minus value to subtract in the total score, and when the opposite is true, with the shoulder of the condyle more sloping, there is on balance a greater minus value in the total score. The lateral epicondyle (measurement 7) also adds a greater plus value when it is high, while the breadth of the medial epicondyle (measurement 6) adds to a plus value (or rather subtracts *least* from a total value) when it is relatively narrow.

Table 1 shows that the above are indeed characteristic human-chimp differences in the averages (though small ones), all of which tend to produce higher score values for the chimpanzee. We note that there is almost no *absolute* difference in measurement 6, the breadth of the medial epicondyle, certainly not a significant one, and yet this measurement is important in discrimination because it is *relatively* narrow in chimpanzees, whose other measurements (in these samples) are larger, on the average, than the human measurements.

We notice also that because the discriminant score is affected by all measurements, it takes account of variation in form toward or away from a basic pattern: if a chimpanzee bone lacks any snubbing of the medial epicondyle, it may exhibit another combination of narrow epicondylar face or high lateral epicondyle, and so it may score in a chimpanzee direction anyhow.

When the discriminant scores were calculated, they produced a far greater separation of human and chimpanzee bones than did any of the measurements singly. Here are the mean score values, and the limits of the individuals in each group:

	Mean	Range
Chimpanzee	99.77	67–130
Man	61.42	40–84

All but two of the chimpanzee values fall between 80 and 120, and all but one of the human values fall between 50 and 75, which are nonoverlapping intervals. So the separation was very good: of 80 specimens, only three overlapped, falling closer to the wrong mean figure than to their own. Unquestionably, this is a successful procedure to distinguish human and chimpanzee humeri by measurement, with a much greater probability of correct assignment than is possible by eye.

Now for the scores of the Kanapoi and *Paranthropus* fragments. These were 59.4 and 63.9, respectively, very close to the human average (almost too good to be true, being closer than most of the known human individuals) and, of course, outside the range of the 40 chimpanzee values entirely. Using statistical theory we compute that, had either bone actually belonged to a chimpanzee, it would have a discriminant score as small as those above (or smaller) with a probability of only about 1 in 500. With so small a probability, we conclude that the two bones did not come from chimpanzees, but from hominids, and that is the answer to the question we framed.

Of course, we must be careful. The real question (because of the material we used) was this: how do the fragments classify themselves when they are asked to choose between *modern* human and *modern* chimpanzee arm bones? These were the only alternatives which we offered to fossil creatures which existed when there were no modern men, and when ancestral chimpanzees might also have differed significantly from those of today. Nevertheless, we have good grounds for inferring from their shape that the arm bones were used, on the whole, like those of men and, at least, not like those of the African apes, terrestrial though they are to a great extent. This takes care of the Kanapoi individual and, as a bonus, says the same thing for the hitherto baffling fragment from Kromdraai.

To review: unable to establish from visual inspection that the Kanapoi fossil did not belong to an animal like a chimpanzee, we turned to measurement and a statistical procedure that could be applied with the help of a computer. (As biologists, we knew from experience how to state the problem and extract a great deal of information, but how to order, analyze, and judge it we learned from statisticians.) Though moderately complex, the discriminant function is well suited to the biological realities of individual variation and group differences in shape and gives an answer that states a numerical probability from the known evidence. So, by the middle of 1966, Professor Patterson and I had concluded that we could rule out the possibility that he had found the bone of an ape and that from what we know about East Africa, the only other possible possessor of the fossil was an early hominid, that is, an australopithecine.

Happy though this made us, the original question since has become partly academic. Patterson went back to the Lake Rudolf region and other teams have been at work in fossil-bearing areas to the north. It is now clear to Patterson (from such evidence as fossil pigs and elephants and more radio-isotope dates) that the Kanapoi formation is over 4 million years old, not 2½ million. Numerous australopithecine fossils—skulls, jaws, teeth, leg bones— have been found in different places with ages of 1 to 4 million years; and at a locality known as Lothagam Hill, near Lake Rudolf, Patterson found a piece of a lower jaw that is clearly hominid, that is, another early australopithecine, and that is over 5 million years old. So it has to be accepted that the Kanapoi bone is that of an australopithecine, agreeing with the result obtained from the discriminant function.

If the original question now has less meaning, our analysis also implies something about the actual form of the bone: it was used in human fashion. This is equally important, if not more so. Several anthropologists and graduate students have recently been using similar, but more complex, analyses to study shape and function of other bones and fragments of our early ancestors, the australopithecines.

PROBLEMS

1. In the hypothetical example of Figure 1, project the values of measurement 2 for groups A and B onto the vertical axis and indicate the interval where the two populations overlap on measurement 2. Using Figure 1 explain why one needs to look at both measurements (or a function of these measurements) to be able to classify an individual as being in A or B.

2. Suppose instead of the two populations in Figure 1 we have the populations in Figure 3 below. It is then very easy to discriminate between the two populations. How would you do it?

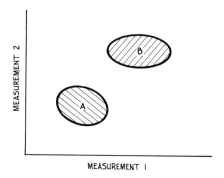

Figure 3

Using this example and the one in Figure 1 explain when discriminant functions are needed.

3. In Table 1 note that there is almost no difference in measurement 6 for humans and chimpanzees, yet it was one of the measurements to help discriminate between the two populations. Explain how measurement 6 might help in discriminating between the two populations.

4. Give a detailed example of your own in which a discriminant function analysis might be appropriate.

Part two

OUR POLITICAL WORLD

HOW WELL DO SOCIAL INNOVATIONS WORK? *

John P. Gilbert   *Harvard University*
Richard J. Light   *Harvard University*
Frederick Mosteller   *Harvard University*

How EFFECTIVE are modern large-scale social action programs? To see how well such programs accomplish their primary mission, we have reviewed the performance of a large number. Our examples are drawn from public and private social action programs, from applied social research, and from studies in medicine and mental illness. We have two purposes in mind: first, to find out how frequently social innovations succeed, and second, to call attention to the importance of evaluations and the need to design them well.

*This paper is adapted from a much fuller report by the same authors entitled "Assessing social innovations: an empirical base for policy" appearing as Chapter 2 in Carl A. Bennett and Arthur A. Lumsdaine (Eds.) *Evaluation and Experiment: Some Critical Issues in Assessing Social Programs*, Academic Press, New York, 1975, pp. 39–193 with the permission of the Academic Press.

How do we choose an innovation to include in this review, when there exist currently hundreds of social and socio-medical programs? The answer is that we have restricted our review to include only those innovations that have been well evaluated. By "well evaluated" we mean that a particular sort of field study was part of the innovation—a randomized, controlled field study. Only a small fraction of evaluations fall into this category. Yet we will find later on that evaluations based on less stringent design criteria lead to less assured conclusions.

The studies we review came to us from several sources: a list compiled by Jack Elinson and Cyrille Gell, a list compiled by Robert Boruch, our own list which we have been collecting for years, and finally Bucknam McPeek using the MEDLARS computer-based reference retrieval system provided us with some surgical studies.

If all innovations worked well, the need for evaluations would be less pressing. And if we forecast nearly perfectly, the suggestion might even be made that we were not trying enough new things. Our findings show that this happy state of affairs does not hold; they show, rather, that only a modest fraction of innovations work well. This finding makes clear that innovations need to have their performance assessed. And if society decides to sponsor such assessments, and acts upon their outcomes, designing evaluations that are clearly reliable becomes crucially important.

MEDICAL AND SOCIAL INNOVATIONS

As we studied many of the evaluations that examine social programs, we were struck by how many of the research designs had been taken directly from the methods of agronomy or the natural sciences. But the idea that the exact techniques that have worked so well in agriculture or physics can be directly applied to the evaluation of social programs is naive. Perhaps a much closer parallel to social research comes from medical research, a field that suffers many of the same difficulties that beset the evaluation of social programs. Physicians, in general, and surgeons, in particular, have been diligent in their attempts to evaluate the effects of their therapies on their patients. Indeed, some techniques for implementing and analyzing randomized controlled field trials have been developed in this context. Some examples of problems faced by medical programs that parallel those often faced by social programs are:

- Multiple outcomes and often negative side effects.
- In multi-institutional trials the actual treatments delivered may differ from place to place, although the original intention was to deliver identical treatments in several places.
- Patients often differ in their general condition, in the state of their disease, and in their response to treatment.

- Ethical tensions may exist between study design and the perceived best interests of the patients.
- Patients may adapt their lives so as to minimize their symptoms and thus prevent their disease from being diagnosed.
- Patients may receive additional treatments, unprescribed and uncontrolled, that are unknown to the physician and perhaps never discovered.

We feel that the social evaluator will find a closer parallel to his own work in medical and health investigations than in those of the laboratory scientist.

OUR RATINGS OF INNOVATIONS

In our full study from which this essay is adapted, we studied in detail 28 well-evaluated innovations. We present here a sample of these well-evaluated innovations to illustrate the thrust of our findings. An overall summary of all 28 follows.

We rate each innovation according to a five-point scale, from double plus (++), meaning a very successful innovation—it does well some of the major things it was supposed to do; to double minus (−−), meaning a definitely harmful innovation. A single minus (−) indicates a slightly harmful or negative innovation. A zero (0) means that the innovation does not seem to have much if any effect in either the positive or negative direction. It could mean also that the innovation has both small positive and small negative effects, neither overwhelming the other. A plus (+) means that the innovation seems to have a somewhat positive effect, but that one would have to weigh particularly carefully whether it was worth its price. We have not carried out detailed cost-benefit analyses here.

The reader might disagree with our ratings, but this worry is not as substantial as one might at first suppose. The reader who wants more information about the individual studies will find it in the full paper [Gilbert et al., 1975] as well as references to the original research reports.

We wish to stress again that *we are not rating the methodology* of the field trial. All of the field trials we discuss were in our judgment of sufficient size and quality to give strong evidence of the efficacy of the innovation. Thus the rating applies to the *innovation* as measured by the field trial.

FIVE EXAMPLES TO ILLUSTRATE THE STUDIES AND THE RATINGS

1. The Salk Vaccine Trials (Francis et al., 1955; Meier, 1972)

The 1954 trial to test a new preventive medication for paralytic polio, the Salk vaccine, is most instructive. First, it exposed children to a new vaccine, and thereby showed that we as a nation have been willing to experiment on people, even our dearest ones, our children. Secondly, the preliminary arguments over the plan instructed us, as did the way it was actually carried out —in two parallel studies.

In the initial design—the *observed control method*—the plan was to give the vaccine to those second graders whose parents volunteered them for the study, to give nothing to the first and third graders, and then to compare the average result for the untreated first and third grade with the treated group in the second grade.

There are troubles here. In the more sanitary neighborhoods, polio occurs more frequently than in unsanitary neighborhoods, and the more sanitary regions are associated with higher income and better education. It is also a social fact that better educated people tend to volunteer more than less well educated ones. Consequently we could expect that the volunteers in the second grade would be more prone to have the diseases in the first place than the average second grader, and than the average of the first and third graders. The comparison might well not be valid because of this bias.

Some state public health officials noticed these difficulties and recommended instead a second design, the *placebo control method*, which randomizes the vaccine among volunteers from all grade groups; that is, these officials recommended a randomized controlled field trial. Half the volunteers got the vaccine and half a salt water injection (placebo), so that the "blindness" of the diagnoses could be protected. Thus the physician could be protected from his expectations for the outcome in making a diagnosis. This meant that the self-selection effects and their associated bias would be balanced between the vaccinated and unvaccinated groups of volunteers, and that the hazards to validity from an epidemic in a grade would be insured against.

In actuality, both methods were used: one in some states and the other in others. The result has been carefully analyzed, and the randomized trial (placebo control) shows conclusively a reduction in paralytic polio rate from about 57 per hundred thousand among the controls to about 16 per hundred thousand in the vaccinated group. (See Table 1.)

In the states where only the second-grade volunteers were vaccinated, the vaccinated volunteers had about the same rate (17 per hundred thousand) as those vaccinated (16 per hundred thousand) in the placebo control areas. The expected bias of an increased rate for volunteers as compared to non-volunteers appeared among the whole group. Among the placebo controls, the volunteers who were not vaccinated had the highest rate (57 per hundred thousand) and those who declined to volunteer had 35 or 36 per hundred thousand. In the states using the observed control method, the first and third graders, who were not asked to volunteer and were not vaccinated, had a rate between the two extremes, 46 per hundred thousand.

Rating: ++

TABLE 1.   Summary of Study Cases by Vaccination Status for Salk Vaccine Experiment.

Study group	Study population (thousands)	Paralytic poliomyelitis cases: rate per hundred thousand
Placebo control areas: Total	749	36
Vaccinated	201	16
Placebo	201	57
Not inoculated*	339	36
Incomplete vaccinations	8	12
Observed control areas: Total	1,081	38
Vaccinated	222	17
Controls**	725	46
Grade 2 not inoculated	124	35
Incomplete vaccinations	10	40

*Includes 8,577 children who received one or two injections of placebo.
**First- and third-grade total population.
Source: Paul Meier, 1972; from Table 1, p. 11.

2. The Gamma Globulin Study (U.S. Public Health Service, 1954)

The general success of the Salk vaccine randomized study can be contrasted with the results of a corresponding earlier study of gamma globulin which was carried out in a nonrandomized trial. In 1953, during the summer, 235,000 children were inoculated in the hope of preventing or modifying the severity of poliomyelitis. "The committee recognized that it would be very difficult to conduct rigidly controlled studies in the United States during 1953" (p. 3). They hoped to use mass inoculation in various places and compare differential attack rates at different sites, as well as to analyze other epidemiological data. In the end this approach turned out to be inconclusive, and the authors of that study describe the need for a more carefully controlled experiment.

No Rating: not a randomized study

Why do we introduce this brief reference to a nonrandomized evaluation amidst all the others that we consider well done? Because we consider the contrast between the Salk and the gamma globulin studies particularly striking, and therefore informative. Both studies had, by most standards, very large sample sizes. Yet putting these studies side by side we see that simply having a large number of participants in an evaluation does not imply the results are capable of reliable interpretation. The Salk vaccine study had two investigations in parallel and they support each other. But our confidence in the findings is based primarily on the randomized component of the overall study. This does not help us in a situation where an observed control study is performed alone.

Uncontrolled biases can make interpretation difficult because it becomes necessary for the interpreter to guess the size of the bias, and, when its size may be comparable to that of the treatment effect, the interpreter is guessing the final result.

This is what happend in the gamma globulin study. We must recognize that whether or not gamma globulin was good for the purpose, the lack of randomization undermined the expert investigators' ability to draw firm conclusions, in spite of the large size of the study. The children were put at risk. Although it must be acknowledged that conducting randomized studies would have been difficult in 1953, those who argue today that randomized trials have ethical problems may wish to think about the problems of studies that put the same or greater numbers of people at risk without being able to generate data that can answer the questions being asked.

We now return to several further examples of *well-done* evaluations.

3. Delinquent Girls (Meyer, Borgatta, & Jones, 1965)

The investigators tried to reduce juvenile delinquency among teen-age girls in two ways: first, by predicting which girls were likely to become delinquent; and secondly, by applying a combination of individual and group treatment to girls who exhibited potential problem behaviors. The population was four cohorts of girls entering a vocational high school; of these, approximately one quarter were screened into this study as indicating potential problems. These girls were randomly assigned to a treatment group which was given treatment at Youth Consultation Services (YCS), a social agency; and to a nontreatment (control) group given no special services. The assignments were 189 to YCS, 192 to the nontreatment group.

The result was that in spite of the group and the individual counseling, delinquency was not reduced. The investigators were successful in identifying girls who were likely to become delinquent, but that was not the primary purpose of the study. They report that, "on all the measures . . . grouped together as school-related behavior . . . none of them supplies conclusive evidence of an effect by the therapeutic program" (p. 176). Similar findings were reported for out-of-school behavior.

We view this innovation as rating a zero, though the detection ability might be of value on another occasion.

Rating: 0

4. Probation for Drunk Arrests (Ditman et al., 1967)

Encouraged by preliminary work on the use of probation with suspended sentence as a way of getting chronic drunk offenders into treatment, Ditman et al. developed with the cooperation of the San Diego Municipal Court a randomized controlled field study. Offenders who had had either two drunk arrests in the previous three months or three in the previous year were fined $25, given a 30-day suspended sentence, and then assigned by judges to one of three groups: 1) no treatment, 2) alcoholic clinic, and 3) Alcoholics Anonymous. The primary payoff variables were number of rearrests and time before first rearrest. The total study included 301 individuals, divided randomly into the three groups. The results, based on the 80% of the subjects for whom good records are available, were that the "no treatment" group did as well as or better than the other two groups, which performed practically identically. Table 2 shows the detailed results. Ditman et al. conclude that the study gives no support to a general policy of forced short term referrals, on the basis of the suggestive evidence contained in the paper.

Table 2 shows that 44% of the "no treatment" group had no rearrests (in the first year) as opposed to about 32% in the other groups. Although the missing 20% of the data might change this picture somewhat, the overall result is compelling. (The missing data arise because of difficulty in getting complete data from two distinct sources of records.) Since the difference, though favoring the control, is well within the range of chance effects, we rate this innovation as zero rather than minus.

Rating: 0

TABLE 2.   Number of Drunk Rearrests Among 241 Offenders in Three Treatment Groups.

| Treatment Group | Rearrests | | | |
	None	One	Two or More	Total
No treatment	32 (44%)	14 (19%)	27 (37%)	73
Alcoholism clinic	26 (32%)	23 (28%)	33 (40%)	82
Alcoholics Anonymous	27 (31%)	19 (22%)	40 (47%)	86
Total	85	56	100	241

Source: Ditman et al., *American Journal of Psychiatry*, 1967, *124*, 160–63. Copyright © the American Psychiatric Association.

5. Psychiatric After-Care (Sheldon, 1964)

Mental hospitals in England had high readmission rates, so a field study was undertaken to see if "after-care" of discharged patients could reduce readmission significantly. Women between the ages of 20 and 59 were randomly assigned to psychiatric after-care treatment or to their general practitioner, the latter being viewed as the standard treatment. The after-care involved 45 women; the standard involved 44. The psychiatric after-care group was further divided into a day center nurse treatment mode and an outpatient clinic with a doctor; this assignment was also random.

After six months, the general practioners returned their patients to the hospital in about 47% of the cases, while the nurse and the MD (the psychiatric team) sent back about 18% of theirs. The psychiatric team kept their group under care for a longer period than the general practitioners, but had less rehospitalization. The investigation found that the better the attendance, the less the readmission in all three groups. But the psychiatric team was more frequently associated with good attendance by the patient than was the general practitioner.

We rate this innovation plus. The lower return-to-hospital data make it seem like a double plus, but since the innovators may be biased in keeping patients out, we think caution is in order. Were the decision for rehospitalization being made by a separate decision group, the research would be tighter.

Rating: +

SUMMARY OF RATINGS

We hope these several examples illustrate our general procedure.

Table 3 summarizes the ratings of the 28 studies we analyzed in our original paper. Since the gamma globulin study was not a randomized trial, we do not include it in the ratings.

Overall, six innovations, or about 21% of the total, were scored double plus. The rate of double pluses does not differ sharply among the three groups. The pile-up at zero, 13 out of 28, or 46%, suggests that innovations that are carefully evaluated yield disappointing results a fair proportion of the time.

We warn that there may be some upward bias owing to selective reporting and selective finding by our searches. Even so, when we consider the high rate of failure of laboratory innovations, we can take pleasure in a success rate as high as the one seen here.

Except for the surgical innovations where we have a rather solid description of our population, the skeptical reader may feel we have no grounds for discussing rates of successful innovations in the absence of a population and in the presence of several possible selection effects. The difficulty is not unique—if one wants to know the percentage of new products that succeed or the percentage of new businesses that succeed, the same problems of definition of population and success arise.

Yet we can make some reasonable guesses about direction of bias. First, evaluations that find a successful innovation are probably more likely to be written up and published than those which find an innovation to be a failure. Further, more successful programs are more likely to have come to our attention in our original review of studies. Consequently, we believe that the

TABLE 3. Summary of Ratings.

Rating		--	-	0	+	++
Total Social Innovations	8	0	0	3 D2. Welfare Workers D3. Girls at Voc. High D7. Pretrial Conf.	2 D4. Cottage Life D6. L.A. Police (or ++)	3 D1. Neg. Income Tax D5. Manhattan Bail D8. ESAP
Total Socio-medical Innovations	8	0	1 F1. Kansas Blue-Cross	4 F3. Comp. Med. Care F4. Drunk Probation F7. Nursing Home F8. Family Medical Care	2 F5. Psychiatric After-care F6. Mental Illness	1 F2. Tonsillectomy
Total Medical	12	1 H8. Everting	1 H10. Yttrium-90	6 H2. Vagotomy (Cox) H3. Vagotomy (Kennedy) H5. Cancer H6. Portacaval Shunt H9. Chlorhexidine H11. *Gastric Freezing	2 H4. Bronchus (possibly ++) H7. Ampicillin	2 H1. Vagotomy (Johnson) H12. *Salk Vaccine
Grand Total	28	1	2	13	6	6

*These two studies did not emerge from the MEDLARS search. All the other medical innovations did.

estimate of about 21% successful innovations is high compared to what a census would yield.

To sum up, the major findings of this table are that 1) among societal innovations studied here, about one in five succeeded; and 2) among those that succeeded, the gain was often small in size, though not in importance.

Both findings have important consequences for the policy maker considering methods of evaluation and for the attitudes toward programs that society needs to develop.

NONRANDOMIZED STUDIES

Now that we have looked at randomized controlled field trials, we explore the results of investigations that for the most part did not use randomization. We will see the consequences of this approach for both weakness of the findings and the ultimate time taken to gather firm information.

We have mentioned earlier the value for social investigators of learning from the medical experience. What we have been observing recently in medicine are systematic attempts to appreciate the interpretative difficulties in ordinary nonrandomized investigations as compared with randomized controlled clinical trials. Much experience is building up, and we can profit from a short review of very extensive work in a few medical areas.

About 1945 an operation called portcaval shunt was introduced to treat bleeding in the esophagus for certain patients, and this operation has been extended for other purposes. After 20 years of experience with this operation and 154 papers on the subject, it still was not clear (Grace, Muench, & Chalmers, 1966) what advice a physician should give to a patient with esophageal varices. Grace, Muench, and Chalmers reviewed the literature to see whether they could resolve such questions as whether the operation would prevent further hemorrhage, what disabling side effects there might be, or what expectation of life went with the operation as compared with not having it.

They rated the investigations on two variables: degree of control in the investigation and degree of enthusiasm for the operation as expressed in the article reporting the trial. The degrees of enthusiasm after the study are: marked, moderate (with some reservations), and no conclusions or enthusiasm. The degrees of control are: 1) well-controlled—random assignment to treatment groups; 2) poorly controlled—selection of patients for treatment (compared with an unselected group or some other experience); and 3) uncontrolled—no comparison with another group of untreated patients.

Table 4 shows clearly that following their uncontrolled studies, investigators almost invariably express some enthusiasm for the shunt, and more than two thirds of the time express marked enthusiasm. Poorly controlled investigations have much the same outcome. On the other hand, in the six instances where the study was well-controlled, three investigators expressed moderate enthusiasm; the rest none. We assume that the investigators using the well-controlled field trials had better grounds for their degree of enthusiasm than did those with no controls or poor controls. (See footnote on Table 4 for more detail on the operations.)

This investigation is informative because we can put the results of uncontrolled, poorly controlled, and well-controlled studies side by side. In all, the results of the investigation show that uncontrolled and poorly controlled studies led to greater enthusiasm than was warranted on the basis of the well-controlled studies. If the poorly controlled studies had suggested conclusions similar to those of the well-controlled trials, we, the surgeons, and policy makers, could be more comfortable with the results of related studies in similar contexts. But we see instead that the results are far from the same. By performing many poorly controlled trials we waste time and human experience and mislead ourselves as well. The argument that sometimes has been advanced that we do not have time to wait for well-controlled trials, because decisions need to be made immediately, does not seem to have been applicable here.

TABLE 4.   Degree of Control Versus Degree of Investigator Enthusiasm for Shunt Operation in 53 Studies Having at Least 10 Patients in the Series.

	Degree of enthusiasm			
Degree of control	Marked	Moderate	None	Totals
Well-controlled	0	3*	3*	6
Poorly controlled	10	3	2	15
Uncontrolled	24	7	1	32
Totals	34	13	6	53

Source: Revised from Grace, Muench, & Chalmers (1966), Table 2, p. 685. Copyright © 1966 The Williams and Wilkins Co., Baltimore.

*In the original source, the cell "well-controlled—moderate enthusiasm" had one entry, but Dr. Chalmers informed us by personal communication that two studies can now be added to that cell. Furthermore, he told us that the "well-controlled—moderate enthusiasm" group is associated with therapeutic shunts, and the "well-controlled—none" with prophylactic shunts.

FINDINGS, INTERPRETATIONS, AND RECOMMENDATIONS

THE RESULTS OF INNOVATIONS. To see how effectively new ideas for helping people worked out in practice, we have collected a series of innovations from social programs and medicine that have been well evaluated. The overall findings are that 1) about a fifth of these programs were clear and substantial successes; 2) a similar number had some small to moderate positive effects; and 3) most of the remaining programs either had no discernible effects or were mixed in their effects, while a few were even found to be harmful rather than beneficial. These proportions do not differ sharply among the social, medical, and socio-medical studies.

How should we interpret these findings? If most innovations had worked, one might well feel that we were not being expansive or broad enough in our attempts to ameliorate social problems. If hardly any had worked, one might conclude that not enough thought and planning were being put into these programs and that they were wasting society's resources. Although our results fall between these two extremes, we would have liked to see a higher proportion of successful innovations, particularly since we feel that the selection biases of our observational study are probably causing the data to overestimate the proportion of successful innovations rather than underestimate them. Thus it seems to us that the more successful innovations would be both more likely to have been well evaluated and more likely to have come to our attention. If it is true that we are less apt to evaluate programs that are feared to have little effect, society should be concerned that so many very large programs have been evaluated with only nonrandomized studies, if at all.

FINDINGS FOR NONRANDOMIZED TRIALS. Although we are often pushed to do them for reasons of expediency, uncontrolled trials and observational studies have frequently been misleading in their results. Such misdirection leads to the evaluations' being ineffective and occasionally even harmful in their role as tools for decision makers. This was well illustrated in the medical studies by Grace et al., and we are concerned because similar troublesome features are present in many evaluations of social programs. Nonrandomized studies may or may not lead to a correct inference, but without other data the suspicion will persist that their results reflect selection effects. This suspicion leads to three difficulties for the decision maker. First, his confidence in the evaluation is limited, and even when he does believe in the result, he may be reluctant or unable to act because others are not convinced by the data. Second, because of this lingering suspicion, observational studies are rarely successful in resolving a controversy about causality. Though controversy about policy implications may of course persist, few controversies about the effects of new programs survive a series of carefully designed randomized controlled trials. Third, nonrandomized studies are not simply neutral in their effects; they may be *harmful* if they result in the *postponement* of randomized studies.

BENEFICIAL SMALL EFFECTS. The observation that few programs have slam-bang effects stresses the importance of measuring small effects reliably. Once small effects are found and documented, it may be possible to build improvements upon them. The banking and insurance businesses have built their fortunes on small effects—effects the size of interest rates. Ten percent per year doubles the principal in a little over seven years. Similarly, a small effect that can be cumulated over several periods—for example, the school life of a student—has the potential of mounting up into a large gain. Naturally, small effects require stronger methods for their accurate detection than do large ones. One must be sure that the observed effects are not due to initial differences between groups, or to other spurious causes. Randomized controlled field trials are virtually essential for controlling these sources of bias, and so are necessary for the accurate measurement of small effects.

CONTROLLED TRIALS VS. FOOLING AROUND. Ethical problems have often been cited as reasons for not carrying out well-controlled studies. This is frequently a false issue. The basic question involves comparing the ethics of gathering information systematically about our large scale programs with the ethics of haphazardly implementing and changing treatments as so routinely happens in education, welfare, and other areas. Since the latter approach generates little reliable information, it is unlikely to provide lasting benefits. Although they must be closely monitored, like all investigations involving human subjects, we believe randomized controlled field trials can give society valuable information about how to improve its programs. Conducting such investigations is far preferable to the current practice of "fooling around with people," without their informed consent.

PROBLEMS

1. Why is it true that forecasting the success of social innovations very well could lead to a reluctance to try new things?

2. The authors point out that their technique for selecting studies to study probably leads, if anything, to an *overestimation* of the success of social innovations. Explain why this is so.

3. (a) Explain the rating system the authors use for social innovations.
   (b) Does the methodology of an experiment affect its rating? Explain.

4. If your class were asked for volunteers to participate in a study of grade averages, how would you expect the grade averages of the volunteer group to compare with those of the class as a whole? Explain.

5. Explain the difference between the *observed control method* and the placebo control method in the Salk vaccine trials. Which method is superior and why?

6. What do the authors mean by "blindness" in their discussion of the Salk vaccine trials?

7. Consider Table 1. In the placebo control areas, approximately how many cases of poliomyelitis were there among those fully vaccinated?

8. Explain in some detail, and in contrast to the Salk vaccine trials, the methodological weaknesses of the gamma globulin study.

9. Consider Table 2.

(a) What percentage of the 241 people in the study had no rearrests?

(b) What percentage of those with 2 or more rearrests had received no treatment?

10. In the psychiatric after-care experiment, the innovators (psychiatric team) also made the decision to return patients to the hospital.

(a) Why does this affect the rating of this experiment?

(b) How does this situation differ from the delinquent girls experiment? (Hint: Who decides whether the girls become delinquent?)

11. Consider Table 3. What percentage of the overall number of studies looked at showed that the innovation had at least a somewhat positive effect? What percentage, of the social innovations only, had a negative effect?

12. How did the outcomes of the controlled and uncontrolled investigations of the portacaval shunt operation differ? What are the implications of this difference for social policy makers?

REFERENCES

K. S. Ditman, G. G. Crawford, E. W. Forgy, H. Moskowitz, and C. MacAndrew. 1967. "A Controlled Experiment on the Use of Court Probation for Drunk Arrests." *American Journal of Psychiatry*. 124, pp. 160–163.

Thomas Francis, Jr., et al. 1955. "An Evaluation of the 1954 Poliomyelitis Vaccine Trials—Summary Report." *American Journal of Public Health*. 45:5, pp. 1–63.

J. P. Gilbert, R. J. Light, and F. Mosteller. 1975. "Assessing Social Innovations: An Empirical Basis for Policy." C. A. Bennett and A. A. Lumsdaine, eds., *Evaluation and Experiment: Some Critical Issues in Assessing Social Programs*. New York: Academic Press.

N. D. Grace, H. Muench, and T. C. Chalmers. 1966. "The Present Status of Shunts for Portal Hypertension in Cirrhosis." *Gastroenterology* 50, pp. 684–691.

Paul Meier. 1972. "The Biggest Health Experiment Ever: The 1954 Field Trial of the Salk Poliomyelitis Vaccine." J. M. Tanur et al., eds., *Statistics: A Guide to the Unknown*. San Francisco: Holden-Day.

Meyer, H. J., Borgatta, E. F., and Jones, W. C. *Girls at vocational high: An experiment in social work intervention*. New York: Russell Sage Foundation, 1965.

Sheldon, A. An evaluation of psychiatric after-care. *British Journal of Psychiatry*, 1964, *110*, 662–667.

PARKING TICKETS AND MISSING WOMEN: Statistics and the Law

Hans Zeisel   *University of Chicago*

Harry Kalven, Jr.   *University of Chicago*

THE LAW's traditional stance toward quantification and statistics was wittily expressed some years ago by one of its great professors, Thomas Reed Powell, who spoke of research in which thinkers don't count and counters don't think. By tradition, the concerns of the law have been qualitative and in large part based on the individual case; therefore, the thinker who would not count has been favored.

In recent decades, however, the law has shown some appetite for quantification, and we shall sketch some ways in which statistics has impinged on contemporary law.

PROOF BY DISPROOF OF COINCIDENCE

Perhaps the most striking use of statistics is to calculate the probability that a given event occurred by chance. The alternative explanation is that the event occurred by intent or another identifiable cause, and the recourse to statistics is to refute or support the contention that the matter was simply a coincidence. Two examples are offered here, one simple, the other somewhat more complex.

Parking Tickets. The simple example comes from a Swedish trial on a charge of overtime parking. A policeman had noted the position of the valves of the front and rear tires on one side of the parked car, in the manner pilots note directions: one valve pointed, say, to one o'clock, the other to six o'clock, in both cases to the closest "hour" (see Figure 1). After the allowed time had run out, the car was still there, with the two valves still pointing toward one and six o'clock. In court, however, the accused denied any violation. He had left the parking place in time, he claimed, but had returned to it later, and the valves just happened to come to rest in the same positions as before. The court had an expert compute the probability of such a coincidence by chance, the answer was that the probability is 1 in 144 (12 × 12), because there are 12 positions for each of two wheels. In acquitting the defendant, the judge remarked that if all *four* wheels had been checked and found to point in the same directions as before, then the coincidence claim would have been rejected as too improbable and the defendant convicted; four wheels with 12 positions each can combine in 20,736 (= 12 × 12 × 12 × 12) different ways, so the probability of a chance repetition of the original position would be only 1 in 20,736. Actually, these formulas probably understate the probability of a chance coincidence because they are based on the assumption that all four wheels rotate independently of each other, which, of course, they do not. On an idealized straight road all rotate together, in principle. It is only in the curves that the outside wheels turn more rapidly than the inside wheels, but even then the front and rear wheels on each side will presumably rotate about the same amount. (See Zeisel 1968.)

FIGURE 1

Schematic diagram of parked car with valves at 1 and 6 o'clock

Missing Women. The second example arose from the 1968 trial of the pediatrician-author Dr. Benjamin Spock and others in the U.S. District Court in Boston for conspiracy to violate the Selective Service Act by encouraging resistance to the war in Vietnam. In that trial, the defense challenged the legality of the jury-selection method. Although more than half of all eligible jurors in Boston were women, there were no women on Dr. Spock's jury. Yet he, more than any defendant, would have wanted some because so many mothers have raised their children "according to Dr. Spock"; moreover, the opinion polls showed women in general to be more opposed to the Vietnam war than men.

The question was whether this total absence of women jurors was an accident of this particular jury or whether it had resulted from systematic discrimination. Statistical reasoning was to provide the answer.

In the Boston District Court, jurors are selected in three stages. The City Directory is used for the first stage; from it, the Clerk of the Court is supposed to select 300 names at random, that is, by a lotterylike method, and put a slip with each of these names into a box. The City Directory is renewed annually by censuslike household visits of the police, and it lists all adult individuals in the Boston area. The Directory lists slightly more women than men. The second selection stage occurs when a trial is about to begin. From the 300 names in the box, the names of 30 or more potential jurors are drawn. These people are ordered to appear in court on the morning of the trial. The subgroup of 30 or more is called a *venire*. In the third stage, the one that most of us think of as jury selection, 12 actual jurors are selected after interrogation by both the prosecutor and the defense counsel. Figure 2 shows the percentages of women in some 46 such venires selected by all seven judges of the Federal District Court in Boston.

The average proportion of women drawn by the six judicial colleagues of the Spock trial judge was 29%, and furthermore, the averages of these six judges bunched closely around the group average. This suggests that the proportion of women among the names in the 300-name panels in the jury box was somewhere close to that 29% mark. But Figure 2 shows also that the Spock judge's venires had consistently lower percentages of women, with an overall average of only 14.6% women, almost exactly half of that of his colleagues.

It is *possible,* of course, that the selection method used by the trial judge was the same as that of his six colleagues. But what is the probability that a difference as large (or larger) as that between 14.6 and 29% could arise by chance? Statistical computation revealed the probability to be 1 in 1,000,000,000,000,000,000 that the "luck of the draw" would yield the distribution of women jurors obtained by the trial judge or a more extreme one. The conclusion, therefore, was virtually inescapable: the venires for the trial judge

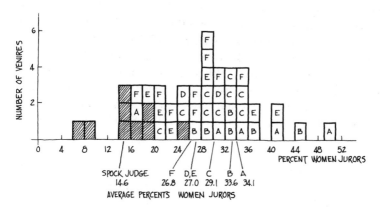

FIGURE 2

Number of venires by proportion of women (shaded blocks are for Spock judge venires; unshaded blocks are for other judges—A, B, C, D, E, F—of the Federal District Court in Boston). Averages are weighted by size of venire (not shown here). Source: Zeisel (1969)

must have been drawn from the central jury lists in a fashion that somehow *systematically* reduced the proportion of women jurors.

Thus the proportion of women among the potential jurors *twice* suffered an improper reduction—first, when the court clerk reduced their share from a majority in the City Directory to 29% in the jury lists and, second, when judge managed to lower the 29% to his private average of 14.6%. In the Spock trial, only one potential woman juror came before the court, and she was easily eliminated in stage 3 by the prosecutor under his quota of peremptory challenges (for which he need not give any reasons). (For further discussion see Zeisel 1969.)

ILLUMINATING DESCRIPTION

A second major use of statistics in the law is careful description. At times, it becomes relevant to measure, or at least to estimate within limits, some frequency, range, ratio, or level. Three examples will be given.

Juries and Judges. There has been perennial debate over the merits of the jury system, in particular over the differences between jury verdicts and the verdicts that would have been arrived at by the judge alone. If judge and jury hardly ever differed, the jury would be a somewhat wasteful institution; if they differed too often, grave questions might be raised about the rationality of the jury, about the judge, and indeed about the meaning of justice under law.

This debate has been advanced by a statistical analysis of 3576 criminal jury trials in which the presiding trial judges reported how they would have decided the case without a jury. Table 1 shows a two-by-two distribution that was a fundamental outcome of the analysis. The boxed percents refer to the 78% (=14% + 64%) of the trials in which jury and judge agree on the verdict, 14% on acquittal and 64% on conviction. The remaining 22% are disagreements: 3% in which the judge would have acquitted but the jury convicted, and 19% in which the jury acquitted but the judge would have convicted. Note, in particular, that for those cases in which the jury finds guilt, the judge agrees nearly all (96%) of the time, but for the cases in which the jury acquits, the judge agrees only about 42% of the time. Thus the jury tends to have a softening effect.

The figures fall into a range that approaches neither of the extremes; that is, the jury is not superfluous, nor do juries and judges disagree intolerably. If the reasons for the jury's disagreement turn out to be understandable, as indeed they do, the statistics offer basic insights into the viability of the jury as an institution. The range of reasons that move a jury to disagree with the judge is wide: a sense of justice concerning the particular crime that does not coincide with the letter of the law, a different view of the weight of the evidence, special attitudes towards the particular defendant, and so forth.

Quality of Counsel. Another such illumination of a heretofore dark corner pertains to how adequately defendants in criminal jury trials are represented

TABLE 1. Percent Agreement* and Disagreement Between Jury and Trial Judge (3576 Cases = 100%)

	JURY	
JUDGE	Acquitted	Convicted
Acquitted	14*	3
Convicted	19	64

Source: Kalven and Zeisel (1966).
* Boxed percents show agreement.

TABLE 2.   Quality of Counsel in Criminal Jury Trials

	PERCENT
Defense counsel superior to prosecutor	11
Abilities equal	76
Prosecutor superior to defense counsel	13
Total	100
Number of Cases	(3576)

Source: Kalven and Zeisel (1966).

in the courts.   In recent years, there has been concern that all criminal defendants be represented at the various stages of the criminal process.   Less attention has been paid to the quality of representation, although systematic disparity would be a disturbing commentary on the basic fairness of our administration of justice.

The statistics in Table 2 summarize assessment by the presiding trial judges mentioned above and are reassuring.

As we shall see below, inequality of counsel does affect the outcome of the trial, so it is comforting to learn that in over three-fourths of the criminal jury trials the ideal of the adversary system is realized: roughly equal champions on both sides.   It is also reassuring to learn that in the remaining trials, the inequality goes in both directions about the same proportion of times.   More complex questions might be raised about characteristics of the defendants who recruit superior or inferior counsel, but we do not treat of them here.

Automobile Injuries.   One more set of descriptive data, despite its simplicity, provides insights and perspective on the functioning of a major part of the legal system, this time the law of torts.   There is much debate these days over how well the law works to compensate the victims of accidents, especially automobile accidents.   A recent elaborate study of auto accidents in Michigan revealed the distribution of reparations shown in Table 3.

The result may surprise those who assume that liability law is overwhelmingly the most important source of compensation for accident victims. As the figures indicate, this is no longer the case, and the importance of compensation from other sources suggests the feasibility of major legal reforms in the direction of insured compensation for damages irrespective of fault on anybody's part.   Several states have adopted, at this point, "no-fault" automobile insurance.

TABLE 3.   Sources of Reparation for Automobile Injuries in Michigan, 1958

	PERCENT OF TOTAL DOLLARS
Liability of third parties who had negligently caused the accident (mostly insured some uninsured)	55
Injured's own insurance	
Accident	22
Hospital and medical	11
Life and burial	5
Social security	2
Employer and Workmen's Compensation	1
Other	4
Total	100% (= $85,196,000)

Source: Conrad et al. (1964, p. 63).

SAMPLING

The above examples depend mostly on sampling rather than censuslike operations, and in the study of legal as well as other institutions, sampling has enormously facilitated the possibilities of quantitative studies.

Sampling also has invaded the very core of the legal system, the fact-finding process in court trials and in hearings before administrative agencies. An important reason is that sampling can replace the cumbersome, and sometimes impossible, complete count or census. For example, a company's share of the consumer market may be determined by auditing a sample of stores, and the degree to which two competing trademarks are confused may be determined by interviewing a sample of consumers.

The rules of evidence that guide the courts and, to a lesser degree, the administrative agencies, have sometimes made it difficult to bring in sampling and survey data. If the data are collected through personal interviews, the rule against hearsay evidence, coupled with the guarantee of anonymity that is obligatory in most interviewing, may stand in the way. Sometimes the court simply distrusts sampling operations altogether. Thus, in a California case, a department store instituted suit for overpayment of local taxes, and submitted an estimate for the amount in question of $27,000, based on a sample of sales slips. The court insisted on the full count, only to discover that the correct amount was $26,750. On the whole, however, sampling operations have become more and more acceptable.

CAUSE AND EFFECT

The most complex use of statistics in law is measurement of the effect of particular rules or institutions. We conclude with two examples of this kind.

Pretrial Hearings. The first example comes from a controlled experiment, designed to reveal whether the worrisome number of cases requiring trial

TABLE 4.   Obligatory Versus Optional Pretrial

	OBLIGATORY	OPTIONAL
Average length of trial time of the cases that reached trial	8¼ hours	7¼ hours
Percent cases not settled, hence reaching trial	24%	22%

Source: Rosenberg (1964).

is being reduced by the procedure known as pretrial hearing.   This is a hearing in which, prior to the trial proper, the litigants and their counsel are requested to appear before a judge to attempt to prepare the case so as to reduce the time required for its trial or to settle it there and then.

To learn whether pretrial hearings accomplish these aims, the state of New Jersey authorized an experiment: a random half of the filed suits were pretried as usual, but the other half were pretried only if one (or both) of the litigants requested it.   This happened only in 48% of this half of the cases.   The results of the experiment are summarized in Table 4. *Obligatory* pretrial achieved neither of its purposes and consumed court time, so the state of New Jersey decided to abolish it.

Effect of Counsel Quality.   As a final example, we shall discuss a survey that gauges the effect of superior counsel in criminal trials.   (A more detailed discussion appears in Kalven and Zeisel 1966.)

The survey data come from the real jury trials mentioned above; after each trial the presiding judge told us in confidence how he would have decided the case if it had been tried without a jury.   We use the judge's private decision as a base line against which to compare the jury, under the working assumption that the judge is far less affected by the skill of counsel than is the jury.   Table 5 summarizes the results of the survey.

The 88% in the upper left-hand corner of Table 5 has the following meaning: in those trials in which the judge would have acquitted and in which

TABLE 5.   Effects of Counsel Ability on Jury Verdicts in Criminal Cases

	DEFENSE COUNSEL SUPERIOR	ABILITIES EQUAL	PROSECUTOR SUPERIOR
Percent of cases where the judge would have *acquitted* that the jury acquitted	88%	82%	76%
Percent of cases where the judge would have *convicted* that the jury convicted	60%	78%	86%

Source: Kalven and Zeisel (1966).

the defense counsel was superior to the prosecutor, the jury also acquitted in 88% of the cases. (In the remaining 12%, the jury convicted, thus acting in disagreement with the judge's private conclusions.) Moving one step to the right, we see that when the judge would have acquitted, but the defense counsel and prosecutor were equal in ability, the jury also acquitted less often, in 82% of the cases. That is in conformance with intuition because for this group of cases we would expect the jury to be less swayed in the direction of acquittal by the skills of the defense counsel. Moving one step more to the right, the 76% in the upper right shows that a superior prosecutor sways the jury away from acquittal by about the same amount that a superior defense counsel sways the jury toward acquittal (for cases in which the judge would have acquitted). The arithmetic is direct: $88 - 82 = 6$, and $82 - 76 = 6$.

The bottom row gives analogous information for those cases in which the judge would have *convicted*. When the defense counsel is superior, the jury also convicts in 60% of the cases; when the abilities are equal, the jury convicts more often, in 78% of the cases; when the prosecutor is superior, the jury convictions rise to 86%. Here the changes in percent agreements with the judge are larger than before and unequal, 18% and 8%.

We note also what we saw earlier, that when the judge would have acquitted, the jury agreed with the judge more often than when the judge would have convicted. For a symmetric comparison with respect to defense or prosecution superiority, it may be seen that 88% is greater than 86%, 82% greater than 78%, and 76% greater than 60%.

To temper the force of these estimates of counsel ability, it is useful to keep in mind Table 2, which showed that in most jury trials (76% of them) the quality of counsel on the two sides is about equal. Hence the *overall* impact of superior counsel on the outcome of jury trials is modest. Still, that is small consolation to the defendant with counsel of inferior ability who loses his particular case.

FINAL REMARKS

The foregoing examples illustrate the many ways in which statistics has begun to illuminate legal problems. Yet this compact recital of examples should not leave the impression that the law is quick to appreciate the power of statistics. On the contrary, statistics is only just beginning to enter the legal realm at rare and selected points. It finds its most ready acceptance in the trial courts and before the administrative agencies, in litigation in which the issues depend on counting and measurement. In constitutional adjudication and legislative action, however, the law typically states its issues in terms of principles that at least superficially appear to be less accessible to a statistical approach, but even here some progress is being made. (See the essay by Alker.)

PROBLEMS

1. Assume the rear wheels of a car rotate independently of each other, while the front wheels rotate together. What is the probability of a person driving away from a space and returning to find all four valves pointing at the same "hours" as before departure?

2. What is meant by "proof by disproof of coincidence"?

3. What is the most frequent percentage of women jurors in a venire? (Give an approximate answer. Use Figure 2.)

4. In *Juries and Judges*, why do the authors say "the jury tends to have a softening effect"?

5. Referring to Table 1, what percentage of all cases were acquitted by the jury? Would have been acquitted by the judge?

6. Refer to Table 2. In how many cases were the lawyers of unequal ability?

7. How does Table 3 support a case for no-fault auto insurance?

8. What is sampling? Why is it sometimes difficult to use sampling and surveys as evidence?

9. Explain the experiment dealing with pretrial hearings. Why is this called a *controlled* experiment?

10. Suppose a second controlled experiment were conducted in which a random half of the filed suits were pretried as usual while the other half were not pretried. Suppose the results of this second experiment are given in the table.

Table. Pretrial Versus no Pretrial

	Pretrial	No Pretrials
Average length of trial time of cases that reached trial	6½ hours	8½ hours

Discuss the difference in design and results of the experiment reported here and the one reported in the text.

11. Refer to Table 5. Assume that the judge would have convicted in 100 cases, and that he felt 60% of the cases had lawyers of equal ability. The defense counsel was superior in half of the remaining cases. In how many cases did the jury acquit when the defense was superior?

REFERENCES

Alfred F. Conard et al. 1964. *Automobile Accident Costs and Payments.* Ann Arbor, Mich.: University of Michigan.

H. Kalven, Jr. and H. Zeisel. 1966. *The American Jury.* Boston: Little, Brown.

M. Rosenberg. 1964. *The Pre-Trial Conference and Effective Justice.* New York: Columbia.

H. Zeisel. 1968. "Statistics as Legal Evidence." D. L. Sills, ed., *International Encyclopedia of the Social Sciences.* New York: Macmillan.

H. Zeisel. 1969. "Dr. Spock and the Case of the Vanishing Women Jurors." *University of Chicago Law Review* 37:1–18.

H. Zeisel, H. Kalven, Jr., and B. Buchholz. 1959. *Delay in the Court.* Boston: Little, Brown.

SIZE OF POLICE FORCE VERSUS CRIME

S. James Press   *University of Chicago*

CRIME AND its prevention are much discussed these days, especially crimes of violence against individuals. Among the ways society moves to prevent or decrease crime, an important and time-honored one is the use of the police.

How effective are the police, and how may they be made more effective? For example, what impact does an increase of police manpower have? Such questions have in the past been remarkably little studied, in part because it is difficult and expensive to do the studies. This essay describes one study that was carried out where statistics played a central role. In this study, police manpower was increased substantially in one precinct of New York City. It was found that the manpower increase was accompanied by decreases in certain kinds of crime, in particular robbery and auto theft, but there were no changes in other kinds of crime, for example, burglary.

THE PROBLEM

To allocate police resources efficiently, it is useful to know whether the amount of crime committed in a given area is affected by increasing police manpower and associated equipment in the area; if crime is affected, by how much; and which crimes are affected the most. Getting at such questions requires careful statistical analysis because of a number of difficulties. Some difficulties are observational, in that data most appropriate for examining these questions have simply not been collected. Other difficulties are inherent in the problem. Some of them are:

(1) The true amount of crime committed in an area is never known precisely since only some fraction of it is actually reported. Further, the proportion reported may itself increase if police manpower increases.

(2) Crime changes must be analyzed in a controlled fashion so that the effects of changed. manpower can be compared with what crime *would* have been had the changes not been carried out. If crime patterns are continually shifting, such controlled study becomes more difficult. Thus, if an area experiences basic sociological or economic changes (such as might happen after an influx of low-income immigrants), patterns of crime might be expected to change independently of changes in police procedure.

(3) If a "crackdown on crime" is attempted in one area of the city and not in others, crime may decrease in that part of the city but increase in other parts as criminals merely relocate their activities. If such increases are spread over many areas in small amounts, their detection would be difficult because the effect would be masked by random fluctuations in reported crime, and by other factors external to the basic problem.

Because of data-gathering problems and inherent problems like those above, it has been difficult to quantify the effects of increased police manpower on crime. We shall see that statistical analysis helped clarify and interpret data from a New York City study in which police manpower was substantially increased in a single police precinct. Reported crimes in that precinct (and others) were recorded before and after the increase in police manpower.

The police made available data on daily reported crimes during the five-year period from January 1, 1963, to December 31, 1967. The crime reports were classified according to crime type (robbery, burglary, etc.), exact date, time, place, and nature of crime, and whether or not the crime was of the "outside" type (visible from the street). An example is shown by calendar quarter for "total robberies" in Figure 1.

On October 18, 1966, the police increased the number of patrolmen assigned to the 20th Precinct, located on the West Side of Manhattan (see

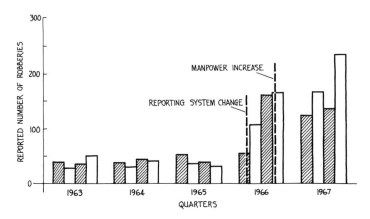

FIGURE 1

*Reported number of robberies
in the 20th Precinct. Source:
Press (1971)*

Figure 2), from an average of 212 to an average of 298 while police manpower
elsewhere in the city remained fairly constant.

METHOD OF STUDY

The fundamental question of the study was whether the increase in police
manpower decreased reported crimes. To answer this question, a number
of analytic decisions had to be made, for example, should crimes be studied

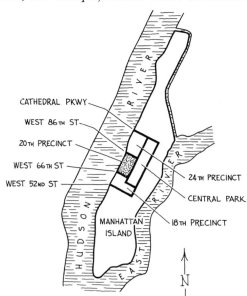

FIGURE 2

*Locations of police precincts
under study. Source: Press
(1971)*

on a daily, weekly, or some other time-period basis? The decision was to study crime on a weekly basis because of unevenness in the incidence of crime over the days of the week; more crime is committed on weekends.

A second analytic problem was that of seasonality; more outside crime is committed during the warmer months than during the colder months; for other crime types the reverse is true. A naive study of crime records might easily mistake an indicated increase for a "real" increase, rather than a seasonal effect. By using a statistical averaging technique this seasonal effect can be minimized.

The total period was divided into a low-manpower period (before the increase in police manpower) and a high-manpower period. The seasonally adjusted average weekly crime volumes during the low- and high-manpower periods could now be compared. Their differences, however, might not reflect the effect of increased police manpower since there was no control. That is, there was no way to tell whether an observed change in reported crime might have occurred anyhow—without a change in police manpower. To provide some measure of control for this effect (since it had not been provided by the design of the study), the following procedure was established. A group of precincts located in other parts of the city were selected as controls for the 20th Precinct, separately for each crime type, on the basis of how "similar" they were to the 20th, from a crime standpoint. Crime changes were then studied for the control precincts in the same way they were studied for the 20th Precinct. The difference in average weekly crime from low- to high-manpower periods in the 20th Precinct was compared with the average difference in the control precincts; the difference between these differences was taken to be the net effect associated with the increase in police manpower in the 20th Precinct.

The net changes in reported crime were evaluated as described above. At this point, decisions were made about which net changes were larger than could be explained by purely chance, or sampling variation. For example, if it were found that the estimated net change in reported crime (over and above the change in the control precincts) was a *decrease* of two crimes per week, was this observed change real, or just an effect explainable by purely chance (sampling) variation? Perhaps if the experiment were carried out again, a net *increase* of two crimes per week would be observed, just because of slightly different conditions at the times of observation. To reduce such uncertainties about interpretation, intervals of credibility, that is, intervals which reflect a high degree of belief in the net change results, were established. The idea behind these intervals is the following.

Prior to examining the data it was believed that almost any net change might be observed. After observing the data, however, it was determined (by means of statistical techniques) that uncertainty about the true net change of crime could then be confined to a rather small interval.

For example, suppose as in the illustration above, it were found that the observed net decrease in crime was about two crimes per week, and further, that the chance was very great (95%) that the *true* decrease was between 1.3 crimes per week and 2.6 crimes per week.   Then, it could be concluded with little chance of error that a real decrease was being observed (since the interval includes only decreases) and that an observed increase would be highly unlikely in a future study.   In many cases it was found that, although a net decrease or increase was observed, in fact, the interval of credibility included both positive and negative numbers and thus the apparent change could be attributable to sampling variation and might not be a real effect at all.

RESULTS

Some major results of the study are given in Table 1.   A percentage change is relative to what would have occurred without the police manpower change. A net change was taken to be "statistically significant" if the 95% credibility interval included only positive or only negative numbers.   In the table, all results are statistically significant unless the contrary is stated.

To understand how the numbers in the table were developed, consider the case of *outside robbery*.   It was found that 4.56 crimes per week were reported on the average, in the 20th Precinct, prior to the substantial change

TABLE 1.   Changes in Crime Rate of Inside and Outside Crimes*

ROBBERY:	A net decrease of 2.6 crimes per week (33%) for crimes visible from the street and a net decrease of about 2 crimes per week (21%) for others
GRAND LARCENY:	A net decrease of 17 crimes per week (49%) for crimes visible from the street and a net decrease of 6.6 crimes per week (29%) for others
BURGLARY:	Changes in reported crime not visible from the street (97% of all burglaries in New York) were not statistically significant
AUTO THEFT:	There was a net decrease of 7.7 crimes per week (49%)
MISCELLANEOUS FELONIES:	There was a net decrease of 1.9 crimes per week (38%) for crimes visible from the street
TOTAL FELONIES:	A net decrease of 23.7 crimes per week (36%) for crimes visible from the street and a net decrease of 4.4 crimes per week (5%) for others
MISCELLANEOUS MISDEMEANORS:	Crimes visible from the street showed a net decrease of 8 crimes per week (15%); other crime changes were not statistically signicant
TOTAL MISDEMEANORS:	Net changes were not statistically significant

* The data classified reported crimes as *outside crimes*, which are visible from the street, and *inside crimes*.

in police manpower (after adjustment for seasonal effects). After the manpower change, an average of 5.01 crimes per week were reported (after seasonal adjustment). In the control precinct, the corresponding numbers were 4.76 and 7.79, respectively. Thus, there was an increase of only 0.45 crimes per week in the 20th Precinct while there was an increase of 3.03 crimes per week in the control precinct (where there was *no* change in police manpower). The implication is that the reported number of outside robberies would have increased by 3.03 crimes per week (instead of only 0.45) during the same time interval had there not been a substantial increase in police manpower. The net effect of the police manpower change is 3.03 minus 0.45, or a decrease of about 2.6 crimes per week, the entry reported in the table. The 95% credibility interval was computed to be from 1.49 to 3.70. Since the interval included only decreases, it was concluded that the measured reduction in reported crime was a real effect.

It may be noted from the table that the increase in police manpower had little if any effect upon burglaries. This is not surprising since almost all burglaries are inside-type crimes. Placing more police in the neighborhood is not likely to have an appreciable effect upon such crimes unless the police manpower increase is accompanied by a change in the pattern of deployment to one which is designed to focus on burglaries.

The police manpower increase was accompanied by appreciable decreases in the violent crime of robbery (a crime that, by definition, requires a confrontation between criminal and victim) and the property crimes of grand larceny and auto theft. These results make sense since these crimes take place in the street, where a police manpower increase is likely to have its greatest effect.

TECHNICAL AND INTERPRETIVE PROBLEMS

Some of the technical difficulties which had to be overcome to arrive at the above conclusions, and some of the difficulties of interpretation of results of any such study, are described below.

Unfortunately, there was no opportunity for the statisticians to help in the design of the study so that, for example, the quality of the additional patrolmen (relative to the others) could not be assessed, the way in which the additional patrolmen were deployed and utilized could not be determined, and it could not be determined whether or not there was a *Hawthorne effect*, that is, a reduction or increase in reported crime simply because the patrolmen and residents knew a change was taking place (and therefore tried hard to effect a change). Also, the experiment was not repeated so that random errors would tend to average out. That is, several precincts in which police manpower was increased by the same percentage might have been, but were not, used. Another possibility might have been to increase police manpower (and then remove the additional force) several times in the 20th Precinct.

Moreover, the 20th Precinct might have peculiarities not common to the other precincts, and had several precincts been selected at random and the results averaged, results would have been more acceptable as representative of the City.

During the period of investigation the method of reporting crime underwent major changes. On March 10, 1966, a central reporting bureau was established for the entire city to replace the earlier precinct-by-precinct reporting system. Moreover, some of the definitions of what constitutes a reportable crime changed. The overall effect was to increase the number of reported crimes, that is, to reduce the number of crimes the police are aware of, but which previously may have gone unreported on official records. As a result, reported incidence of crime showed a substantial fictitious increase after the change in the reporting system. From an analysis standpoint, the data in the two periods (before and after the reporting system change) are not directly comparable so that reporting comparisons without statistical analysis would be misleading. (Data collected *after* the change in reporting system but *prior* to the increase in police manpower could be and were compared directly with data collected after the police manpower increase.) It was decided that although data collected before and after the reporting system change were somewhat different, the "early" data could at least be used to estimate the pattern of seasonal variation since that factor is likely to be least affected by changes in the reporting system. To evaluate seasonal variation, the early data were averaged in a special way designed to eliminate all effects except the seasonal component. Since the seasonal variation has a period of one year, the average was taken for one year periods over the approximately three years of data available prior to the change in reporting system. The results of this procedure were then used to eliminate the seasonal component from all the data collected subsequent to the change in reporting system.

It was also reasoned that when a new system of reporting crime is instituted it takes some time for the police personnel to adapt to the new system. That is, if crime records were used from the instant of adoption of the new system, part of any observed change in reported crime might be attributable to this transient effect. It was decided that about a month of adaptation time would be adequate, since over this period mistakes could be made one week, discovered during the next week or two, and corrected thereafter. After a month, most adaptation errors should have been eliminated.

In an analogous way, it was reasoned that after a substantial change in police manpower in an area the residents might change their rate of reporting crime, and therefore the reported incidence of crime. This might occur because of their increased awareness of the presence of police in the area, and therefore a greater feeling that something might be done about the crimes they report. To allow, at least in part, for the short-run aspect of this adapta-

tion effect, the first month of data after the manpower increase began was
not used in the detailed analysis.

SUMMARY

The problem of main interest was whether increased police manpower is effective as a deterrent to crime.   In a New York City police precinct, police
manpower was increased by 40%.   Large volumes of data were collected
for reported crime, over a five-year period, for all police precincts in the
city.   The simplest methods of summarizing the data could not, by themselves,
be used to answer the question posed.   By using techniques of seasonal adjustment, the method of comparison with control precincts, and credibility intervals, modern statistics was used to shed more light on a difficult problem.
That is, by using statistical techniques, results were obtained, suggesting that,
while increased police manpower is probably not very effective against certain
types of crimes such as burglary and misdemeanors, it may be effective against
other crimes such as robbery, grand larceny, and auto theft.

PROBLEMS

1.   What is the common property or characteristic of the crimes against
which an increase in police manpower seemed to be effective?

2.   Does Figure 1 show that the 20th precinct had about 155 robberies
in the third quarter of 1966? Explain your answer. (Hint: the answer is
no.)

3.   Figure 1 shows approximately 35 reported robberies in the second
quarter of 1965 and 105 in the same quarter of 1966. Are these figures
comparable? Explain.

4.   If auto theft were the only crime under consideration, what seasonal
variation, if any, would you expect?

5.   Explain the system of control used in this study. Was it originally
part of the design of the study? What is the weakness of this system of
control?

6.   What is an interval of credibility?

7.   Consider the following table.

Changes in Crime Rate for Auto Theft

	20th Precinct	"Control" Precincts
Rate Before Manpower Increase	5.7/week	5.5/week
Rate After Manpower Increase	5.9/week	6.7/week

(a) What is the estimated net effect of the manpower increase on the auto theft rate in the 20th precinct?

(b) If the interval of credibility were −1.2 to .5 crimes per week with 95% credibility would we conclude we had a true change?

(c) If the interval of credibility were −1.2 to −.5 crimes per week with 95% credibility would we conclude we had a true change?

Explain your answers.

8. Consider Table 1 (and the accompanying text). Which crime(s) showed "statistically significant" changes?

9. What is the meaning of the term *Hawthorne effect*?

10. How did the study's analysis take account of the change in the system of reporting crimes?

REFERENCES

A. D. Biderman. 1967. "Surveys of Population Samples for Estimating Crime Incidence." *The Annals of the American Academy of Political and Social Science* 374: Nov.

"The Challenge of Crime in a Free Society." *A Report by the President's Commission on Law Enforcement and the Administration of Justice.* Washington: U.S. Government Printing Office. Feb. 1967.

"Criminal Victimization in the United States: A Report of a National Survey." *Field Survey II,* The President's Commission on Law Enforcement and the Administration of Justice, National Opinion Research Center, Chicago. May 1967.

"Operation 25." 1955. City of New York Police Department.

S. J. Press. 1971. "Some Effects of an Increase in Police Manpower in the 20th Precinct of New York City." R-704-NYC. New York: The New York City RAND Institute.

MEASURING THE EFFECTS OF SOCIAL INNOVATIONS BY MEANS OF TIME SERIES*

Donald T. Campbell   *Northwestern University*

WE LIVE in an age of social reforms, of large-scale efforts to correct specific social problems. In the past most such efforts have not been adequately evaluated: usually there has been no scientifically valid evidence as to whether the problem was alleviated or not. Since there are always a variety of proposed solutions for any one problem, as well as numerous other problems calling for funds and attention, it becomes important that society be able to learn how effective any specific innovation has been.

From the statistician's point of view, the best designed experiments, whether in the laboratory or out in the community, involve setting up an *experimental group* and a *control group* similar in every way possible to the experimental

* Supported in part by NSF grant GS 1309X.

group except that it does not receive the same experimental treatment. The statistician's way of achieving this all-purpose equivalence of experimental and control groups is randomization. Persons (or plots of land or other units) are assigned at random (as by the roll of dice) to either an experimental or a control group. After the treatment, the two groups are compared, and the differences that are larger than chance would explain are attributed to the experimental treatment. This ideal procedure is beginning to be used in pilot tests of social policy, as in the current New Jersey negative income-tax experiment where several hundred low income employed families who agreed to cooperate have been randomly assigned to experimental groups that receive income supplements of differing sizes and a control group that receives no financial aid. The effects of this aid on the amounts of other earnings, on health, family stability, and the like are being studied.

While such experimental designs are ideal, they are not often feasible. They are impossible to use, for example, in evaluating any new program which is applied to all citizens at once, as most legal changes are. In these more common situations much less satisfactory modes of experimental inference must suffice. The *interrupted time series design,* on which this paper will concentrate, is one of the most useful of these quasi-experimental designs. The proper interpretation of such data presents complex statistical problems, some of which are not yet adequately solved. This paper will touch upon a number of these, in terms of words and graphs rather than mathematical symbols. The discussion will start off by considering two actual cases.

THE CONNECTICUT CRACKDOWN ON SPEEDING

On December 23, 1955, Connecticut instituted an exceptionally severe and prolonged crackdown on speeding. Like most public reporting of program effectiveness, the results were reported in terms of simple before-and-after measures: a comparison of this year's figures with those of a year ago. The 1956 total of 284 traffic deaths was compared with the 1955 total of 324, and the governor stated, "With a saving of 40 lives in 1956 . . . we can say the program is definitely worthwhile." Figure 1 presents his data graphically. This simple quasi-experimental design is very weak and deceptive. There are so many other possible explanations for the change from 324 to 284 highway fatalities. In attributing all of this change to his crackdown, the governor is making the implicit assumption that without the crackdown there would have been no change at all. A time series presentation, using the fatality records of several prior and subsequent years, adds greatly to the strength of the analysis. Figure 2 shows such data for the Connecticut crackdown. In this larger context the 1955–56 drop looks trivial. We can see that the implicit assumption underlying the governor's statement was almost certainly wrong.

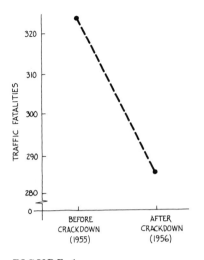

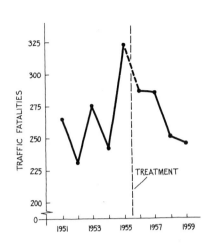

FIGURE 1

Connecticut traffic fatalities, 1955–56. Source: Campbell and Ross (1968)

FIGURE 2

Connecticut traffic fatalities, 1951–59. Source: Campbell and Ross (1968)

To explore this more fully, turn to Figure 3, which presents in a stylized manner how an identical shift in values before and after a treatment can in some instances be clearcut evidence of an effect, and in other cases no evidence at all of a change. Thus with only 1955 and 1956 data to go on, that drop of 40 traffic fatalities shown in Figure 1 might have been a part of a steady annual drop already in progress (the reverse of the steady rise in line F of Figure 3), or of an unstable zigzag (as in line G of Figure 3), etc. Figure 2 shows that in Connecticut, the unstable zigzag is the case. The 1955–56 drop is about the same size as the drops of 1951–52, 1953–54, and 1957–58, times when no crackdowns were present to explain them. Furthermore, the 1955–56 drop is only half the size of the 1954–55 rise. Thus with all this previous instability in full graphic view, one would be unlikely to claim all of any year-to-year change as due to a crackdown, as the governor seemed to do.

Later, we shall examine Figure 2 again to raise a more difficult problem of inference. But before we do this, let's spend more time on the stability issue, with the help of an illustration from a reform that even a skeptical methodologist can believe was successful.

THE BRITISH "BREATHALYSER" CRACKDOWN OF 1967

In September 1967, the British government started a new program of enforcement with regard to drunken driving. It took its popular name from a device for ascertaining the degree of intoxication from a sample of a person's breath.

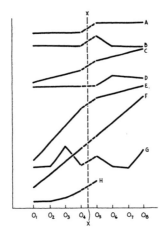

FIGURE 3

Some possible outcome patterns from the introduction of a treatment at point X into a time series of measurements, O_1 to O_8. The $O_4 - O_5$ gain is the same for all time series, expect for D, while the legitimacy of inferring an effect varies widely, being strongest in A and B, and totally unjustified in F, G, and H. Source: Campbell and Stanley (1963)

Police administered this simple test to drivers stopped on suspicion, and if it showed intoxication, then took them into the police station for more thorough tests. This new testing procedure was accompanied by more stringent punishment, including suspension of license. Figure 4 shows the effect of this crackdown on Friday and Saturday night casualties (fatalities plus serious injuries). The effect is dramatically clear. There is an immediate drop of around 40% and a leveling off at a level that seems some 30% below the precrackdown rate, although this is hard to tell for sure since we don't know what changes time would have brought in the casualty rate without the crackdown.

Does the effect show up when casualties at all hours of all days are totaled? Figure 5 shows such data. While the effect is probably still there, it is cer-

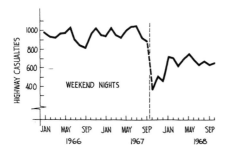

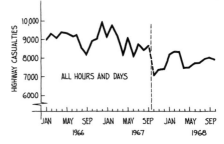

FIGURE 4

Effects of the September 1967 English "Breathalyser" crackdown on drunken driving. Fatalities plus serious injuries, Fridays and Saturdays, 10:00 PM to 4:00 AM, by month. Source: Ross et al. (1970)

FIGURE 5

Effects of the September 1967 English "Breathalyser" crackdown. Fatalities plus serious injuries, all hours and days, by month. Source: Ross et al. (1970)

tainly less clear, the crackdown drop being not much larger than the unexplained instability of other time periods. (The crackdown drop is, however, the largest month-to-month change, not only during the plotted period, but also for a longer period going back to 1961, for these data from which the seasonal fluctuations have been removed.)

THE STATISTICAL ANALYSIS OF INSTABILITY

The problem of the statistician is to formalize the grounds for inference that we have used informally or intuitively in our judgments from these graphs. It is clear that the more unstable the line is before the policy change or treatment point, the bigger the difference has to be to impress us as a real effect. One approach of statisticians is to assume that the time series is a result of a general trend plus specific random deviations at each time period. The theory of this type of analysis is well worked out for the case in which the random deviations at each point are completely independent of deviations at other points. But in real-life situations the sources of deviation or perturbation at any one point are apt to be similar for adjacent and near points in time, and dissimilar for more remote points. This creates deceptive situations both for statistical tests of significance and for visual interpretation. Figures 6 and 7 illustrate this with computer simulated time series. For each point in time, times 1 to 40, there is a *true score*. These true scores, if plotted,

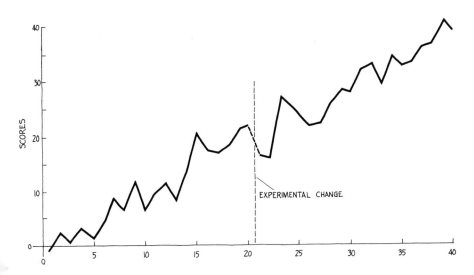

FIGURE 6
Simulated time series with independent error. Source: Ross et al. (1970)

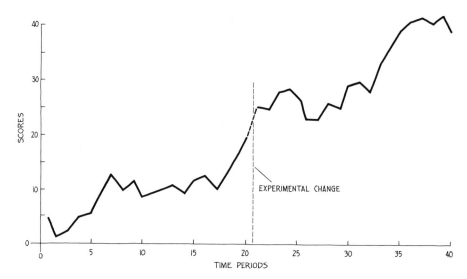

FIGURE 7

Simulated time series with correlated (lagged) error. Source: Campbell (1969)

would make a straight diagonal line from a lower left score of zero to an upper right score of 40, with no bump whatsoever at the hypothetical treatment point. These true scores are the same for Figures 6 and 7. To each true score has been added or subtracted a randomly chosen deviation. In Figure 6, the deviation at each time point was drawn independently of every other deviation. This is a simulation of the case in which the hypothetical treatment introduced between time periods 20 and 21 had no effect at all. Occasionally, by chance, random deviations will occur in such a pattern so as to make it look as though the treatment had an effect, as perhaps in Figure 6. It is the task of *tests of significance* to estimate when the difference from before treatment to after treatment is more than such random deviations could account for. Statistical formulas have been worked out which do this well in the case of independent deviations such as Figure 6 illustrates.

Figure 7 is based upon the same straight diagonal line as Figure 6. It has the same magnitude of deviations added. But the deviations are no longer independent. Instead, four smaller deviations have been added at each point, in a staggered or lagged pattern. A new deviation is introduced at each time period and persists for three subsequent periods. As a result, each point shares three such deviations with the period immediately prior and three with the period immediately following. It shares two deviations with periods 2 steps away in either direction, and one deviation with periods 3 steps away. For periods 4 or more steps away, the deviations are independent. While

Figure 7, like Figure 6, is a straight line distorted by random error, note how much more dynamic and cyclical it seems. Such nonindependent deviations mislead both visual judgments of effect and tests of significance which assume independence, through producing judgments of statistically significant effect much too frequently. (To emphasize the lack of any true or systematic departures, let it be emphasized that were one to repeat each simulation 1000 times, and to average the results, each average would approximate the perfectly straight diagonal line of the true scores.)   There is a variety of ways in which statisticians are attempting to get appropriate tests of significance for the real-life situations in which nonindependent deviations are characteristic. While none is completely satisfactory as yet, great strides are being made on the problem.

REGRESSION ARTIFACTS

We have moved from simple problems of inference to more complex ones. We will return soon to some more easily understood problems. But before doing that, let us attempt to understand a final difficult problem, known in one statistical tradition as *regression artifacts*. If we can be sure that the policy change took place independently of the ups and downs of the previous time periods, there is no worry. But if the timing of the policy change was chosen just because of an extreme value immediately prior, then a "regression artifact" will be sufficient to explain the occurrence of subsequent less extreme values. To see if a regression artifact might be at work in the Connecticut case, let us return to Figure 2. Here we can note that the most dramatic change in the whole series is the 1954–55 increase. By studying the newspapers and the governor's pronouncements, we can tell that it was this striking increase which caused him to initiate the crackdown. Thus the treatment came when it did because of the 1955 high point.

In any unstable time series, after any point which is an extreme departure from the general trend, the subsequent points will on the average be nearer the general trend. Try this out on Figure 6. Move your eye from left to right, noting each point that is "the highest so far." For most of these, the next point is lower, or has *regressed* toward the general trend. Such regression subsequent to points selected for their extremity is an automatic feature of the very fact of instability and should not be given a causal interpretation. Applied to Figure 2, this means that even with no true effect from the crackdown at all, we would expect 1956 to be lower than the extreme of 1955.

OTHER REASONS FOR SHIFTS IN TIME SERIES

It is going to be important for administrators, legislators, the voting public, and other groups of nonstatisticians to be able to draw conclusions from time series data on important public programs.   For this reason, two further points will be made that are less directly statistical.   First note that there are many reasons for abrupt shifts in time series other than the introduction of a program change.   One very deceptive reason is a shift in record keeping procedures. Such shifts are apt to be made at the same time as other policy changes.   For example, a major change in Chicago's police system came in 1959 when Professor Orlando Wilson was brought in from the University of California to reform a corrupt police department.   Figure 8 shows his apparent effect on thefts—a dramatic *increase*.   This turns out to be due to his reform of the record keeping system, and the rise was anticipated for that reason.

In a real situation, unlike in an insulated laboratory, many other causes may be operating at the same time as the experimental policy change.   Thus in Connecticut or in England, a drop in traffic casualties might have been due to especially dry weather, or fewer cars on the road, or to new safety devices, or to a multitude of other factors.   If one had been able to design an experiment with the randomized control groups discussed at the beginning of this essay, such explanations would have been ruled out statistically.   However, setting up such an experiment would have been impossible in these two situations.   We must instead try to rule out these rival explanations of an effect in other ways.   One useful approach is to look at newspaper records of rainfall, changes in traffic density, and other possible causes of the shift.

Another approach is to look for some control comparison that should show the effects of these other causes, if they are operating, but where the specific reform treatment was not applied.   For Connecticut, the data from four nearby states are relevant, as shown in Figure 9.   All of these states should have been affected by changes in weather, new safety features in cars, etc. While these data support the notion that the 1956 Connecticut fatalities would

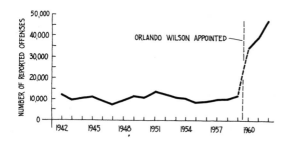

FIGURE 8

Reported larcenies under $50 in Chicago from 1942 to 1962. Source: Campbell   (1969). Data from Uniform Crime Reports for the United States, 1942–62.

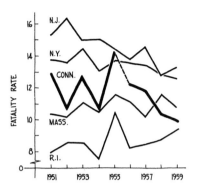

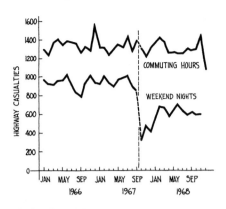

FIGURE 9

Traffic fatalities for Connecticut, New York, New Jersey, Rhode Island, and Massachusetts (per 100,000 persons). Source: Campbell (1969)

FIGURE 10

A comparison of casualties during closed hours (commuting hours) and weekend nights in the English "Breathalyser" crackdown. Source: Ross et al. (1970)

have been lower than 1955 even without the crackdown, the persisting decline throughout 1957, 1958, and 1959 is much the steepest in Connecticut, and may well indicate a genuine effect of the prolonged crackdown. While visually we have little difficulty in using these supplementary data, the statistician has many real problems in combining them all in an appropriate test of significance.

For England, there were no appropriate comparison nations available. But British pubs are closed before and during commuting hours, so casualties from such hours provide a kind of comparison base, as shown in Figure 10. Unfortunately there is a lot of instability in these data so that they do not enable us to estimate with much confidence the degree to which the initial crackdown effects are persisting.

A FINAL NOTE

There are several general lessons to be learned from these brief illustrations. The first involves the distinction between *true experiments,* in which experimental and control groups are assigned by randomization, and *quasi experiments.* True experiments, when they are possible, offer much greater power and precision of inference than do quasi experiments. The administrators of social innovations, in consultation with statisticians, should attempt to use such designs where possible. Where true experiments are not possible, or have not been used, there are some quasi-experimental designs such as the interrupted time series which can be very useful in evaluating policy changes. These too require statistical skill to avoid misleading conclusions. Evaluation of social innovations is an important and challenging area of application for modern statistics.

PROBLEMS

1. Why is the design on which the paper concentrates called an *interrupted* time series design?

2. How does Figure 2 change your perception of Figure 1?

3. Consider Figure 3. Suppose that X was a currency devaluation which resulted in a price increase between O_4 and O_5. What effect did X have on the price of product F? Product A? Product C?

4. Consider Figure 3. Why does the author say that one can most legitimately infer an effect in A and B while totally unjustified in inferring one in F, G, and H?

5. Consider Figures 4 and 5. Is the effect of the breathalyser crackdown more pronounced in one of the figures? If so, which one?

6. (a) Explain the difference between independent and lagged error.
    (b) How would you characterize the effects of the two types of error on the plots of data in Figures 6 and 7?

7. What does the author mean when he says Figure 7 is more "dynamic and cyclical" than Figure 6?

8. Explain what the author means by the term *regression artifact*.

9. Refer to Figure 8. Explain the sharp increase in the number of reported offenses when Orlando Wilson was appointed.

10. Should the 1955–56 traffic fatality decline in Rhode Island (Figure 9) be attributed to the speeding crackdown in neighboring Connecticut? Explain your answer.

11. What is a *true experiment*? What is a *quasi-experiment*?

12. Comment on this statement: Quasi-experiments are used in evaluating policy changes because it is virtually impossible to apply true experimental design to social situations.

REFERENCES

D. T. Campbell. 1969. "Reforms as Experiments." *American Psychologist* 24:4, pp. 409–429.

D. T. Campbell and H. L. Ross. 1968. "The Connecticut Crackdown on Speeding: Time-Series Data in Quasi-Experimental Analysis." *Law & Society Review* 3:1, pp. 33–53.

D. T. Campbell and J. C. Stanley. 1963. "Experimental and Quasi-Experimental Designs for Research on Teaching." N. L. Gage, ed., *Handbook of Research on Teaching*. Chicago: Rand-McNally. Pp. 171–246. Reprinted as *Experimental and Quasi-Experimental Designs for Research*. 1966. Chicago: Rand-McNally.

H. L. Ross, D. T. Campbell, and G. V. Glass. 1970. "Determining the Social Effects of a Legal Reform: The British 'Breathalyser' Crackdown of 1967." *American Behavioral Scientist* 15:1, pp. 110–113.

J. S. Wholey, J. W. Scanlon, H. G. Duffy, J. S. Fukamato, and L. M. Vogt. 1970. *Federal Evaluation Policy*. 9-121-21. Washington: The Urban Institute.

DO SPEED LIMITS REDUCE TRAFFIC ACCIDENTS?

Frank A. Haight   *Pennsylvania Transportation and Traffic Safety Center*

Iᴛ ɪs surprisingly difficult to discover what factors cause increases or decreases in traffic accidents. Sometimes changes are made in speed limits, in highway design, in driver-licensing standards, or in vehicle specifications in an attempt to decrease accidents. Yet, it can sometimes happen that the number of accidents will actually increase. These surprising increases may come merely from increases in population, or from increases in number of vehicles on the road, or perhaps from greater distances traveled.

MEASURING ACCIDENT RATES

So it is sensible to measure accident *rates* rather than the raw number of accidents. Of course, there is a variety of such rates; among the most common are *accidents per person in the population, accidents per registered vehicle,* and *accidents per vehicle mile traveled.* For example, if along a certain highway there were 500 accidents in 1970 and 10 million vehicle miles, then the last rate would be 50 accidents per million vehicle miles. Often these rates are measured in terms of fatalities rather than in terms of all accidents.

The behavior of these rates in the U.S. since the late thirties has been approximately as follows: (1) the fatality rate per person is increasing slightly; (2) the fatality rate per registered vehicle is decreasing; and (3) the fatality rate per vehicle mile is decreasing, rather quickly during the forties and fifties and less quickly later. From 1963 to 1970 it may even have been stationary. This combination of rates would be consistent with a trend of proportionately more cars but relatively safer ones. This is roughly true also for other places in the world where road traffic plays an important part in the society: Western Europe, Australia, and New Zealand.

The total size of the population considered affects the stability of all such accident rates. For a large country like the United States there is very little fluctuation from year to year or even from month to month because we are dealing with averages based upon large numbers of cars and large numbers of accidents. In a small town, to take the other extreme, it may be nearly impossible to "see the forest for the trees" since a very few accidents may have the effect of changing the average greatly.

This tendency is shown clearly by the data of Figure 1 from the National

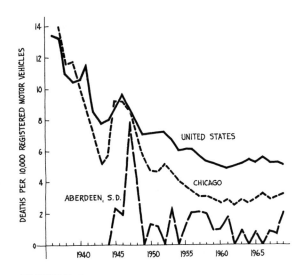

FIGURE 1

Deaths per 10,000 registered motor vehicles, 1936–69, U.S. and selected localities. Source: Data from National Safety Council

Safety Council. It should be noted that the rate given in Figure 1 is *fatalities per ten thousand registered vehicles.* A similar graph based on the rate of *fatalities per hundred million vehicle miles driven* would be even more informative, but this rate is not known for subdivisions smaller than whole states. The reason is that the estimate of vehicle miles traveled is based on gasoline consumption and records of this type are not usually kept by municipalities. (Thus, if you buy a gallon of gasoline and throw it away without using it in your car, you will actually lower the fatality rate per vehicle mile for that year as much as if you had used the gas safely in your car!)   Because the averages for the country are so stable but those of small towns so unstable, predictions of the number of fatalities expected over holiday weekends are usually given for the country as a whole rather than for smaller geographical divisions.

EVALUATING CHANGES IN ACCIDENT RATES

From the point of view of accident prevention there are two important consequences of this variability. The first is that since the most commonly used rate is generally going down, its decrease after some specific change is made does not by itself prove that the change was beneficial. Second, if the change takes place over a small population—even as small as several million—it may be well nigh impossible to separate the effect of this change from the general chance fluctuation. The statistical point is not really the population size, but the general level of the numbers of accidents. The numbers of accidents do not vary much percentagewise when the general level is large, but they do when it is small (see the essay by Campbell).

An added difficulty is that most changes in traffic patterns, laws, or law enforcement are expensive: rebuilding highways, eliminating grade crossings, building safer cars, and enforcing safety rules are typically costly items. Some of these factors are, in addition, nearly permanent; once done it is not easy to undo them. For example, if an expensive bridge is built, it cannot readily be relocated. Hence, there is a tendency to introduce improvements in the road system on the basis of good sense and experience rather than on empirically demonstrated improvements in highway safety; many claims of benefit are to be taken in a general rather than specific sense—the accident rate *is* decreasing although it is hard to pinpoint the exact causes.

This may in some ways be a gratifying situation, but it is not helpful in planning new measures. Given a fixed budget, should investment be made in driver training, in better enforcement of traffic regulations, in alcohol testing, in road reconstruction, in speed limit changes, or in some completely different approach such as subsidizing public transit? Furthermore, safety and budget are not the only considerations; convenience counts too. How can we find out if each of these proposed measures is effective?

EXPERIMENTS TO EVALUATE PROPOSED CHANGES

It would seem that a logical method would be to set up an experiment. But, in field after field of public policy, responsible authorities object to carefully controlled experimentation because they feel they cannot tamper with such important and expensive arrangements. Although there is much to discuss on this point, we shall not debate the rights and wrongs of it here, but report only that road authorities are usually unwilling to have the road transport system play the role of a guinea pig. They regard it as expensive and confusing to the traveling public and likely to be inconclusive.

Nevertheless, let us think about how such an experiment to test the effect of an improvement in traffic control on accident fatalities should be designed. Because fatal accidents are rare (less than six fatalities per 100 million vehicle miles in the U.S., for example), a large sample of driving is needed for such an experiment to give a clear answer through the mist of chance fluctuation. Furthermore, to allow for a general decreasing trend in rates the statistician would like to have the proposed improvement operate only on alternate days of the week or be "turned on and off" according to some reliable, recognizable program. The policeman will usually not agree to have traffic regulations change so often, although he may be willing to experiment for a week or two along a few miles of highway.

THE SCANDINAVIAN EXPERIMENT

An exception to this usual official reluctance to experiment has been in progress in Denmark and Sweden for nearly ten years. The governments of these countries have introduced *periodic speed limits* (speed limits that stay the same for long periods of time over long stretches of road, and then are changed for other long periods).

We must realize that speed limits of any sort are bitterly opposed in many European countries and are considered to be justified only if their reduction of accident rates can be strictly demonstrated. The American idea that speed limits are favorable for all aspects of road usage (for example, maximum flow), not merely safety, is in general not accepted.

In Sweden, a royal commission was established in 1961 to investigate various aspects of speed control; the commission members were statisticians and other scientists from the country's foremost technical university. The experiment for the first year involved the comparison of a 90-kilometer-per-hour limit (about 56.7 miles per hour) against no speed limit over 71 different road sections, chosen to represent a variety of road conditions. In that first year the speed limitation was in force from Friday, May 19, through Wednesday, May 24, from Thursday, June 22, through Wednesday, July 12, and from Friday, September 1, through Monday, September 11.

In subsequent years, the experimental program has been more ambitious and more complicated, involving speed limits of 90 kph, 110 kph, and 130 kph, as well as larger road networks; in 1968, all the roads of the European highway system in Sweden were included, together with some of the principal national main roads. Also the speed limits (or lack of limits) were maintained for longer periods of time, sometimes several months or a year, in order to obtain larger samples of accidents.

ACCIDENT RATES AND THE POISSON PROCESS

With the immense quantity of data obtained (7000 accidents during the 1968 experiment, for example) many different types of statistical analysis were performed. This article deals with one analysis which evoked some theoretical problems: before-and-after comparisons of accident rates.

Long study of the traffic accident phenomenon shows that a simple random process which well describes the instant at which accidents occur is the so-called *Poisson process*. A Poisson process is a quantitative way of expressing the fact that the accidents are indeed *accidental,* that is, that each accident occurs at a moment in time which is completely independent of the moments when other accidents occur—the times of the accidents are perfectly scrambled. This specification is incomplete, however, unless it also includes a so-called mean value that expresses how often accidents occur *on the average.* There can be a Poisson accident process on a main highway with, for example, 100 accidents per year on the average, or on a small rural road with only one accident per year on the average. In such a case the accident record for the rural road for 100 years would, if compressed into a year, duplicate the statistical properties of the single year on the main road. Thus, if we let the letter m denote the mean number of accidents observed for a section of road, we find that m will depend (among other things) on the traffic volume characteristic of the road sections. In the Swedish experiments, the volumes ranged from less than 150 vehicles per day up to values in excess of 5000 per day.

This variability in the traffic volume typical of a road section was complicated further by variability in volume over the days of the year. Some roads have heavy traffic during the summer, others in the winter. On certain days there are special events such as football games or national holidays to increase traffic flow. The weather influences both the volume of traffic and the risk of an accident, and even if the record is limited to the same day in consecutive years, there may have been a blizzard on one of these days and not on the other.

Therefore, the problem of comparing values of m without the speed limit to values of m with a speed limit seems to break down into thousands of special tests, one for each bit of road and bit of time. Fragmentation of the problem in this way reduces the sample sizes and seems to nullify the whole purpose of the experiment. Also, the conclusions would be far less reliable because of the small samples, and in the majority of cases, it might be impossible to come to any conclusion whatsoever.

It is clearly desirable to group the road segments and days into larger groups that will be homogeneous, but the factors involved are so varied as to make this nearly impossible. How can we compare the effect of a rain-shower with that of a construction project, or of a national holiday with an extra-wide shoulder? Is there any quantitative equivalence between a road in the far north experiencing wintry conditions for long periods and one near a ferry terminal serving many foreign tourists?

The solution to this problem was found to lie in grouping together road segments having the same number of accidents during the total period of the study, and for each group examining the proportion of accidents that occurred in the "before" period, and the complementary proportion that occurred during the "after" period; these time periods are of equal length. Thus, for example, we group together all road segments having ten accidents over the total experimental period; these road segments might be very different, yet their qualitative differences balance out to the extent that they experienced the same overall number of accidents.

To get an intuitive idea about the reasoning, let us suppose that there are but two conditions: having a speed limit and not having one. If we suppose that having a speed limit is effective in improving safety, then when we collect all the road segments we should find that more accidents occur under the "no limit" condition than under the "speed limit" condition. We have many segments, so we can look at the results for many similar segments, and thus pile up a considerable record. Furthermore, in some segments, the "speed limit" condition would have come first, and in others second, and we can check whether the order mattered. We can also see whether "speed limits" matter more to safety in segments with high accident rates or with low, and so on.

In any case, the key statistical technique was to put together those road segments having the same total number of accidents in the "before" and "after" periods, even though the segments grouped together might have nothing in common beyond their accident experience. With this approach, it was found that deciding whether the Poisson mean value m had changed as a result of the speed limit could be reduced to a simpler statistical problem involving the ratio of the m value "before" to the m value "after," given that the total accidents "before-and-after" was the same.

SOME RESULTS

One general result of the Swedish and Danish analyses is especially interesting. It appears that speed limits were more effective in Sweden than in Denmark! The reason behind this is not at all clear.

We can only speculate on the value of a similar experiment in the U.S. Probably it would not be useful to choose "speed limit" as the experimental variable because most of our communities have roughly similar attitudes towards speed restriction, with relatively small local variations. A more interesting variable might be vehicle inspection systems, which vary greatly from no compulsory inspection at all in California to very rigorous periodic checks in Pennsylvania. There has been recent discussion in technical journals about the effect of inspection laws on accident experience. How could we design an experiment that would test this factor in isolation from others? We could, perhaps, trace the accidents which Pennsylvania vehicles have on California roads and vice versa. Or we might begin an inspection system which applies only to blue cars and trace the proportion of blue cars in accidents. Either of these, or some other design, would give some clear indication of the usefulness of various inspection systems.

PROBLEMS

1. Why is it more reasonable to consider accident rates, rather than the actual number of accidents?

2. How has the fatality rate per vehicle mile changed between the 1930's and 1970? Can you speculate on some explanations? (Hint: How large were the road systems and the number of automobiles in the 1930's?)

3. In Figure 1, why does Aberdeen, S.D., show the sharpest increases and decreases in deaths?

4. Considering Figure 1, was the death rate per 10,000 registered vehicles larger for Chicago or the United States as a whole in 1949? In 1945?

5. Why does the author say when commenting on the rate given in Figure 1: "A similar graph based on the rate of *fatalities per hundred million vehicle miles driven* would be even more informative"?

6. Explain how you think number of vehicle miles travelled is estimated from gasoline consumption.

7. If, after a specific safety measure is instituted, the accident rate continues to decline, can this decline be attributed to the safety measure? Explain.

8. The advantages of experimenting to identify measures which reliably reduce the fatality rates are clear. Why haven't more such experiments been done?

9. Why is the Poisson process important for the study of accident rates?

10. In the Scandinavian experiment, what was m? On what did it depend?

11. (a) In the Scandinavian experiment, why couldn't the road segments be grouped by traffic volume?

(b) How were the road segments finally grouped? How did this treat their differences?

12. Can you think of any explanations of why speed limits in Sweden might be more effective than in Denmark?

ELECTION NIGHT ON TELEVISION

Richard F. Link   *Artronic Information Systems, Inc.*

DURING THE evening of the first Tuesday in November in even-numbered years, millions of people all over the U.S. watch the election shows provided by the three major networks. The viewers see a rapid tabulation of the votes cast for the major state offices of senator and governor, and in years when a president is elected, a rapid tabulation of the presidential vote by state and for the nation. They also see a tabulation of the vote for the members of the House of Representatives. They usually hear an announcement of the winner after only a few percent of the vote has been reported, often within minutes of the closing of the polls. As the evening progresses they are treated to analyses that explain how a given candidate won, that is, where his strength and weakness lay, and why it appeared that he won.

Massive machinery operates behind this effort. This machinery is physical in the sense that it requires a very elaborate communications network and extensive usage of computers, but it is also statistical and mathematical in the sense that it requires rapid summaries and interpretations so that the findings can quickly be passed to the viewing public.

We shall not attempt to describe the complete organization necessary to produce the election night show, but we shall describe the three parts of the show that lean most heavily upon computer and statistical technology: vote tabulation, projection of winners, and detailed analysis of the vote. The vote tabulation system is basically the same for all networks, but they differ in their methods for projecting winners and in their analysis of the vote. We shall describe in this paper only the method of projection used by one network (NBC).

Before discussing the procedures and methods used today, we shall give a brief history of the reporting of election night results to give a feel for why and how today's shows came about.

A BRIEF HISTORY OF ELECTION REPORTING

Persons living in the United States, as in other free societies that hold elections, have always had an intense interest in the outcome of elections. Most intense for the elections that involve the presidency, it is reasonably high for gubernatorial and senatorial elections, and at least the numbers of Republicans and Democrats composing the House of Representatives are of concern, even though the election of a particular member usually does not have national significance.

Thus election results have always been news of great interest. Until about 1928, this news reached the public via their newspapers. In general the coverage was relatively slow and incomplete. Radio changed this situation, and election reporting was speeded up. For example, radio reported the upset victory of Harry S. Truman in the early hours following election day in November 1948. Television began to report elections on a national scale in 1952 and has increased its scope and coverage and speed of gathering the vote since then. Extensive primary coverage was introduced during the presidential year of 1964 and continues to be a feature of television reporting today even in "off years".

Speed of coverage is influenced by two factors: the speed with which the vote is obtained from its source (basically a precinct), and then the speed with which it is reported. The speed of reporting the vote, once collected, was greatly increased by reporting via radio as opposed to reporting via newspaper. This reporting speed has not been particularly increased by television, since both radio and television are capable of essentially instantaneous reporting. The speed of vote collection, however, has been greatly increased by the television networks. It is worth reviewing the collection procedures utilized in the past and today.

The U.S. has approximately 175,000 precincts. In the official electoral machinery, the precinct vote is usually forwarded to a county collection center, and then to a state center, often to the Secretary of State there, who then certifies the official vote. Final official collection and certification often take several months. The precinct vote, however, is forwarded to the county level fairly rapidly, perhaps by phone or courier, and the vote at the county level is often quickly available on an unofficial basis. The job of collecting the vote at the county level is much less arduous than that of collecting at the precinct level since there are only about 3000 counties in the country. The vote can be collected faster, nonetheless, if it is collected at the precinct level, and this is the basic innovation that television introduced to vote collection. The networks with their large economic resources were instrumental in establishing a mechanism for obtaining the vote at the precinct level, and communicating it by phone to a central location where it could be processed by a computer.

Competition by television networks in the area of extensive vote collection became very intense by the primary elections of 1964. That year, the New Hampshire primary saw all three of the major television networks collecting and reporting the vote at the precinct level. In fact, some wags have said that there were more television workers in New Hampshire during the 1964 primary than voters, or to put it another way, that it would have been cheaper to bring the New Hampshire voters to New York to vote at a central location, than to collect the vote in New Hampshire. Needless to say, these remarks are exaggerated, but they do indicate the magnitude of the expense involved. The competition in collection became more intense that spring, and culminated in the reporting of the California presidential primary where each of the three networks collected the vote in the more than 30,000 precincts in California. This enormous expense brought only a mixed blessing. The newspaper wire services continued to collect and report the vote in the traditional manner, from complete county returns, so that on the day after the election, after the television networks had reported Goldwater the winner, the newspapers all showed Rockefeller with a substantial lead. The reason for this disparity was that Los Angeles county, with approximately a third of the precincts in the state, did not have a complete county report until too late to meet the newspaper deadlines, and Goldwater ran very strongly in Los Angeles.

This confusion was coupled with another, arising from the fact that each network would report its own vote totals at any instant of time, and, since they were being collected independently, at any given moment their totals were all different. All this led to the formation of an organization called the News Election Service (NES) whose sole purpose is to collect the vote and report it to its members. This service was formed by a cooperative effort of the three television networks (ABC, CBS, NBC) and the two wire services (AP, UPI).

We next examine briefly the operation of the News Election Service.

THE NEWS ELECTION SERVICE

The news election service was formed to provide a uniform and consistent report of the vote to the American public. The figures obtained by this organization are released to its members simultaneously, so that at any instant all networks and news services are able to report the same basic data to the public.

The massive operation functions in the following manner. Reporters, called stringers, are on duty at more than 100,000 of the largest of the 175,000 precincts in the country and at each of the 3000 county reporting centers. These reporters collect the vote at the precinct and county levels and then phone this vote to a central location, adding enough information to identify the source of the report. The vote information is then put into a computer, which checks to see that it appears valid; for example, if it is a precinct report, it checks to see that the vote does not exceed registration in that precinct, or if a county report, that the number of precincts in the county has not been exceeded. Because registration figures and data on number of precincts are not exactly accurate at this time the check depends upon a statistical tolerance rather than an absolute cut-off. Once the report has been checked, if it is a precinct report it is added to the precinct results already reported for that county. If it is a county report, it replaces the previous county report (the county reports are made on a cumulative basis). At regular intervals the computer generates a vote report for each election race for each county in the state and also provides a state total for the presidential, senatorial, and gubernatorial races and a national total for the presidential race. The summary report is generated by comparing the county votes from the county reports with the votes in the county calculated from the precinct reports. It uses the larger figure for the county figure, and then sums over the counties in the state to obtain a state figure. In presidential years an additional summation is made over the states to obtain a national vote figure for the presidency. In addition to providing summary vote totals, percentages for each candidate are reported, as is the fraction of precincts reporting. Similar accumulations are made for house races, but they are organized on a congressional district basis.

Once the information has been calculated, the computer releases the information to its clients. It provides this information in printed form both at the computer location and at the television studio, and it also makes the information available via telephone lines that can be used for input into the various network computer systems.

Extensive research must be done before the election not only to train the stringers, but also to gather the registration figures for the precincts, to find the number of precincts in each county, and to collect other basic data.

This information is essential for the checking process, and for accurate reporting of the fraction of the vote that has been counted.

Extensive preparation must also go into the operation that gets the vote into the computer and into the preparation of the computer to accept the vote, add it properly, and report it correctly. The general name of the operation that prepares the computer to work properly is called coding. Its importance cannot be overestimated. In 1968, mistakes made in the instructions for the computer caused the computer to malfunction, and the vast stream of votes from NES dried up to a trickle shortly after midnight (EST) election night. This malfunction was partly responsible for the uncertainty about the winner in the presidential race, which was not reported by the television networks until Wednesday morning.

This brings us to an area which is still the subject of intensive competition among the three networks. Although they are all constrained to report the same vote totals, they are not constrained in the interpretation of these vote totals; for although the vote total at any instant in time may be interesting in itself, the real interest in an election lies in who wins, in how much he wins by, and in why he wins.

PROJECTING ELECTION WINNERS

The rapid collection and reporting of the vote requires a great deal of organization, computer capability, and communication equipment. All that activity, nonetheless, goes simply to adding the vote up. The question for the election forecaster always remains "When can I be reasonably sure I have tabulated enough of the vote to decide who will be the ultimate winner?"

An easy answer to that question is "Wait until all the votes are counted," but this may take days. Statistical theory, however, sometimes allows us to give an answer earlier. Sometimes it allows us to determine the winner of an election when only a fraction of one percent of the vote has been reported to the analyst. It happens frequently that projections can be made on the basis of information collected by the network and available to the analysts in the television studio before a single vote has been posted for the television audience, because NES has not yet produced vote totals (which, by agreement are the only ones that can be released to the public).

The projection of election night winners requires a combination of historical information, statistical theory for the construction of an appropriate mathematical model of the vote and for deciding when one is sure enough to make a projection, and the actual election night vote. The networks have different schemes for projecting winners, but all of these schemes have the basic elements we have described. We next describe the general scheme used by one network, NBC. We first discuss the projection of the winner in a state race, then the projection of a winner in a presidential election, and finally the projection of the composition of the House of Representatives.

STATE RACES

The information for projecting the winner of a state race comes from three separate sources. First, there is an estimate of the percentage each candidate will get, which is available before election day. This information comes from public opinion polls, from newspaper reporters, from politicians in the state, and similar sources. This initial estimate is often quite accurate and may give a definite indication of how the race is likely to turn out. A second source of information comes from the vote of specially selected precincts the network collects in addition to the vote that NES collects. Typically, there will be 50 to 150 such precincts for each state. Thus nationally a network may have a precinct collection system which has reporters in over 5000 precincts and is completely independent of the NES effort. These precincts are carefully investigated and their voting behavior in past elections is carefully analyzed. Finally, the information from NES is available at a county level.

The information from these three sources is ordered in time. The initial estimate is obviously available first, since it is available before election day. The vote of the special precincts, called *key precincts,* which the networks collect themselves is usually the first vote information available to the network. By the agreement forming NES, this information cannot be used to tabulate the vote, that is, to show to the public, but it can be used to project winners. Often if a race is one-sided, the vote in the initial precincts to come in from the network collection system is sufficient to allow the projection of a winner.

If the race is close, however, more of the special precincts are needed before a winner may be projected, and often it is necessary to use the county information available from NES.

To use the information from the counties, it is necessary to develop a mathematical model. The reason is that different counties have different voting behaviors. For example, New York City is always more Democratic in its vote than the rest of the State of New York. The difference in voting behavior in terms of relative Democratic or Republican leanings can be incorporated into a mathematical model.

The statistical model uses the voting patterns from the recent past. For example, the fraction of the New York State vote in New York City is typically 0.4, and the fraction in the rest of the state 0.6. In a typical past election for governor the Democratic candidate got 50% of the vote in New York City and 40% of the vote in the rest of the state. His statewide vote, then, was $0.4 \ (50\%) + 0.6 \ (40\%) = 44\%$. Thus New York City was 6% more Democratic than the state average, and the rest of the state was 4% less

Democratic than the state average. This fact can be incorporated into a model so that in this simplest instance, if in a new election the early returns from New York City show 54% for the Democratic candidate and the rest of the state shows 48% for the Democratic candidate, the state projection in percent would be $0.4(54\%) + 0.6(48\%) = 50.4\%$. This indicates that the Democratic candidate would win, although if the returns were very early, this projection would not be considered sufficiently accurate to make an announcement of a victory.

Another factor considered in the statistical model is whether the fraction of the vote assigned to the various parts of the state is accurate for this election. If it snowed heavily in upstate New York, but not in New York City, and cut the vote upstate, but not in New York City, the fraction of the vote in the election might be 0.5 for New York City and 0.5 for the rest of the state; that is, the relative voter turnout in New York City would be higher than normal. In this case the projection would be $0.5(54\%) + 0.5(48\%) = 51\%$, indicating a better chance for Democratic victory. Thus the differential turnout must also be considered in the model in order to make vote projections.

The use of computers allows such a model to be constructed using detailed information for all the counties of a state rather than just the two regions in our example, to provide not only a projection, but also an indication of the accuracy of the projection, so that one can decide when a projection may safely be announced.

It is useful for the model to include the prior estimate available to the network, and results for the special key precincts, so that all information available to the network is effectively utilized. Such a model sometimes allows the results of a race to be called with near certainty, even though only a small fraction of the vote is reported, and the race is relatively close.

It is network policy not to predict the winner unless it is almost a certainty that the predicted winner will actually win. The accuracy of the predictions can be gauged by the fact that, during a given evening when over a hundred predictions may be made, there is usually at most one mistake. The use of such models, developed by statistical theory, allows the networks to enforce their policy, and at the same time "call" close races, because the precision of the estimates developed by the models is always known. One of the most important outputs of the statistical model in this decision problem, as in many others, is the estimated precision of the result.

PRESIDENTIAL RACE

Part of the output of the models used in the calling of state races is the percentage of the vote each candidate is likely to get, as well as an indication as to the accuracy of that percentage. This information is also available for all states for presidential races. (Note that for some states early in the evening the only information available is the preelection estimate.) In this case, however, the percentages are not the final output of the model, but only necessary information to feed into another model whose purpose is to project the winner of the presidential race. The outcome for each state must be utilized in an electoral college model to project the winner of the presidency. This model also provides an estimate of precision, so that the accuracy of the estimate can be assessed, and again the correctness of the prediction can be evaluated.

The loss of the NES vote totals in 1968 held up the projection of the presidential winner because the race was very close in several important states, and the prior and precinct information was not sufficient to make an early responsible projection of the winner.

HOUSE RACES

A projection of the composition of the House of Representatives requires a model similar to the one used for projection of presidential races, the main difference being that each house seat counts as 1, the prior estimates are ordinarily less reliable than those for states in presidential elections, and the vote is only reported by house district. In addition to the projection and vote information, the networks also provide an analysis of the vote. This is the next topic that we shall consider.

NEWS ANALYSIS

We have seen that there is a vast reservoir of information available to the network on election night. The networks utilize this information for vote tabulation and to project winners, but they also make more detailed breakdowns of the information to analyze the election results in depth.

The networks often organize the vote by areas in the state; these areas may be geographical or they may be demographic, for example, showing separately the urban, suburban, and rural vote. The percentage for each candidate can be calculated for each of these categories.

In addition to this detailed inspection of the vote, the networks often conduct election day polls. At randomly selected precincts, network employees ask randomly selected voters to fill in a questionnaire after they have voted. The results are then phoned to a central location and processed by a computer. Analysis of these questionnaires gives insight on the voters' attitudes towards issues and how these may have affected their voting, as well as allowing an estimate of the voting patterns for ethnic groups, racial groups, and groups defined by other demographic characteristics. Analysis of these data helps in gaining a rapid understanding of many of the implications of the vote.

CONCLUSION

The reporting effort of the television networks represents an area of activity that could not exist without the computer and without modern statistics. It represents a blend of modern technology and the traditional skills of the reporter.

The statistical techniques of vote projection may have other applications. For example, it might be possible by similar methods to establish the pattern of yields of corn county by county in Iowa from historical records, and accurately to estimate the state yield from the yields of only a few early harvesting counties.

PROBLEMS

1. What are the three parts of the election night show which rely most heavily on computer and statistical technology?

2. What are the advantages of precinct level vote collection by the media? The disadvantages?

3. Why was the NES formed? Does this mean that the only data available to the networks is from the NES?

4. Statistical theory enters into winner projection in two ways. Describe them.

5. What is a "key precinct"? Are they the same for all networks?

6. The gubernatorial candidate in New York is assured of 60% of the New York City vote and 50% in the rest of the state. As stated in this article, the New York City vote usually represents 40% of the statewide total. What percentage of the total vote can our candidate expect?

7. Our candidate is dismayed. A sudden blizzard has hit New York City on the first Tuesday in November, cutting the city's voter turnout to 30% of the state total. Can the candidate still win?

8. Why is the precision of an estimate important in winner projection?

9. Besides the estimated voting percentages themselves, what is an equally important output of the projection models discussed in the article?

OPINION POLLING IN A DEMOCRACY

George Gallup   *Chairman, American Institute of Public Opinion*

IN ONE form or another, the public opinion poll has been part of the American scene for well over 100 years. As early as July 24, 1824, a report in the Harrisburg *Pennsylvania* told of a "straw vote taken without discrimination of parties" which indicated Jackson to be the popular presidential choice over Adams. Polls of a more careful nature than "straws" were occasionally undertaken in the early part of the current century, but these did not generally deal with issues of the day. It was not until the mid-thirties that polls based on carefully drawn samples were undertaken on a continuing basis.

The superiority of polls based on scientific sampling procedures over those which relied for their validity only on the size of an unscientifically chosen group of persons was demonstrated in dramatic fashion by the 1936 presidential election when Franklin D. Roosevelt defeated Alfred E. Landon in a landslide vote. A Landon victory had been predicted by the *Literary Digest,* a magazine which ran the oldest, largest, and most widely publicized of the polls at that time.   The *Digest*'s final prediction was based on 2,376,523 questionnaires by mail. Yet despite the massive size of this sample, it failed to predict a Roosevelt victory, being off the mark by 19 percentage points.   The Gallup Poll and the Roper Poll, on the other hand, predicted a Roosevelt victory.

The failure of the *Literary Digest*'s polling approach can be explained rather simply. The *Digest*'s sample of voters was drawn from lists of automobile and telephone owners.   This sampling system produced accurate results so long as voters in average and above-average income groups were as likely to vote Democratic as Republican; and conversely those in the lower income brackets— the have-nots—were as likely to vote for either party candidate.   With the advent of the New Deal, however, the American electorate became sharply stratified along income lines, with many persons in the above-average income groups gravitating to the Republican party and many of those in the below-average income groups moving to the Democratic side.

Obviously, a sampling system that reached only telephone subscribers and automobile owners—who were largely among the better-off in that era—was certain to overestimate Republican strength in the 1936 election.   And that is precisely what did happen.

In contrast, the scientific sampling methods which were employed by Gallup, Elmo Roper, and Archibald Crossley for the first time in this election were designed to include the proper proportion of voters from each economic stratum–not just those who owned automobiles and telephones.   These samples much more accurately reflected the proportion of Democrats and Republicans in the population.   And the findings produced by the three organizations, therefore, were closer to the actual election results.

The 1936 election experience provides an excellent example of how election polling serves the science of opinion measurement by providing a kind of "acid test" of statistical methods.   An election represents one of the few situations in which the figures produced by survey organizations can be compared to the actual voting results.

The progress made in polling techniques since 1935 is revealed by examining the error between the Gallup Poll's final election figures and the actual vote. For the seven national elections between 1936 and 1948, the average error recorded for the Gallup Poll was 4.0 percentage points.   For the 14 national elections since 1948, the average error is 1.5 points.

The Gallup Poll and other survey organizations have demonstrated that when scientific methods, rather than procedures relying heavily on subjective judgment, are employed, the prediction of aggregate human behavior can be closely approximated.

The American Institute of Public Opinion was founded in the fall of 1935 for the purpose of determining the public's views on the important political, social, and economic issues of the day.   The operation, as planned, was to be continuous, with survey reports prepared for distribution at regular intervals. The press and the press services traditionally had confined their efforts largely to reporting events—*what people do.* This new effort was designed to deal with a different aspect of life—*what people think.*

The need for a way to measure public opinion had been suggested near the end of the last century by James Bryce, an Englishman who had established himself as a leading authority on the American government. In his book *The American Commonwealth,* which was widely used in American universities, Bryce observed, "The obvious weakness of government by public opinion is the difficulty of ascertaining it." He predicted that the next and final stage in the development of democracies would be reached when the will of the people could be known at all times.

This final stage as predicted by Bryce is close at hand. With developments of recent years, it is now possible to poll a sample of the entire U.S. in a matter of hours. In fact, there is little difference today in the speed with which the media of communications cover major events and the speed with which opinions can be gathered regarding these same events. National surveys have been conducted in a matter of hours to measure first reactions to occurrences such as the nationwide postal strike in 1970 and the Calley verdict in 1971. Also, Gallup affiliates in countries around the world have frequently measured multinational opinion in as little time as 72 hours about such events as the launching of the first Sputnik in 1957 and the visit of Soviet Premier Khrushchev to the U.S. in 1959.

In 1922, Walter Lippmann, in a prophetic statement in his widely read and quoted book *Public Opinion* said "The social scientist will acquire his dignity and his strength when he has worked out his method. He will do that by turning into opportunity the need of the great society for *instruments of analysis* by which an invisible and most stupendously difficult environment can be made intelligible."

The environment has not become any less complex in the half century since Lippmann wrote these words. And the modern poll is at least one instrument of analysis that can and does help to make the environment more intelligible.

DETERMINING AREAS OF IGNORANCE

Some critics have questioned the value of opinion polling, saying that the great mass of people are uninformed on most issues of the day, and therefore, their views have little significance. If persons in a survey feel that they are not competent to answer certain questions or have no opinion because they lack information, they will usually say they don't know or have no opinion. Moreover, opinions that have most significance concern issues or problems that touch the daily lives of the general public. And the range of these is great; in fact, it covers most of the vital issues of the day.

On some issues, it is important to separate informed opinion from the uninformed. This can be accomplished by a simple survey procedure devised by the Gallup Poll. It is a series of questions that begins: "Have you heard or read about X issue?" The respondent can answer either "yes" or "no." If the answer is "yes," the respondent is asked "Please tell me in your own words what you consider the chief issue to be." And the interviewer writes the respondent's exact words on his interviewing form. The next question seeks to discover the extent, or level, of the respondent's knowledge of the subject. The respondent may be asked to state the positions held by various people or countries involved in a controversy, for example. The next question asks "How do you think this issue should be resolved?" or, depending on the nature of the controversy, a variation of this question. The respondent is permitted to explain his views with as many qualifications as he wishes.

The next in the series poses specific questions that can be answered "yes" or "no." Often it is possible to explain the issue in a few sentences (in effect, to inform the person being interviewed) and then to record his opinions. Individuals who say they have not heard or read about the issue are eligible at this point to answer both this and the last question.

The last question is intended to establish the "intensity" with which the respondent holds his views. How strongly does he feel that he is right? What steps would he be willing to take to implement his opinions?

Thus we have seen that not all questions asked in a public opinion poll are of the yes-no variety. Complex problems, as pointed out, typically require a series of questions. But eventually all issues, especially those dealt with by legislative bodies, sooner or later have to be resolved and the legislator must, whether he likes it or not, vote "yes" or "no." In similar fashion, the ordinary voter, whether he is voting on candidates in a presidential election or on a state referendum issue, must eventually cast a simple "yes" or "no" vote. There is no provision on the ballot or voting machine for qualifications or modifications. He can't put his X in a box marked "no opinion," though, of course, he can skip voting on the issue.

In the process of discovering what the public knows in certain areas, it is possible to shed light on the strengths and weaknesses of the educational system which has brought the public to its present level of knowledge. The best way to judge the quality of the product, and one of the key functions of a polling operation, is to determine levels of knowledge and "areas of ignorance."

The survey approach to social problems has been widely accepted. As Professor Kenneth Boulding has written

> Perhaps the most important single development pointing towards more scientific images of social systems is the improvement in the collection and processing of social information. The method of sample surveys is the telescope of the social sciences. It enables us to scan the social universe, at some small cost in statistical error, in ways we have never been able to do before.

SOME EXAMPLES OF SURVEY RESULTS

Since 1935, the Gallup Poll has published over 6500 reports covering a wide range of subjects.   Following are some of the questions asked in recent months and the national findings.

What is the SMALLEST amount of money a family of four (husband, wife, and two children) needs each week to get along in this community? (Reported: Feb. 29, 1976)

Median of responses (Nonfarm families): $177 per week

In politics, as of today, do you consider yourself a Republican, Democrat, or Independent? (Reported: December, 1976)

        Democrats .... .48%
        Republicans .... .23%
        Independents .... .29%

If your party nominated a woman for President, would you vote for her if she were qualified for the job?

Reported:	Aug. 5, 1971	Sept. 18, 1975
Yes	66%	73%
No	29%	23%
No opinion	5%	4%

Let us look briefly at some other issues and what the public has to say about them.   (Of course the wording of questions makes a difference in how people answer, but the following will serve to give some idea of public sentiment as determined by the Gallup Poll.)

A majority of Americans would like to overhaul the whole process of electing a president; they favor nationwide primaries, making the conventions, if held, more dignified, shortening the campaign, and abandoning the present electoral college system.   Long before such legislation was passed, polls showed that our fellow citizens wanted the voting age lowered to 18.   Americans favor stiffer laws on drinking and driving, tougher gun laws, less leniency toward criminals on the part of courts, compulsory arbitration in the case of strikes (particularly those strikes affecting the public welfare), tougher laws on pornography, guaranteed work rather than a guaranteed annual income.   Americans think all young men should be required to give one year's service to their country, either in the armed forces or in some nonmilitary work, such as VISTA or the Peace Corps.

OTHER CONTRIBUTIONS OF SAMPLE SURVEYS

Even more important are the contributions that sample surveys of the population can make in the improvement of government. The modern poll can, and to a certain extent does, function as a *creative arm of government*. It can discover the likely response of the public to any new proposal, law, or innovation. It can do this by presenting ideas to the public for their appraisal and judgment—ideas that range from specific proposals for dealing with strikes and racial problems to proposals for ending the war in Vietnam.

More and more, the modern poll is dealing with new ideas or proposals for dealing in new ways with current problems. The poll in this respect has a natural advantage over legislators. It can go directly to the people without fear of political repercussions. It can determine the degree of acceptance of or resistance to any proposal—its appeal or lack of appeal, at least in its early stages of acceptance or rejection. It is this creative function that may, in the years ahead, offer the public opinion poll its greatest opportunity for service to the nation.

In many ways it is unfortunate that modern polls should be closely identified in the minds of so many persons with elections and election predictions. Although election polling is an important part of the work of survey organizations, providing important evidence of the accuracy of polling methods and of progress in the technology of this field, the prominence given election polling frequently tends to obscure the many other functions that modern polls can perform to make the political environment more intelligible. In fact, polls can do things that were scarcely dreamed of in earlier days. For example, the modern poll can simulate a national election by determining the relative strength of candidates, pitting leading contenders against each other, in any combination. It can also simulate a nationwide referendum on any issue of current importance. And the results arrived at through polling can be expected to differ little from a national election or referendum held at the same time.

The modern poll can provide a continuous check on the popularity of the president—a sort of American equivalent of a vote of confidence in the government such as that found in those nations with a parliamentary form of government. The Gallup Poll's measurement of presidential popularity has been used at regular intervals during the administrations of those presidents beginning with Franklin D. Roosevelt. It has proved to be a sensitive barometer of public attitudes regarding the president, with wide fluctuations recorded in approval and disapproval.

The highest approval rating recorded for a president was the 87 percent registered for Truman when the nation rallied around him in the period immediately following Roosevelt's death. It is a curious fact that the lowest rating obtained was also recorded for President Truman, this occurred in 1951 when his approval rating dropped to 23 percent during a difficult period in the Korean war. Nixon's approval rating reached a low point of 24 percent in July and August of 1974 just prior to his resignation. His highest rating, 68 percent, was recorded following a nationwide televised speech in November, 1969. President Johnson's highest rating was 80 percent, and this was reached soon after he took office in November, 1963, following the death of President Kennedy. His low point, 35 percent approval, came in August, 1968, when Gallup surveys showed disillusionment over our involvement in Vietnam to be at a peak. The Gallup Poll's first measurement of the public's response to President Carter about two weeks after his inauguration produced an approval rating of 66 percent.

The modern poll can beam a bright and devastating light on the gap which too often exists between the will of the people and the translation of this will into law by legislators.   From years of measuring how the average citizen reacts to a wide range of ideas, it is clear that his thinking is sound, his common-sense quotient high.   Congressional action, as a matter of fact, supports this belief; historically, it has been true that sooner or later the public's will is translated into law.

PROBLEMS

1. Describe the *Literary Digest's* sampling system for its 1936 poll. Was this "scientific sampling"? Why or why not? Why was the poll so far off in predicting the election result?

2. Letters to the editor also reflect public opinion. In what ways is polling a better way of soliciting this opinion?

3. What might a pollster mean by "area of ignorance"?

4. Explain the procedure the Gallup Poll uses to separate "informed" and "uninformed" opinion on an issue.

5. Comment on the following statement: The polling process ultimately elicits a simple yes or no from the respondent.

6. What *percentage* of registerable voters were estimated to be Democrats in November 1971?

7. In the poll on party affiliation, does an answer of "undecided" mean that the respondent can't decide between "Democrat" and "Republican"? What about other parties?

8. (a) The final pre-election poll shows a neck-and-neck race between presidential candidates. What effect would you expect this to have on voter turnout? (See the article by Tufte.)

(b) If the poll showed the following:

Candidate A	55%
Candidate B	35%
Undecided	10%

would you expect the winner prediction to be correct? What level of voter turnout would you expect?

9. How does polling serve as a "creative arm of government"?

REFERENCES

James Bryce.   1888.   *The American Commonwealth,* vol. 2.   New York: AMS Press.

Walter Lippmann.   1965.   *Public Opinion.*   New York: Free Press.

REGISTRATION
EVERY SUNDAY
2 AM to 4 AM
GET OUT AND VOTE!
COURTESY OF
MAYOR
JOE TRICKY
"THE BEST MAN
FOR THE JOB!"

REGISTRATION AND VOTING

Edward R. Tufte   *Princeton University*

THE QUESTIONS

If citizens are to express their preferences on election day in the U.S., they must register to vote some weeks or even months prior to the election. The inability or failure to do so deprives the citizens of their votes. Many other democracies have automatic registration of all voters or no registration procedures at all. What are the consequences, if any, of the more elaborate voter registration practices in the U.S.? And furthermore, since the particular rules for registration differ greatly from place to place throughout the country, what are the consequences of such differences?

Recently several students of politics sought to answer these questions by analyzing quantitative data on registration and voting. In particular, they discovered answers to questions such as "Why does the proportion of citizens registered to vote differ so much in different cities? Why are, for example, 96% of all eligible citizens registered in South Bend, Indiana, about 65% in New York City and Dallas, and only 34% in Atlanta? Why is voter turnout on election day in the U.S. lower than in other democracies, such as England, France, Norway, and Canada?"

Many have suggested answers to these questions. For example, some have felt that the low voter turnout in the U.S. occurs because American citizens are more apathetic than their counterparts in other democracies. But when this suggestion and others are investigated more precisely and deeply with careful statistical procedures, not all of the old speculations prove to be correct. And the new answers have important consequences for increasing political participation in the U.S.

THE IDEAS BEHIND THE STUDY OF REGISTRATION AND VOTING

A key part of the idea of democracy is that citizens participate in the choice of their leaders. Constraints on the ability of citizens to participate in politics restrict voting to those who have the resources and energy to overcome such obstacles. In theory, citizens will generally make greater efforts to overcome limitations on their ability to participate if they feel their efforts will amount to something; that is, if a citizen feels his vote will make a difference, he may be willing to stand in line to register weeks before the election and then, on election day, walk through the rain in order to cast a ballot. This reasoning, which suggests that people assess (whether consciously or unconsciously) the costs and potential benefits of registering and voting, implies that citizens will be more likely to register and then later vote if the costs of registering to vote are low and the election is thought to be closely contested. In other words, citizens may attach more value to their votes, and therefore be more likely to vote, if they think the election is going to be close simply because they believe their votes might make a difference, other things being equal. Now it is important to note at this point that these assertions have not been proven; they are only a plausible theory. Three scholars at Princeton University—Stanley Kelley, Jr., Richard E. Ayres, and William G. Bowen—set out to test these ideas. Let us now see what they learned about registration and voting in their study.

THE STUDY AND THE RESULTS

Their first question was "Do rates of registration vary in different parts of the country and, if so, are the differences in registration rates important in a political sense?" In studying 104 of the nation's largest cities, they found that voter registration rates ranged from a high of 96.4% of those of voting age who were registered in South Bend, Indiana, to a low of 32.1% in Columbus, Georgia.

Table 1 shows both the registration and the voting rates for all 104 cities. The rates varied a great deal from city to city and, moreover, registration

TABLE 1. Registration and Voting Rates in 104 Cities, 1960*

CITY	REGISTRATION RATE AS PERCENT OF VOTING AGE POPULATION	TURNOUT RATE AS PERCENT OF VOTING AGE POPULATION
South Bend, Ind.	96.4	85.2
Des Moines, Iowa	92.6	71.6
Minneapolis, Minn.	92.5	58.5
Detroit, Mich.	92.0	70.0
Seattle, Wash.	92.0	70.8
Lansing, Mich.	91.9	72.4
St. Paul, Minn.	91.2	72.1
Berkeley, Calif.	90.5	70.4
Scranton, Pa.	90.4	80.3
Spokane, Wash.	89.4	67.0
Dearborn, Mich.	89.3	81.2
Albany, N. Y.	88.4	87.2
Torrance, Calif.	87.7	76.5
Peoria, Ill.	87.4	64.9
Gary, Ind.	87.3	72.5
Tacoma, Wash.	87.3	67.8
Salt Lake City, Utah	87.0	76.6
Portland, Ore.	85.8	74.1
Duluth, Minn.	85.1	74.9
Glendale, Calif.	84.9	73.1
Memphis, Tenn.	84.7	50.1
Hammond, Ind.	84.0	71.3
Pasadena, Calif.	83.2	69.2
Grand Rapids, Mich.	83.2	72.9
Buffalo, N. Y.	83.0	69.7
New Bedford, Mass.	82.4	74.6
Tulsa, Okla.	82.4	69.4
Rockford, Ill.	82.0	75.1
Topeka, Kans.	81.9	69.3
Fort Wayne, Ind.	81.7	71.1
Waterbury, Conn.	81.4	77.4
Camden, N. J.	81.3	69.0
Pittsburgh, Pa.	81.2	68.3
Fresno, Calif.	81.1	44.6
Jersey City, N. J.	81.1	72.8
Worcester, Mass.	81.0	74.0
Youngstown, Ohio	81.0	71.6
Canton, Ohio	80.9	73.2
Oklahoma City, Okla.	80.4	62.7
Omaha, Neb.	79.8	66.8
Flint, Mich.	79.6	69.4
Lincoln, Neb.	79.4	64.5
Cincinnati, Ohio	79.4	67.9

TABLE 1.   Registration and Voting Rates in 104 Cities, 1960* *(Continued)*

CITY	REGISTRATION RATE AS PERCENT OF VOTING AGE POPULATION	TURNOUT RATE AS PERCENT OF VOTING AGE PGPULATION
Syracuse, N. Y.	79.3	72.3
New Haven, Conn.	79.2	72.1
Kansas City, Kans.	78.9	66.4
Erie, Pa.	78.8	68.3
Philadelphia, Pa.	77.6	69.8
Sacramento, Calif.	77.3	66.4
Springfield, Mass.	77.1	67.0
Utica, N. Y.	77.1	76.1
Los Angeles, Calif.	77.0	64.2
Akron, Ohio	77.0	70.5
Toledo, Ohio	76.9	69.4
Trenton, N. J.	75.8	63.8
Elizabeth, N. J.	75.6	68.0
Santa Ana, Calif.	75.1	60.1
Rochester, N. Y.	74.9	72.2
Boston, Mass.	74.0	63.3
San Diego, Calif.	73.9	61.4
Cambridge, Mass.	73.8	65.9
Dayton, Ohio	73.6	62.5
Columbus, Ohio	72.4	63.1
Oakland, Calif.	71.9	66.2
Cleveland, Ohio	71.5	61.4
Winston-Salem, N. C.	71.2	50.5
Hartford, Conn.	70.7	34.1
Chattanooga, Tenn.	70.7	46.6
Bridgeport, Conn.	70.6	67.5
Charlotte, N. C.	69.9	54.5
St. Petersburg, Fla.	69.7	59.5
Tampa, Fla.	68.8	63.6
St. Louis, Mo.	68.5	62.0
Patterson, N. J.	68.4	55.4
Baltimore, Md.	68.1	54.0
San Francisco, Calif.	68.0	64.4
Niagara Falls, N. Y.	67.7	55.4
Allentown, Pa.	67.7	60.2
Greensboro, N. C.	66.6	52.6
Kansas City, Mo.	65.8	59.8
New York, N. Y.	65.7	58.8
Dallas, Texas	65.0	57.3
Baton Rouge, La.	64.7	47.8
Wichita, Kans.	62.2	43.0
Corpus Christi, Texas	61.8	53.9

TABLE 1.   Registration and Voting Rates in 104 Cities, 1960* *(Continued)*

CITY	REGISTRATION RATE AS PERCENT OF VOTING AGE POPULATION	TURNOUT RATE AS PERCENT OF VOTING AGE POPULATION
Newark, N. J.	61.4	50.4
Little Rock, Ark.	61.2	46.9
Honolulu, Hawaii	60.0	54.7
Houston, Texas	60.0	57.2
Miami, Fla.	59.2	43.7
Louisville, Ky.	59.0	32.4
Nashville, Tenn.	55.9	38.0
New Orleans, La.	55.6	45.9
Jacksonville, Fla.	54.9	46.9
Fort Worth, Texas	48.4	23.9
Austin, Texas	48.3	28.0
Richmond, Va.	46.5	31.2
Norfolk, Va.	43.6	22.4
San Antonio, Texas	42.6	31.4
Birmingham, Ala.	39.1	13.8
Portsmouth, Va.	38.0	25.7
Newport News, Va.	35.0	28.8
Atlanta, Ga.	33.8	25.6
Columbus, Ga.	32.1	24.2

* The list consists of all the cities in the U.S. with populations greater than 100,000 in 1960, with the following exceptions: registration figures were not available from 16 cities; similarly it was impossible to get accurate information concerning registration procedures for eight other cities.   For further discussion, see the original study.

rates were closely related to the turnout on election days in all the cities.   The table shows the cities ordered from highest to lowest registration rate.   As the table shows, the turnout figures generally followed the pattern of the registration figures quite closely, and the correlation between registration and voting was 0.88 (where 0.00 represents no linear association and 1.00 represents a perfect linear association).   Note also the pattern of the *differences* between cities; the authors report "if the percentage of the population of voting age registered to vote in city A was one percent higher than in city B, then the percentage of the population of voting age actually voting in city A was, on the average, almost exactly one percent higher than in city B."   This relationship held quite reliably for almost all of the 104 cities.   (See Figure 1, in which the data of Table 1 are plotted.)

Plainly those factors related to low turnout in cities might also be related to low registration rates.   But, in addition, does the nature of the registration procedure itself operate to limit the number of voters who register and, conse-

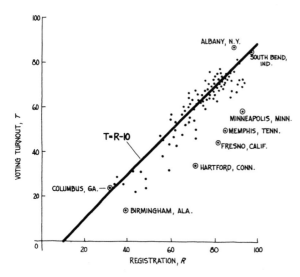

FIGURE 1

City voting registration and
election-turnout percentages

quently, to limit turnout? The strong relationship between registration and turnout naturally suggests taking a careful look at factors that may produce the wide range of differences among cities.

The preceding discussion suggests three sets of forces determining the registration rate:

Factors affecting the value of the vote, measured by
  Closeness of recent elections
Factors affecting the costs of registration, measured by
  Closing date for registration
  Provisions regarding literacy tests
  Times and places of registration
Socioeconomic characteristics, measured by
  Age (percent who are 20 to 34)
  Race (percent white)
  Education (median school years completed by persons over 25 years of age)

Each of the 104 cities was measured on each of these seven variables. Census data were used to determine the age, racial, and educational distribution of each city; the results of recent elections for President and Governor assessed the competiveness or closeness of elections; and information on the differing registration procedures provided data on the "costs of registration" in each city.

The statistical technique of multiple regression was used to estimate the impact of each variable on the registration rate. Simply, the multiple regression equation provides, under some assumptions, estimates of the effect of each of the variables in determining the registration rate. These estimated weights of impact, combined with each city's score on each of the seven variables, generate a predicted value for the registration rate of each of the 104 cities. This predicted registration rate can then be compared with the city's actual registration rate in order to assess the accuracy of the prediction equation. In this study, the predictions were generally quite accurate and most (80%) of the variation in the registration rate from city to city was described by, or in the statistical jargon, "explained by," the weighted combination of the seven variables. By using these statistical procedures, the authors concluded that

(1) "Extending the closing date for registration from, say, one month to one week prior to election day would tend to increase the percentage of the population registered by about 3.6 percent." Thus the *convenience* of registration for the potential voter was strongly related to the rate of registration. It appears that in many cities political parties and politicians have manipulated the convenience of registration in order to decrease or increase the size of the electorate. This is, of course, a familiar story in some southern cities where Blacks have been prevented from registering to vote by means of violence, poll taxes, literacy tests, and other cumbersome and expensive registration procedures. Politicians and parties in the North, however, have also not been immune from designing registration procedures that have effectively prevented many citizens from voting. For example, one study found an almost perfect correlation between the ease of registration in different wards in a major city and the proportion of votes for that city's long-time incumbent mayor in the different wards.

(2) The closeness or competitiveness of past elections was also strongly related to the registration rate: the closer the previous elections in the state, the more people registered to vote in the next election. Competitive elections not only offer the voter a greater range of plausible choices, but they also probably lead some voters to value their vote more than they would if elections were not close.

(3) The various socioeconomic variables entered into the regression equation indicating that they, too, were associated with the registration rate. If a relatively large proportion of young people, Blacks, and families with less than average education lived in a city, then the registration rate in that city tended to be low.

CONCLUSIONS AND IMPLICATIONS

The results of the statistical analysis point to a number of important conclusions. The authors report that differences from city to city in participation in elections by the citizens were to a large extent related to registration rates, which, in turn, strongly reflected the local laws and practices regulating registration. Registration in some cities was made so difficult and so costly that fewer than one-third of the eligible voters ever registered to vote. In other cities, however, the ease of registration procedures, the competitiveness of elections, and certain socioeconomic factors resulted in more than 95% of the eligible citizens registering to vote. It is clear that many citizens were excluded from the polls because the costs of registering were too high for them to overcome. The elimination of restrictive and difficult registration requirements, both in the North and in the South, would increase political participation in the U.S. by reducing the costs of voting. Successful efforts to reduce the stringency of registration requirements would be far more effective in increasing voter turnout than exhortations by the mass media for citizens to go out and vote.

These suggestions, which grew out of the statistical study of registration and voting reported here, receive some further support from the experiences following the passage of the 1965 Voting Rights Act. Since the passage of the law, designed to simplify registration procedures and to reduce intimidation and other obstacles associated with registering in areas with low rates of registration, millions of citizens (both black and white) in the South now are registered and voting for the first time. In this political "experiment" the sudden reduction in the cost of registration led to sharp increases in political participation: prior to the law, less than 10% of the eligible citizens voted in some areas; now, in those same areas, 60% to 70% vote. No increases in registration and voting have been observed in areas not covered by the Voting Rights Act. Note that the theory that Americans are uniquely apathetic compared to the citizens of other democracies neither suggests remedies such as the Voting Rights Act nor explains the sudden upsurge in registration and voting in those areas covered by the Act. The experience with the new law, although certainly not representing a carefully controlled experiment, does at least provide some further independent evidence consistent with the results of the multiple regression study of registration and voting reported here.

The results also help to explain why turnout in elections in the U.S. is lower than in many other countries. Many democracies simply do not have voter registration procedures. (A few countries, in fact, even seek to increase the cost of *nonvoting* by means of compulsory voting.) Because a potential voter has to expend less time and energy to vote in democracies other than the U.S., it is not surprising that a great fraction of the citizens of other countries votes.   Roughly 80% of all potentially eligible voters register to vote in the U.S.   And about 80% of those registered actually do vote on election day—resulting in an overall turnout of around 64% of all potentially eligible voters.   In those countries (Canada, France, and Great Britain) with automatic registration requiring no effort on the part of the citizen, typically about 75% to 80% of all eligible citizens vote in national elections. Thus the persistently lower turnout in the U.S. is more likely due to inconvenient registration procedures than to any lack of civic virtue unique to Americans.

Finally, the findings in this statistical study of registration and voting suggest that nonvoting results from *political* factors as well as from socioeconomic factors.   Kelley, Ayres, and Bowen found evidence to support the conclusion that registration rules are manipulated by the party in power in order to make it easier for that party to continue to rule.   Thus, although nonvoting is related, in part, to persistent social conditions, it also often occurs because the dominant party has simply raised the inconvenience of voting to a high enough level so as to exclude many voters from the polls.

PROBLEMS

1.   What are the assertions that Kelley, Ayres, and Bowen set out to test in their study?

2.   Find Minneapolis on Table 1. What is its rank in voter registration? In voter turnout? What percentage of those registered actually voted?

3.   Suppose the voter registration and voter turnout rates in city i are R_i and T_i respectively. What is the range of *possible* values for T_i? [Answer: $0 \leqslant T_i \leqslant R_i$.]

4.   Consider Figure 1 where the relationship of registration and turnout is estimated. Suppose a city's registration was 75%. What is the predicted estimate of the city's turnout? Answer the same question for a city with 10% registration.

5.   What is the implicit procedure used by the author in deciding which cities to label in Figure 1?

6.   Can you think of other factors than those listed which might influence the voter registration rate? Explain why each of the factors you list might have an influence.

7.   For a given city in the sample, how do you assess the accuracy of prediction of the regression equation?

8.   Why does the author refer to the 1965 Voting Rights Act as a "political 'experiment' "?

9. True or false: The proportion of those who are registered to vote and actually do is roughly the same in the U. S. and Canada. Explain your answer.

REFERENCE

Stanley Kelly, Jr., Richard E. Ayres, and William G. Bowen. 1967. "Registration and Voting: Putting First Things First." *American Political Science Review* 61: June, pp. 359–379. Reprinted in Edward R. Tufte, ed. 1970. *The Quantitative Analysis of Social Problems.* Reading, Mass.: Addison-Wesley. Pp. 250–283.

Part three

OUR SOCIAL WORLD

Communicating with others
Deciding authorship
Adverbs multiply adjectives
The meaning of words
The sizes of things

People at work
How accountants save money by sampling
The use of subjective probability methods
in estimating demand
Preliminary evaluation of a new food product
Making things right

People at school and play
Calibrating college board scores
Statistics, sports, and some other things
Varieties of military leadership

Counting people and their goods
The Consumer Price Index
How to count better: Using statistics to
improve the Census
Information for the nation from a sample survey

Forecasting population and the economy
How crowded will we become?
Early warning signals for the economy
Statistics for public financial policy

Measuring segregation and inequality
Measuring racial integration potentials
Census statistics in the public service
Measuring sociopolitical inequality

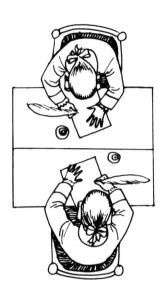

DECIDING AUTHORSHIP

Frederick Mosteller   *Harvard University*

David L. Wallace   *University of Chicago*

ART, MUSIC, literature, the social, biological, and physical sciences share a common need to classify things: What artist painted the picture? Who composed the piece? Who wrote the document? If paroled, will the prisoner repeat his crime? What disease does the patient have? What trace chemical is damaging the process? In the field of statistics, we call these questions classification or discrimination problems.

Questions of authorship are frequent and sometimes important. Most people have heard of the Shakespeare-Bacon-Marlowe controversy over who wrote the great plays usually attributed to Shakespeare. A less well-known but carefully studied question deals with the authorship of a number of Christian religious writings called the Paulines, some being books in the New Testament: Which ones were written by Paul and which by others? In many authorship questions the solution is easy once we set about counting something systematically. But we treat here an especially difficult problem from American history, the controversy over the authorship of the 12 *Federalist* papers claimed by both Alexander Hamilton and James Madison, and we show how a statistical analysis can contribute to the resolution of historical questions.

The Federalist papers were published anonymously in 1787–88 by Alexander Hamilton, John Jay, and James Madison to persuade the citizens of the State of New York to ratify the Constitution. Seventy-seven papers appeared as letters in New York newspapers over the pseudonym "Publius." Together with eight more essays, they were published in book form in 1788 and have been republished repeatedly both in the U.S. and abroad. *The Federalist* remains today an important work in political philosophy. It is also the leading source of information for studying the intent of the framers of the Constitution, as, for example, in recent decisions on congressional reapportionment, since Madison had taken copious notes at the Constitutional Convention.

It was generally known who had written *The Federalist*, but no public assignment of specific papers to authors occurred until 1807, three years after Hamilton's death as a result of his duel with Aaron Burr. Madison made his listing of authors only in 1818 after he had retired from the Presidency. A variety of lists with conflicting claims have been disputed for a century and a half. There is general agreement on the authorship of 70 papers—5 by Jay, 14 by Madison, and 51 by Hamilton. Of the remaining 15, 12 are in dispute between Hamilton and Madison, and 3 are joint works to a disputed extent. No doubt the primary reason the dispute exists is that Madison and Hamilton did not hurry to enter their claims. Within a few years after writing the essays, they had become bitter political enemies and each occasionally took positions opposing some of his own *Federalist* writings.

The political content of the essays has never provided convincing evidence for authorship. Since Hamilton and Madison were writing a brief in favor of ratification, they were like lawyers working for a client; they did not need to believe or endorse every argument they put forward favoring the new Constitution. While this does not mean that they would go out of their way to misrepresent their personal positions, it does mean that we cannot argue "Hamilton wouldn't have said that because he believed otherwise." And, as we have often seen, personal political positions change. Thus the political content of a disputed essay cannot give strong evidence in favor of Hamilton's or of Madison's having written it.

The acceptance of the various claims by historians has tended to change with political climate. Hamilton's claims were favored during the last half of the 19th century, Madison's since then. While the thorough historical studies of the historian Douglass Adair over the past several decades support the Madison claims, the total historical evidence is today not much different from that which historians like the elder Henry Cabot Lodge interpreted as favoring Hamilton. New evidence was needed to obtain definite attributions, and internal statistical stylistic evidence provides one possibility; developing that evidence and the methodology for interpreting it was the heart of our work.

The writings of Hamilton and Madison are difficult to tell apart because both authors were masters of the popular *Spectator* style of writing—complicated and oratorical. To illustrate the difficulty, in 1941 Frederick Williams and Frederick Mosteller counted sentence lengths for the undisputed papers and got averages of 34.5 and 34.6 words respectively for Hamilton and Madison. For sentence length, a measure used successfully to distinguish other authors, Hamilton and Madison are practically twins.

MARKER WORDS

Although sentence length does measure complexity (and an average of 35 words shows that the material is very complex), sentence length is not sensitive enough to distinguish reliably between authors writing in similar styles. The variables used in several recent studies of disputed authorship are the rates of occurrence of specific individual words. Our study was stimulated by Adair's discovery—or rediscovery as it turned out—that Madison and Hamilton differ consistently in their choice between the alternative words *while* and *whilst*. In the 14 *Federalist* essays acknowledged to be written by Madison, *while* never occurs whereas *whilst* occurs in eight of them. *While* occurs in 15 of 48 Hamilton essays, but never a *whilst*. We have here an instance of what are called markers—items whose presence provides a strong indication of authorship for one of the men. Thus the presence of *whilst* in five of the disputed papers points toward Madison's authorship of those five.

Markers contribute a lot to discrimination when they can be found, but they also present difficulties. First, *while* or *whilst* occurs in less than half of the papers. They are absent from the other half, and hence give no evidence either way. We might hope to surmount this by finding enough different marker words or constructions so that one or more will always be present. A second and more serious difficulty is that from the evidence in 14 essays by Madison, we cannot be sure that he would never use *while*. Other writings of Madison were examined and, indeed, he did lapse on two occasions. The presence of *while* then is a good but not sure indication of Hamilton's authorship; the presence of *whilst* is a better, but still imperfect, indicator of Madison's authorship, for Hamilton too might lapse.

A central task of statistics is making inferences in the presence of uncertainty. Giving up the notion of perfect markers leads us to a statistical problem. We must find evidence, assess its strength, and combine it into a composite conclusion. Although the theoretical and practical problems may be difficult, the opportunity exists to assemble far more compelling evidence than even a few nearly perfect markers could provide.

RATES OF WORD USE

Instead of thinking of a word as a marker whose presence or absence settles the authorship of an essay, we can take the rate or relative frequency of use of each word as a measure pointing toward one or the other author. Of course, most words won't help because they were used at about the same rate by both authors. But since we have thousands of words available, some may help. Words form a huge pool of possible discriminators. From a systematic exploration of this pool of words, we found no more pairs like *while–whilst,* but we did find single words used by one author regularly but rarely by the other.

Table 1 shows the behavior of three words: *commonly, innovation, war.* The table summarizes data from 48 political essays known to be written by

TABLE 1. Frequency Distributions of Rate per 1000 Words in 48 Hamilton and 50 Madison Papers for *Commonly, Innovation,* and *War*

COMMONLY			INNOVATION			WAR		
Rate per 1000 Words	H	M	Rate per 1000 Words	H	M	Rate per 1000 Words	H	M
0 (exactly) *	31	49	0 (exactly) *	47	34	0 (exactly) *	23	15
0⁺–0.2	cannot occur †		0⁺–0.2	cannot occur †		0⁺–2	16	13
0.2–0.4	3	1	0.2–0.4		6	2– 4	4	5
0.4–0.6	6		0.4–0.6	1	6	4– 6	2	4
0.6–0.8	3		0.6–0.8		1	6– 8	1	3
0.8–1.0	2		0.8–1.0		2	8–10	1	3
1.0–1.2	2		1.0–1.2		1	10–12		3
1.2–1.4	1					12–14		2
						14–16	1	2
Totals	48	50	Totals	48	50	Totals	48	50

Source: Mosteller and Wallace (1964).

* Each interval, except 0 (exactly), excludes its upper end point. Thus a 2000-word paper in which *commonly* appears twice gives rise to a rate of 1.0 per 1000 exactly, and the paper appears in the count for the 1.0–1.2 interval.

† With the given lengths of the papers used, it accidentally happens that a rate in this interval cannot occur. For example, if a paper has 2000 words, a rate of 1 per 1000 means 2 words, and a single occurrence means a rate of 0.5 per 1000. Hence a 2000-word paper cannot lead to a rate per thousand greater than 0 and less than 0.5.

Hamilton and 50 known to be by Madison. (Some political essays from outside the *The Federalist,* but known to be by Hamilton or Madison, have been included in this study to give a broader base for the inference. Not all Hamilton's later *Federalist* papers have been included. We gathered more papers from outside *The Federalist* for Madison.)

Neither Hamilton nor Madison used *commonly* much, but Hamilton's use is much more frequent than Madison's. The table shows that in 31 of 48 Hamilton papers, the word *commonly* never occurred, but that in the other 17 it occurred one or more times. Madison used it only once in the 50 papers in our study. The papers vary in length from 900 to 3500 words, with 2000 about average. Even one occurrence in 900 words is a heavier usage than two occurrences in 3500 words, so instead of working with the number of occurrences in a paper, we use the rate of occurrence, with 1000 words as a convenient base. Thus, for example, the paper with the highest rate (1.33 per 1000 words) for *commonly* is a paper of 1500 words with 2 occurrences. *Innovation* behaves similarly, but it is a marker for Madison. For each of these two words, the highest rates are a little over 1 per 1000.

The word *war* has a spectacularly different behavior. Although absent from half of Hamilton's papers, when present it is used frequently—in one paper at a rate of 14 per 1000 words. *The Federalist* papers deal with specific topics in the Constitution and huge variations in the rates of such words as *war, law, executive, liberty,* and *trade* can be expected according to the context of the paper. Even though Madison uses *war* considerably more often then Hamilton in the undisputed papers, we explain this more by the division of tasks than by predilections of Madison for using *war.* Data from a word like *war* would give the same troublesome sort of evidence that historians have disagreed about over the last 100 years. Indeed, the dispute has continued because evidence from subject and content has been hopelessly inconclusive.

USE OF NON-CONTEXTUAL WORDS

For the statistical arguments to be valid, information from meaningful, contextual words must be largely discarded. Such a study of authorship will not then contribute directly to any understanding of the greatness of the papers, but the evidence of authorship can be both strengthened and made independent of evidence provided by historical analysis.

Avoidance of judgments about meaningfulness or importance is common in classification and identification procedures. When art critics try to authenticate a picture, in addition to the historical record, they consider little things: how fingernails and ears are painted, what kind of paint and canvas were used.

Relatively little of the final judgment is based upon the painting's artistic excellence. In the same way, police often identify people by their fingerprints, dental records, and scars, without reference to their personality, occupation, or position in society. For literary identification, we need not necessarily be clever about the appraisal of literary style, although it helps in some problems. To identify an object, we need not appreciate its full value or meaning.

What non-contextual words are good candidates for discriminating between authors? Most attractive are the filler words of the language: prepositions, conjunctions, articles. Many other more meaningful words also seem relatively free from context: adverbs such as *commonly, consequently, particularly,* or even abstract nouns like *vigor* or *innovation.* We want words whose use is unrelated to the topic and may be regarded as reflecting minor or perhaps unconscious preferences of the author.

Consider what can be done with filler words. Some of these are the most used words in the language: *the, and, of, to,* and so on. No one writes without them, but we may find that their rates of use differ from author to author. Table 2 shows the distribution of rates for three prepositions—*by, from,* and *to.* First, note the variation from paper to paper. Madison uses *by* typically about 12 times per 1000 words, but sometimes has rates as high as 18 or as low as 6. Even on inspection though, the variation does not obscure

TABLE 2. Frequency Distribution of Rate per Thousand Words in 48 Hamilton and 50 Madison Papers for *By, From,* and *To*

BY Rate per 1000 Words	H	M	FROM Rate per 1000 Words	H	M	TO Rate per 1000 Words	H	M
1– 3*	2		1– 3*	3	3	20–25*		3
3– 5	7		3– 5	15	19	25–30	2	5
5– 7	12	5	5– 7	21	17	30–35	6	19
7– 9	18	7	7– 9	9	6	35–40	14	12
9–11	4	8	9–11		1	40–45	15	9
11–13	5	16	11–13		3	45–50	8	2
13–15		6	13–15		1	50–55	2	
15–17		5				55–60	1	
17–19		3						
Totals	48	50	Totals	48	50	Totals	48	50

Source: Mosteller and Wallace (1964).
* Each interval excludes its upper end point. Thus a paper with a rate of exactly 3 per 1000 words would appear in the count for the 3–5 interval.

Madison's systematic tendency to use *by* more often than Hamilton. Thus low rates for *by* suggest Hamilton's authorship, and high rates Madison's. Rates for *to* run in the opposite direction. Very high rates for *from* point to Madison but low rates give practically no information. The more widely the distributions are separated, the stronger the discriminating power of the word. Here, *by* discriminates better than *to,* which in turn is better than *from.*

PROBABILITY MODELS

To apply any of the theory of statistical inference to evidence from word rates, we must construct an acceptable probability model to represent the variability in word rate from paper to paper. Setting up a complete model for the occurrence of even a single word would be a hopeless task, for the fine structure within a sentence is determined in large measure by nonrandom elements of grammar, meaning, and style. But if our interest is restricted to the rates of use of one or more words in blocks of text of at least 100 or 200 words, we expect that detailed structure of phrases and sentences ought not to be very important. The simplest model can be described in the language of balls in an urn, so common in classical probability. To represent Madison's usage of the word *by,* we suppose there is a typical Madison rate, which would be somewhere near 12 per 1000, and we imagine an urn filled with many thousands of red and black balls, with the red occurring in the proportion 12 per 1000. Our probability model for the occurrence of *by* is the same as the probability model for successive draws from the urn, with a red ball corresponding to *by,* a black ball corresponding to all other words. To extend the model to simultaneous study of two or more words, we would need balls of three or more colors. No grammatical structure or meaning is a part of this model, and it is not intended to represent behavior within sentences. What is desired is that it explain the variation in rates— in counts of occurrences in long blocks of words, corresponding to the essays.

We tested the model by comparing its predictions with actual counts of word frequencies in the papers. We found that while this urn scheme reproduced variability well for many words, for other words additional variability was required. The random variation of the urn scheme represented most of the variation in counts from one essay to another, but in some essays authors change their basic rates a bit. We had to complicate the theoretical model to allow for this, and the model we used is called the negative binomial distribution.

The test showed also that pronouns like *his* and *her* are exceedingly unreliable authorship indicators, worse even than words like *war.*

INFERENCE AND RESULTS

Each possible route from construction of models to quantitative assessment of, say, Madison's authorship of some disputed paper, required solutions of serious theoretical statistical problems, and new mathematics had to be developed. A chief motivation for us was to use the *Federalist* problem as a case study for comparing several different statistical approaches, with special attention to one, called the Bayesian method, that expresses its final results in terms of probabilities, or odds, of authorship.

By whatever methods are used, the results are the same: overwhelming evidence for Madison's authorship of the disputed papers. For only one paper is the evidence more modest, and even there the most thorough study leads to odds of 80 to 1 in favor of Madison.

Figures 1 and 2 illustrate how the 12 disputed papers fit the distributions of Hamilton's and Madison's rates for two of the words finally chosen as discriminators. In Figure 1 the top two histograms portray the data for *by* that was given eariler in Table 2. Madison's rate runs higher on the average. Compare the bottom histogram for the disputed papers first with the top histogram for Hamilton papers, then with the second one for Madison papers. The rates in the disputed papers are, taken as a whole, very Madisonian, though 3 of the 12 papers by themselves are slightly on the Hamilton

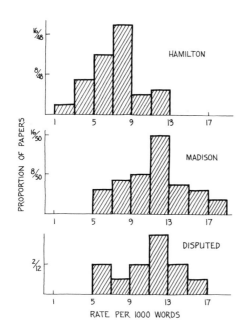

FIGURE 1

Distribution of rates of occurrence of by *in 48 Hamilton papers, 50 Madison papers, 12 disputed papers*

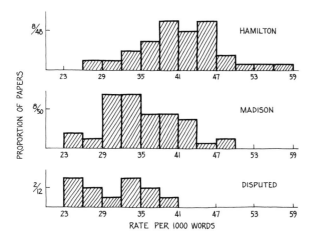

FIGURE 2

Distribution of rates of occur-
rence of to *in 48 Hamilton*
papers, 50 Madison papers, 12
disputed papers

side of the typical rates. Figure 2 shows the corresponding facts for *to*. Here again the disputed papers are consistent with Madison's distribution, but further away from the Hamilton behavior than are the known Madison papers.

Table 3 shows the 30 words used in the final inference, along with the estimated mean rates per thousand in Hamilton's and Madison's writings. The groups are based upon the degree of contextuality anticipated by Mosteller and Wallace prior to the analysis.

The combined evidence from nine common filler words shown as group B was huge—much more important than the combined evidence from 20 low-frequency marker words like *while-whilst* and shown as groups C, D, and E.

There remains one word that showed up early as a powerful discriminator, sufficient almost by itself. When should one write *upon* instead of *on?* Even authoritative books on English usage don't provide good rules. Hamilton and Madison differ tremendously. Hamilton writes *on.* and *upon* almost equally, about 3 times per 1000 words. Madison, on the other hand, rarely uses *upon.* Table 4 shows the distributions for *upon.* In 48 papers Hamilton never failed to use *upon;* indeed, he never used it less than twice. Madison only used it in 9 of 50 papers, and then only with low rates. The disputed papers are clearly Madisonian with *upon* occurring in only one paper. That paper, fortunately, is strongly classified by the other words. It is not the paper with modest overall odds.

TABLE 3.   Words Used in Final Discrimination and Adjusted Rates of Use in Text by Madison and Hamilton

WORD	RATE PER 1000 WORDS		WORD	RATE PER 1000 WORDS	
	Hamilton	Madison		Hamilton	Madison
Group A			**Group D**		
Upon	3.24	0.23	Commonly	0.17	0.05
Group B			Consequently	0.10	0.42
Also	0.32	0.67	Considerable(ly)	0.37	0.17
An	5.95	4.58	According	0.17	0.54
By	7.32	11.43	Apt	0.27	0.08
Of	64.51	57.89	**Group E**		
On	3.38	7.75	Direction	0.17	0.08
There	3.20	1.33	Innovation(s)	0.06	0.15
This	7.77	6.00	Language	0.08	0.18
To	40.79	35.21	Vigor(ous)	0.18	0.08
Group C			Kind	0.69	0.17
Although	0.06	0.17	Matter(s)	0.36	0.09
Both	0.52	1.04	Particularly	0.15	0.37
Enough	0.25	0.10	Probability	0.27	0.09
While	0.21	0.07	Work(s)	0.13	0.27
Whilst	0.08	0.42			
Always	0.58	0.20			
Though	0.91	0.51			

Source: Mosteller and Wallace (1964).

Of course, combining and assessing the total evidence is a large statistical and computational task. High speed computers were employed for many hours in making the calculations, both mathematical calculations for the theory, and empirical ones for the data.

You may have wondered about John Jay. Might he not have taken a hand in the disputed papers? Table 5 shows the rates per thousand for nine words of highest frequency in the English language measured in the writings of Hamilton, Madison, Jay, and, for a change of pace, in James Joyce's *Ulysses*. The table supported the repeated assertion that Madison and Hamilton are similar. Joyce is much different, but so is John Jay. The words *of* and *to* with rate comparisons 65/58 and 41/35 were among the final discriminators between Hamilton and Madison. See how much more easily Jay could be discriminated from either Hamilton or Madison by using *the, of, and, a,* and *that.* The disputed papers are not at all consistent with Jay's rates, and there is no reason to question his omission from the dispute.

TABLE 4. Frequency Distribution of Rate per Thousand Words in 48 Hamilton, 50 Madison, and 12 Disputed Papers for *Upon*

RATE PER 1000 WORDS	HAMILTON	MADISON	DISPUTED
0 (exactly)*		41	11
0+ –0.4		2	
0.4–0.8		4	
0.8–1.2	2	1	1
1.2–1.6	3	2	
1.6–2.0	6		
2.0–3.0	11		
3.0–4.0	11		
4.0–5.0	10		
5.0–6.0	3		
6.0–7.0	1		
7.0–8.0	1		
Totals	48	50	12

Source: Mosteller and Wallace (1964).
* Each interval, except 0 (exactly), excludes its upper endpoint. Thus a paper with a rate of exactly 3 per 1000 words would appear in the count for the 3.0–4.0 interval.

TABLE 5. Word Rates for High-Frequency Words (Rates per 1000 Words)

	HAMILTON (94,000)*	MADISON (114,000)*	JAY (5000)*	JOYCE (ULYSSES) (260,000)*
The	91	94	67	57
Of	65	58	44	30
To	41	35	36	18
And	25	28	45	28
In	24	23	21	19
A	23	20	14	25
Be	20	16	19	3
That	15	14	20	12
It	14	13	17	9

Sources: Hanley (1937), Mosteller and Wallace (1964).
* The number of words of text counted to determine rates.

SUMMARY OF RESULTS

Our data independently supplement the evidence of the historians. Madison is extremely likely, in the sense of degree of belief, to have written the disputed *Federalist* papers, with the possible exception of paper number 55, and there our evidence yields odds of 80 to 1 for Madison—strong, but not overwhelming. Paper 56, next weakest, is a very strong 800 to 1 for Madison. The data are overwhelming for all the rest, including the two papers historians feel weakest about, papers 62 and 63.

For a more extensive discussion of this problem, including historical details, discussion of actual techniques, and a variety of alternative analyses, see Mosteller and Wallace (1964).

PROBLEMS

1. Why can't the authorship of the disputed papers be determined by literary style or political philosophy?

2. (a) What is a discriminator?
    (b) Distinguish at least two categories of discriminators.
    (c) Why is *by* a good discriminator? (Refer to Table 2.)

3. What is a "noncontextual word"?

4. Why do the authors use word frequency per thousand words instead of just the number of occurrences?

5. Refer to Table 1. In how many of the Hamilton papers studied does the word *commonly* appear at least once?

6. Refer to Table 2. In what percentage of the Madison papers studied does *from* occur 3–7 times per 1000 words? (Note: The interval 3–7 uses the authors' convention on intervals.)

7. Consider the "balls in an urn" model. How many colors of balls would we need to extend the model to the simultaneous study of 5 words? Of n words?

8. Consider Figure 1. True or False: More than 1/3 of the Hamilton papers studied use *by* 3–7 times per 1000 words.

9. Study Figure 2. Does the graph for the disputed papers look more like the graph for the Hamilton or the Madison papers?

10. Consider Table 3. Looking at group B, which word would you say was the best Hamilton/Madison discriminator? What was your word-selection criterion? Answer the same questions for group D.

11. Table 1 shows the relative frequency of *war*. Why doesn't *war* appear in Table 3?

REFERENCES

Miles L. Hanley. 1937. *Word Index to James Joyce's Ulysses.* Madison, Wis.: University of Wisconsin.

F. Mosteller and D. L. Wallace. 1964. *Inference and Disputed Authorship: The Federalist.* Reading, Mass.: Addison-Wesley.

ADVERBS MULTIPLY ADJECTIVES

Norman Cliff   *University of Southern California*

Is IT possible that the difference in the meanings of the adjectives *nice* and *unpleasant* is a simple numerical one? If so, what happens when the words are modified by an adverb: *very nice, very unpleasant, somewhat nice, somewhat unpleasant?* Are the meanings of the combinations changed by arbitrary amounts or are the changes systematically related to the meanings of the unmodified words? For a number of years, but especially since the early fifties, there have been studies of questions like this by statistically minded psychologists and linguists. In this essay, we describe one such study and its results. It showed that "adverbs multiply adjectives" in a very literal sense.

There does seem to be an analogy between adverbial modification and multiplication. Compare the meanings of *very nice* and *very unpleasant* to *nice* and *unpleasant* by themselves. The modified pair are more different from each other than the individual adjectives. This could be explained by assuming the *nice* represents a positive number (+2, for example), *un-*

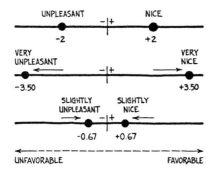

FIGURE 1

The upper part is the hypothetical position of two adjectives on a continuum of favorableness-unfavorableness when they are not modified. The middle portion shows the effect of modification by very *and the lower the effect of modification by* slightly

pleasant a negative one (say, —2), and *very* a positive number greater than one (say, +1.75); then the meaning of the combination is the product of the adverb and adjective numbers. (For example, the meaning of *very nice* would be 1.75 × 2 = 3.50.) The effect is illustrated in Figure 1, where the top portion represents the hypothetical positions of two adjectives on a continuum when unmodified, and the middle their positions when modified by *very*. The effect of adverbs such as *slightly* and *somewhat* that reduce the intensity of the adjectives they modify can be explained by assuming that the numbers they represent are less than unity, as illustrated in the bottom portion of Figure 1, where we assume *slightly* has a value of 0.33, so that *slightly unpleasant* has a value of 0.33 × (—2) = —0.67. Various adjectives occupy different positions on the continuum, and various adverbs have different "multiplying values," so that the function of the adverb is to move adjectives up or down the continuum in a regular way.

Our idea is translatable into a simple mathematical formula:

$$f = ia.$$

It states that the favorableness f of a combination is the product of the intensifying effect i of the adverb times the number a representing the adjective.

QUANTIFYING FAVORABLENESS

The first problem in testing such a theory is quantifying favorableness. Here, as in other similar studies, we started with a fairly straightforward approach based on the judgment of native speakers of the language, but judgments gathered in a formalized way. We defined a scale of favorableness, and 11 categories on this scale. Words and combinations of words were given to a set of judges, who in this case were college students, and each judge was instructed to indicate the category which seemed appropriate to him for each word. The categories were described as running from "most favorable" through "neutral" to "most unfavorable." The setup is illustrated in Figure

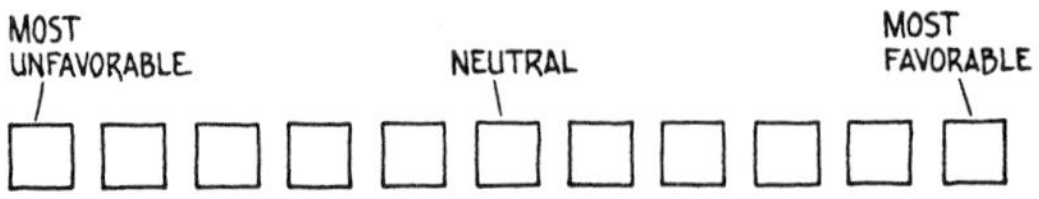

FIGURE 2

Format for judging the combinations. The pattern was repeated down each page, with adjectives or adverb-adjective combinations listed at the left and the boxes printed on the right. The headings "most unfavorable," etc. were given only at the top of the page

2. Then, after the data had been gathered, the numbers —5, —4, . . . , 0, . . . , +5 were assigned to the categories, and the average of the numbers assigned by the judges was calculated.[1] We found that there was good agreement between judges on the rating for each word or combination of words.

In the study of adverb-adjective combinations there were 150 words and combinations of words judged in this way. These were all combinations of the 9 adverbs in Table 1 with the 15 adjectives in the table, plus the adjectives alone. Average ratings of the kind described in the previous paragraph were calculated for all these 150 words and combinations.

DERIVING ADJECTIVE AND ADVERB NUMBERS

The 150 average ratings from the study could be used to test the multiplicative formulation and to derive the adverb and adjective numbers. In the case of the ratings of the adjectives by themselves, we assume that $f = a$ since there is no adverb involved. In order for the idea of multiplicative combination to have much use, we need to assume further that the number which the adverb represents does not change from adjective to adjective, and vice versa, so that we should be able to find a number for each adverb and one for each adjective.

People are, of course, not entirely consistent either with themselves or with other people in the way they rate the adjectives; thus no such mathemati-

---

[1] In the initial study of this kind, the statistical analysis that was used was much more elaborate than simply taking averages and required several weeks of computation using the equipment then available. Subsequent events showed that virtually the same results would have been achieved by using the simple averaging process. For the sake of simplicity and because the results would be equivalent, the study is described as if the simple averaging process had been used.

TABLE 1. Adverbs and Adjectives Used in the Adverb-Adjective Combinations

ADVERBS	ADJECTIVES
Slightly	Evil
Somewhat	Wicked
Rather	Contemptible
Pretty	Immoral
Quite	Disgusting
Decidedly	Bad
Unusually	Inferior
Very	Ordinary
Extremely	Average
	Nice
	Good
	Pleasant
	Charming
	Admirable
	Lovable

cal formulation can be expected to fit the data exactly, so we need to have procedures for the statistical estimation of the adverb and adjective numbers that are most consistent with the data. We also need statistical means of seeing the degree to which the formulation is consistent with the data: is it way off the mark, a rough approximation, or a very accurate description?

The method of estimating the numbers revolved around comparing the favorableness of unmodified adjectives to their favorablenesses when modified by a particular adverb. In Table 2, we have the favorablenesses of five adjectives when by themselves and when modified by *decidedly*. In theory, all we have to do is divide the favorableness of, for example, *decidedly admirable* by the favorableness of *admirable* to get the intensifying value of

TABLE 2. Average Ratings of Five Adjectives

ADJECTIVE	UNMODIFIED (a)	MODIFIED BY DECIDEDLY (ia)	ESTIMATE OF INTENSIFYING VALUE (i)
Disgusting	−3.10	−3.42	1.10
Inferior	−1.94	−2.70	1.39
Ordinary	−0.35	−0.63	1.80
Pleasant	2.18	2.75	1.26
Admirable	2.92	3.33	1.14

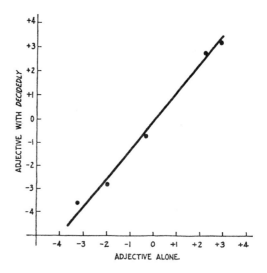

FIGURE 3

Plot of the data in Table 2 in which the mean judgment of each adjective used alone is the horizontal coordinate of each point and its vertical coordinate is the rating of the same adjective modified by decidedly. *Thus, the lowest left point is for* disgusting *by itself* (−3.10) *and* decidedly   disgusting (−3.42)

decidedly because this would be *ia/a*; the *a*'s would cancel out, leaving us with an estimate of *i* for *decidedly*. In this case, that would be 1.14. The trouble with this approach is that we get different values for *i* depending on which adjective we pick to work with, as we see in the last column of Table 2.

Statistical methodology gives us a way out of this inconsistency. It provides a way of arriving at an overall best estimate of the *i* for any adverb. To illustrate the way it does this, the data for the five adjectives are plotted in Figure 3, where the unmodified rating (horizontal axis) of each is plotted against the rating when modified by *decidedly* (vertical axis).

Another way to state the multiplicative principle is to say that the favorableness when modified should be proportional to the favorableness when unmodified, with the intensifying effect of the adverb as the constant of proportionality. This means that the points in the figure should lie along a straight line through the origin, and that the numerical value of its steepness (the tangent of the angle it makes with the *x* axis) is the adverb number. It can be seen that the points in the figure lie pretty close to a straight line. The statistical method called *least squares* allowed us to define the line that is *closest to all the points,* and we used the slope of this line as the most representative value of the intensifying effect of *decidedly*.

This is an example typical of a number of applications of statistical methodology. Whenever, as here, we are trying to fit a model to some observations, there are discrepancies between the model and the data. The model usually states the general form of a relationship but leaves one or more constants (or parameters) to be determined from the data. In this case, our model tells us that the relation should be linear and through the origin, but it doesn't

TABLE 3.   Intensifying Values of Adverbs and
Scale Values of Adjectives

ADVERB	i	ADJECTIVE	a
Slightly	0.54	Evil	−2.64
Somewhat	0.66	Wicked	−2.54
Rather	0.84	Contemptible	−2.20
Pretty	0.88	Immoral	−2.48
Quite	1.05	Disgusting	−2.14
Decidedly	1.16	Bad	−2.59
Unusually	1.28	Inferior	−2.46
Very	1.25	Ordinary	−0.67
Extremely	1.45	Average	−0.79
		Nice	2.62
		Good	3.09
		Pleasant	2.80
		Charming	2.39
		Admirable	3.12
		Lovable	2.43

tell what the slope of the line should be.   Among all the lines that could be used, it is natural to pick the one that gives the least discrepancy between the model and the data; geometrically, that means drawing the line that is the closest to the points.   There are various ways of mathematically defining "closest," and to say we used least squares means we used a particular one of these.   In general, it means picking the value of the parameter that gives the smallest total *squared* deviations between the model and the data.   Here, that means finding the line for which the sum of the squared distances from the points to the line is the smallest.[1]   Just how this is done depends on the exact nature of the application, and, while the basic aspects of the method require only elementary calculus, there are a number of mathematical and computing tricks that have been developed.   As applied here, it allowed us to find "best possible" estimates of the multiplying value of the adverbs.

The process was repeated using each adverb; the slope of the line that most nearly related the favorableness of adjectives combined with it to their favorableness when used alone was determined.   These numbers are the adverbial multiplying values presented in Table 3.   The numbers indicate that *slightly* and *somewhat* diminish the effect of an adjective to about half its original value; *quite* has almost no effect at all (in contrast to its dictionary

---

[1] To a large extent, the reason for using the sum of squared discrepancies rather than some other criterion such as the sum of absolute discrepancies has to do with mathematical convenience.

meaning) ; *extremely* has the greatest intensifying effect, making an adjective about one and a half times as strong as it would be alone.

We could use the average ratings of the unmodified adjectives as their *a* values. These are not the best estimates, however, since they place too much reliance on a single set of data, whereas, according to our formula, the number represented by an adjective enters into the ratings of *all* its combinations. Therefore, once we have determined the adverb numbers we can use them in a statistical procedure analogous to that used in fitting the straight lines to arrive at optimum or best-fitting estimates of the adjective numbers. These are the ones given in Table 3, where we see that the unfavorable adjectives represent negative numbers, and the favorable ones positive numbers. Two adjectives, *ordinary* and *average*, have numbers near 0. Perhaps this is why it seems odd or funny-sounding when they are modified by one of our adverbs: zero multiplied by anything is still zero, so why bother?

CHECKING THE FIT OF THE MODEL

There were statistical checks on the accuracy of our formulation in all of the steps performed to this point (the closeness of points to the lines, for example), but one final, overall evaluation was made. This was to multiply each adverb number by each adjective number and compare the result to the average rating for that combination. [For example, the rating of *very contemptible* should be $(1.25) \times (-2.20) = -2.75$.] This tells us the accuracy with which the 24 derived numbers (9 adverbs plus 15 adjectives) could be used in conjunction with our multiplicative theory to account for the 150 data numbers. When this was done, it was found that the average (unsigned) discrepancy was about 0.15 of a category, so that a combination with actual average rating of say 3.00 might well come out as about 3.15 or 2.85 when we multiply its adjective number by its adverb number. In some cases it would come out closer, and in some cases somewhat farther away, but on the average the unsigned discrepancy would be about 0.15. Since the data we started with were the average categories in which word combinations were placed—a task that many people might think cannot be done consistently with single words much less with combinations—and since the theory tested is that the use of words in communication is really a numerical process, the closeness with which the data fit the theory was surprising.

One additional point may be worth noting. This has to do with the assignment of the numbers -5, -4, etc. to the categories. This might seem very reasonable intuitively, but it could be that the lowest category rather than the middle one should be zero and the categories should be numbered 0, 1, 2, up to 10 if we are to interpret the labels as favorableness numbers. This would be analogous to having used degrees Centigrade when we should be using degrees Absolute in measuring temperature in a gas-law experiment.

The fact is that an aspect of the statistical analysis validated our locating the zero point in the middle category rather than at the end of the scale or some other place. In effect, we were able to use the analysis to locate the *absolute zero point,* and it turned out to be in the middle category.

REPEATING THE STUDY

The same experiment was repeated with two more groups of judges in other parts of the country with an equivalent degree of verification of the principle of multiplication. There was some variation from group to group in the actual adverb and adjective numbers obtained, presumably as a reflection of slightly different local usages. The study was further repeated in Australia and Great Britain, again with very similar results. Since then, the word lists have been translated into French, Polish, Spanish, Norwegian, German, Japanese, and other languages, and in all cases the principle that adverbs multiply adjectives has been verified, although the words' numerical values sometimes surprise the translators. The same sort of principle has also been shown to apply to other classes of words such as adverbs denoting probability, and the tenses of the verb *to be.* Two adjectives, on the other hand, combine in a way that is more like averaging than multiplication. Each adjective retains its own characteristic effect when used in combination with another regardless of what it is combined with, but two adjectives together do not have anywhere near the product or the sum of the effects of each separately.

The idea that words combine numerically, that they even multiply together, is now well established, but it originally seemed rather farfetched. To find that people agree fairly well on the numerical value to assign to a particular word was in itself surprising the first time or two it was done. To find that they did the same thing with combinations of words was an additional surprise. To find that the numbers they assign to the combinations conform to the multiplicative rule initially seemed outlandish. In fact, data that could have been used to establish the multiplicative rule was published some 15 years previous to the study described here, but no one thought of looking at it that way, probably in part because the relevant statistical methods were not widely enough known.

The final conclusion reached in the studies described here was established as the end result of a large amount of statistical methodology employed at numerous stages in the process. It started with the tabulation of the original responses by the judges, and ended with the overall estimate of how well the multiplicative hypothesis agreed with the data. It may be worth noting in passing that the process used here is completely analogous to verification of the laws interrelating the temperature, pressure, and volume of gases, including a check on the validity of the numbers assigned to the categories (analogous to estimating absolute zero on the temperature scale). The inexact nature of the data used here makes statistical methodology play a larger part in this study than it does in most physical science problems, and the use of statistical methods was essential in this work.

PROBLEMS

1.   Consider Figure 3. Suppose "pretty" has a rating of about 2.5. What then is the predicted rating of "decidedly pretty"?

2.   Consider ratings based on the estimates of Table 3 for "slightly inferior" and "somewhat disgusting." Which of the ratings is less favorable?

3.   Suppose "happy" has an average scale value of 3.2 and "completely happy" has an average scale value of 4.8. What then would be the estimate of the intensifying value of "completely"?

4.   What is the slope of the line in Figure 3? (Hint: Use Table 3.)

5.   Refer to Table 2. Why is the third column only an estimate? How were the other figures obtained?

6.   Why was a statistical procedure for estimating adverb and adjective numbers necessary?

7.   There are 9 adverbs and 15 adjectives used in this study. Why are there 150 average ratings?

8.   Reconstruct Figure 1 assuming the effect of adverbs on adjectives is additive.

REFERENCES

The study described here is also reported informally in N. Cliff. 1958. "Intensive Adverbs from a Quantitative Point of View." *College Composition and Communication* 9:20–22. A more technical report is "Adverbs as Multipliers." 1959. *Psychological Review* 66:27–44.

THE MEANING OF WORDS

Joseph B. Kruskal   *Bell Telephone Laboratories, Inc.*

How CAN we possibly use statistics to capture that elusive thing called "meaning"? Even granting the possibility, won't it take all the romance out of poetry and the charm out of graceful speech?

Well, we *can* study the meaning of words by the orderly methods of statistics. I shall discuss two interesting studies of this kind (see also the essay by Cliff). But the romance of words, you will discover, is safe from science.

Practical-minded people may ask *why* we should work so hard to pin down meaning. Doesn't language, after thousands of years of natural evolution, serve well enough? Quite simply, no. Any college admissions officer will testify to the difficulty of interpreting teachers' written recommendations, and many colleges use a standard form to reduce it. Studies like those we discuss may one day improve communication among teachers, students, and colleges. They may even contribute to peace among nations.

Numbers, in other situations, have supplemented descriptive words for a long time, often very helpfully. The use of *inches, pounds,* and *degrees Fahrenheit* certainly has advantages over *fairly short, very light,* and *rather hot.* Once upon a time, *feverish* carried a meaning as vague as *charming* or *surly.* The introduction of the fever thermometer has been a boon to human health.

In subtler areas, such as musical tones, speech sounds, and intelligence, numbers have wide use today. Color provides a striking example. Many people imagine colors so subtle as to elude numerical description, yet several numerical systems enjoy routine use. (In most of them, three numbers describe each color.) These systems played a vital role in developing color television and improving color photography, and have also aided the paint industry, stage makeup, and other fields. Some artists, too, find this scientific way of describing colors helpful. Nothing about this science reduces the scope of their art, nor your pleasure in looking at their paintings, seeing a color movie, or watching the sun go down.

These older examples, however, differ from the studies to be described here in one important way: the manner in which measurements are taken. Physical devices can measure height, weight, and temperature. Colors are handled by elaborate devices for mixing and dimming light, and a standard test measures intelligence. In the work described below, however, a person acts as a measuring instrument, rather than as an object of study. His subjective impression, obtained and analyzed in a very careful way, constitutes the measurement.

PERSONALITY TRAITS

A study by Rosenberg, Nelson, and Vivekananthan (1968) deals with the meaning of 64 personality trait words such as *impulsive, sincere, cautious, irritable,* and *happy,* words that describe people. By using a novel statistical technique it helps to provide order where order is difficult to find.

Everyone uses these words, and they carry meaning to us all. Furthermore, they have importance. Imagine overhearing someone describe you as "humorless" or instead as "good-natured," and think how different these two descriptions would make you feel! When these words are used to arrange blind dates, or by teachers writing recommendations, it makes a big difference just which ones are chosen.

Nevertheless, people use these words differently. If two people both describe the same friend, or the same movie character, we would be surprised if they used *exactly* the same words. In psychiatry, the same uncertainty of meaning exists, not only for common words but also for technical terms such as *schizoid* and *autistic.* Long articles are written about what these words should or do mean.

This creates the problem. With words that are used so frequently and carry so much importance, clarification of their meaning may well prove useful—to psychiatrists perhaps, to college admissions officers possibly, to computer dating organizations, and who knows to whom else.

A DETOUR ON MULTIDIMENSIONAL SCALING

Clarify their meaning—but how? Statistics first enters here, providing an approach to the problem, as well as a method of analysis. First, however, let us take stock of the problem more clearly.

> (1) What do we *want?* An orderly description of the personality trait words, according to their meaning. With success, we will obtain a description as helpful as a city map, which displays roads, bus routes, parks, etc. according to their location.
>
> (2) What *means* do we have for getting this description? Nothing but the way people use the words. We have nothing to measure with a ruler or a thermometer; we can only observe how people use or respond to the words.

A recently developed statistical method called "multidimensional scaling" has particular value in problems like this one. The very fact that Rosenberg and his coauthors were aware of this method helped stimulate their experiment. Statistics provided the approach.

What does multidimensional scaling consist of? It can best be described in three parts: what goes in, what comes out, and what happens in between. What goes in to multidimensional scaling are *similarities* or *dissimilarities* between various objects of one kind. For each pair of objects (in this case, for each pair of words from among the 64 used), a number describes how much alike, or how different, the two objects are. Many methods can provide such numbers. For example, the similarity between two colors may be provided by a person who looks at the two colors and describes, on a scale from 1 to 5, how alike they are. (If a larger number indicates greater likeness we call the numbers "similarities"; in the opposite case, "dissimilarities.") A more elaborate method was used in this study; we'll describe it later.

What comes out of multidimensional scaling is a *map* or *picture,* such as Figure 1. (For the moment ignore the lines in the figure and consider only the configuration of the points.) Briefly, the method places each object in a particular position. The map produced by scaling, though not a real map, shares many characteristics of real ones. For example, we are free to turn it and look at it from any direction we please; it has no particular direction that is truly up. Also we are free to enlarge (or diminish) the map, change the scale, so to speak, though our maps do not really have any scale. Convenience dictates how big we make the map.

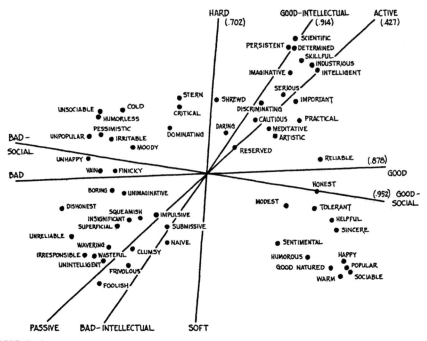

FIGURE 1

Multidimensional scaling map with trend axes. Source: Rosenberg et al. (1968)

What happens between the input and the output of multidimensional scaling? It is simple to say, though hard to do. We place objects on a map with the goal of having objects that are close together be very much alike, and having objects that are far apart be very different. In other words, small distances should correspond to small dissimilarities, and large distances to large dissimilarities; or vice versa for similarities. In brief, the goal while constructing the map is a good relationship between the map distances and the input (dissimilarities or similarities).

How good a relationship can be achieved depends on the input. The quality of the relationship has great importance in multidimensional scaling. Where it is good, the map may tell us something useful; where it is very bad, we may as well throw the map away.

The complex process of constructing the map is almost always carried out by an automatic computer. The user of the computer program does not need to understand this process, any more than the driver of a car needs to understand how its motor works. Though there is great variation according to circumstances, a typical calculation might require a minute of computer time and cost several dollars.

PERSONALITY TRAITS RESUMED

To obtain the similarities, Rosenberg and his coauthors presented 69 subjects (college students) with slips of paper containing 64 personality trait words. Each slip contained one word, and each subject received a complete pack of 64 slips. The subjects were asked to put the slips in their packs into roughly ten groups, so that the words in a group could plausibly describe a single person. (A slip could only be put in one group.) Of several procedures which psychologists use to obtain similarities, this is the most rapid.

By counting how many subjects put two words together, the investigators got an "agreement score" for the two words. For example, 37 of the subjects put *reliable* and *honest* in the same group, so these two words have an agreement score of 37. Several examples are:

(1) *reliable* and *honest,* score = 37
(2) *clumsy* and *naive,* score = 19
(3) *sentimental* and *finicky,* score = 2
(4) *good-natured* and *irritable,* score = 1
(5) *popular* and *unsociable,* score = 0

The agreement scores form a 64-by-64 square array. The 64 agreement scores of a word with itself are of no interest because all equal 69, the number of subjects. The remaining $64 \times 64 - 64 = 4032$ values come in pairs of equal values because the agreement score of *warm* with *intelligent,* for example, is just the same as the agreement score of *intelligent* with *warm.* Thus only $4032/2 = 2016$ distinct agreement scores really matter.

The agreement scores are similarities and can be used as the input for multidimensional scaling. In fact, the authors tried this in the earlier stages of their analysis, but the results were not clear. Fortunately the authors spotted the trouble: many of the agreement scores are so small that comparisons among them have little reliability. The scores given above shout that pair (2) is more alike than pair (3), but they only whisper that pair (3) is more alike than pair (4). With more than half the scores under 10, greater accuracy could help a lot. (Incidentally, this sort of trial and improvement often occurs when data is analyzed in a novel way.)

Fortunately, greater accuracy could be found by using the data differently. Consider any two words, say *sentimental* and *finicky.* Use two rows to list their agreement scores with all 64 words:

	Foolish	Inventive	Wavering	Submissive	Cold	Tolerant. . .
Sentimental	2	6	5	15	3	22 . . .
Finicky	6	2	18	10	9	2 . . .

If two words have similar meanings, then not only should they have a large agreement score, but also *their agreement scores with a third word should be about the same.* Thus for two similar words, the corresponding numbers in the two rows should not differ greatly. If we measure the difference between these two rows of numbers in some reasonable way, then this secondary (or derived) measure should indicate the difference in meaning of the two words. Following this idea, the dissimilarity between *sentimental* and *finicky* is formed by adding up the squares of the differences,

$$(2-6)^2 + (6-2)^2 + (5-18)^2 + (15-10)^2 \\ + (3-9)^2 + (22-2)^2 + \cdots .$$

The larger this dissimilarity, the greater the difference between the words.

The application of multidimensional scaling to the dissimilarities[1] gave the map in Figure 1. This map must be examined, interpreted, and assessed before we can feel we have found something useful or illuminating. The reader will find it profitable to examine Figure 1, to see what groups of words occur together, and to see how meanings change systematically across the figure. If an investigator finds anything sensible at this point, then a preliminary judgment of success may be entered. If the map provides an orderly picture which we did not know about before, that is progress.

PERSONALITY TRAITS—A SECOND STEP

The authors found a good deal of structure in Figure 1:

> Going from the upper-right corner to the lower-left corner, the desirability of traits for intellectual activities appears to decrease systematically. Another systematic change appears to take place as one goes from the upper-left corner to the lower-right corner; in this case the social desirability of the traits increases.

Notice that the upper-right corner includes *intelligent, skillful,* and *scientific,* while the opposite corner includes *unintelligent, foolish,* and *frivolous.* The upper-left corner includes *unsociable, humorless,* and *unpopular,* while the opposite corner includes *sociable, popular,* and *warm.* A good way to indicate these trends is by axes, as drawn in Figure 1. For each trait, its position *along* the axis indicates where it stands with regard to that trend. (To get the position of a word along the axis we drop a perpendicular onto the axis from the plotted position of the word. It makes no difference how far from the axis the word lies.)

This interpretation does not exclude other interpretations. A given direction (such as upper right to lower left) may have several interpretations,

[1] Due to limitations (no longer present) on the computer program, only 60 words were actually included in the multidimensional scaling.

and other directions may have meaning also. Not everybody agrees on the best way to interpret the figure.

The authors did not stop with their own subjective interpretation. Subjects different from those who provided the agreement scores rated the 64 traits on five different scales. Each subject dealt only with a single scale. For example, 34 subjects rated each of the traits on a 7-point scale "according to whether a person who exhibited each of the traits would be good or bad in his intellectual activities." A second scale dealt with social activities. Three other scales, namely, good-bad, hard-soft, and active-passive, were also used. (These were chosen from among the semantic differential scales in a book by C.E. Osgood et al., 1967.)

For each scale, the authors took several steps. They found the median of the subject ratings for each trait. (The median of the ratings is the middle value, smaller than half the values and larger than half the values.) By using a statistical method called *linear regression,* they found the axis that best matches the median ratings. The five axes shown in Figure 1 resulted from this procedure. The second experiment clearly verifies the two trends described by the authors, as well as displaying some other trends. This does not rule out the possibility of still other valid trends.

It is instructive to check the strength of each trend. In other words, how well do the median ratings on each scale match the positions of the words along the corresponding axis? To make this comparison, we calculate the *correlation coefficient* between the median scores for the word on the scale, and their positions on the axis. This widely used statistical measurement indicates how closely two sets of numbers vary with each other. It always lies between -1 and $+1$, and $+1$ indicates perfect agreement in the way the numbers vary (see the essay by Whitney for further explanation). The correlation coefficients, shown in Figure 1, are:

Social good-bad	0.95
Intellectual good-bad	0.91
Good-bad	0.88
Hard-soft	0.70
Active-passive	0.43

These indicate that the first three scales match the map really quite well, the fourth only fairly well, and the last in a very mediocre way (although 0.43 is high enough to prove conclusively that there is *some* connection between this scale and the map).

In conclusion, this analysis makes a significant start towards providing a systematic explanation of the meanings of a considerable set of words for personality traits. The authors have constructed a single map that explains how subjects use and understand these words in several different tasks. Of course, the map only explains *part* of the meaning of these words. Other aspects are entirely ignored. Thus *skillful, industrious,* and *intelligent* are very close on the map, because they often describe the same person, even though their meanings differ greatly in other respects. Also, the map only explains *aggregate* data, based on many subjects. This washes out and ignores individual variation in the use of these words.

Nevertheless, it is striking and illuminating that a simple map captures an important part of the meaning of a wide variety of words for personality traits. Real progress has been made.

MULTIDIMENSIONAL SCALING—A FURTHER COMMENT

The maps produced by multidimensional scaling need not be ordinary two-dimensional maps. They may be (and often are) three-dimensional, four-dimensional, or even higher-dimensional. As a matter of fact, Rosenberg and his coauthors actually constructed a one-dimensional map, a two-dimensional map (Figure 1), and a three-dimensional map, before deciding which one was the right one. (To explain how the choice is made would take us too far afield.)

We mentioned earlier that the quality of the relationship between map distances and dissimilarities has great importance. If the relationship is very bad, the map might as well be thrown away. A picture of this relationship (for the map of Figure 1) appears as the central plot of Figure 2. This plot contains $(60 \times 59)/2 = 1770$ points, one for each pair of words. For each pair, the dissimilarity gives the vertical coordinate of a point, and the distance between the two words (taken from Figure 1, just as you would measure it with a ruler) gives the horizontal coordinate. Clearly quite a

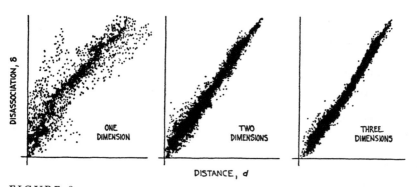

FIGURE 2

Relationships between map distances and word dissimilarities.
Source: Rosenberg et al.
(1968)

good relationship connects dissimilarity and distance. The other two plots of Figure 2 give the corresponding pictures, based on one-dimensional and three-dimensional maps.

NATIONS

The names of nations, like other words, bring to mind many associations. Two experiments [one by Wish (1970) using 12 nations, another by Wish, Deutsch, and Biener (1970, 1972) using 21 nations] investigated how people perceive nations and their interrelationships. The most important questions addressed to the subjects concerned how *similar* various nations are. By methods somewhat like those above, and using students at Columbia University as subjects, the authors discovered several interesting facts.

Of the many characteristics of nations which influence their perceived similarity, it is not surprising that the two which emerge as most important (in both experiments) are political alignment (ranging from "aligned with U.S.A." to "aligned with Russia") and economic development. A more complex characteristic which also displayed importance in the second experiment is "geography, race, and culture," under which the nations break down into four groups: European, Spanish, African, and Oriental.

The importance of these characteristics to different subjects did yield surprises, however. In the first experiment, the 18 subjects were divided into doves, moderates, and hawks according to which of five recommendations each selected concerning the war in Vietnam. All six hawks attached greater importance (in making their similarity judgments) to political alignment, while all seven doves attached greater importance to economic development. Thus political attitude affects even a question like how similar two nations are.

In the second experiment, 75 subjects from eight countries displayed the same systematic effect. Also, females as compared with males and students from underdeveloped countries as compared with students from developed countries attached higher importance to political alignment. The students were also tested for their knowledge about the 21 nations, and divided into better informed and less informed groups. Within each of these groups, the same effects were displayed, though much more strongly among the less informed group.

PROBLEMS

1. In what way are the measurements used in the personality traits study different from pounds or inches?

2. What quality was being quantified in preparing the input for multidimensional scaling?

3. Suppose Rosenberg and his coauthors had been studying the meaning of 50 (rather than 64) personality trait words. How many distinct agreement scores would have then mattered?

4. Consider a Rosenberg-type experiment using only four words with the following agreement scores resulting:

friendly	mean	perfect	
1	4	2	faulty
	0	4	friendly
0		1	mean

What is the dissimilarity between "friendly" and "perfect"?

5. The agreement score for *clumsy* and *naive* is 19. What numerical contribution will this make to the measure of dissimilarity for *clumsy* and *naive*?

6. Refer to Figure 1.
    (a) What five words form an extreme good-soft group?
    (b) How would you characterize *stern* and *critical*?

7. Consider the locations of *meditative* and *impulsive* in Figure 1.
    (a) How do the two words compare on the active/passive axis?
    (b) Does the answer in (a) seem counterintuitive? Explain your answer.

8. Were the axes in Figure 1 produced by the multidimensional scaling procedure? Explain your answer.

9. The number .914 (under Good-Intellectual) in Figure 1 is a correlation coefficient. What two measures are being correlated? What can you conclude about these two measures?

10. Examine Figure 2. How is δmeasured? What about d?

REFERENCES

Charles E. Osgood, George J. Suci, and Percy Tannenbaum. 1967 (1957). *The Measurement of Meaning*. Urbana, Ill.: University of Illinois.

S. Rosenberg, C. Nelson, and P. S. Vivekananthan.   1968.   "A Multidimensional Approach to the Structure of Personality Impressions."   *Journal of Personality and Social Psychology* 9:283–294.

Myron Wish.   1970.   "Comparisons Among Multidimensional Structures of Nations Based on Different Measures of Subjective Similarity."   A. Rapoport and Bentalanffy, eds., *General Systems* 15:55–65.

Myron Wish, M. Deutsch, and L. Biener.   1970.   "Differences in Conceptual Structures of Nations: An Exploratory Study."   *Journal of Personality and Social Psychology* 16:361–373.

Myron Wish, M. Deutsch, and L. Biener.   1972.   "Differences in Perceived Similarity of Nations."   A. K. Romney and S. Nerlove, eds., *Multidimensional Scaling: Theory and Applications in the Behavioral Sciences.*   New York: Academic.

THE SIZES OF THINGS

Herbert A. Simon   *Carnegie-Mellon University*

On Figure 1 are drawn four lines. The lowest one, a simple straight line inclined at a 45° angle, serves merely for purposes of comparison in describing the three slightly wavy lines. The three wavy lines—and particularly the two just above the straight line—depict some curious facts about the world. Whether they are significant facts as well as curious facts is a question we examine.

The lower broken line relates, on a logarithmic scale,[1] the 1965 populations of the twenty largest cities in the U.S. to the ranks by size of the cities,

---

[1] The common logarithm is familiar to many as a tricky device for multiplying numbers through a process of addition. Another way of looking at the logarithm is that taking the logarithm compresses the scale of numbers so as to create a new scale, one that makes multiplying the old number by 10 equivalent to adding one unit to the new number. For examples, the logarithm of 10 is 1, of 100 is 2, of 1000 is 3, and so on. The logarithm of 2000 is about 3.300 and that of 20,000 is about 4.300. If a city has a population of 5,000,000, then the logarithm of its population is about 6.70. The compression achieved by a logarithm scale increases as the numbers do.

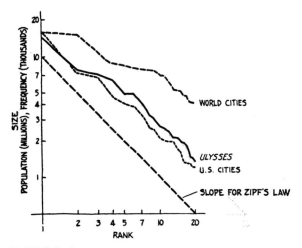

FIGURE 1

Logarithm of size plotted against logarithm of rank for frequencies of words and for populations of cities

arranged with New York, ranked 1 down to Kansas City, ranked 20. The population of each metropolitan area was used, not just that within the city limits. The horizontal axis shows, also on a logarithmic scale, the city ranks, from 1 through 20; on the vertical scale are shown the corresponding logarithms of populations in millions of persons. Ignoring the two largest cities (New York and Chicago), we can see that the rest of the line is nearly straight and inclined nearly at a 45° angle, parallel to the straight line below. Straightening out the left end of the curve would involve raising New York from about 16 million people to about 25 million and Chicago to 12 million (a heavy price to pay for a straight line), but the remaining 18 cities would require very little adjustment—generally less than 10% up or down.

The solid line, just above and very close to the line for cities shows (again on logarithmic scales) the number of occurrences of each of the twenty words most frequently used in James Joyce's *Ulysses,* when the words are arranged in descending order of frequency of occurrence. For this line, the ordinates show the frequencies of occurrences in thousands. The most frequent word in *Ulysses, the,* occurred 14,887 times; the twentieth most frequent, *all,* occurred 1311 times. As with the city sizes, the word frequencies lie almost on a straight line, although straightening the line would again require adjustment of the first few words; *the* would have to be increased to about 26,000 occurrences, *of* to 13,000, and *and* to about 8700. The remaining seventeen frequencies are extremely close to a straight line inclined at 45°.

Observe that in these distributions the product of the rank of each item by its size remains constant over the whole scale. If the first item (rank 1) has size 1,000,000, the tenth item will have size about 100,000 ($10 \times 100,000 = 1,000,000$), and the twentieth item will have size 50,000 ($20 \times 50,000 = 1,000,000$). The task before us is to explain why these regularities hold, why the product of number and rank in these distributions is almost constant, and—even more mysterious—why the size distribution of American cities should obey the same law as the frequency distribution of words in a stylistically unusual book like *Ulysses* (or in any book for that matter). Let's begin with the words.

WORDS: COMMON AND RARE

In the late nineteenth century, several linguists (among them de Saussurre in France) discovered the surprising rank-frequency regularity in the relative contributions of different words to any body of text. Obviously, certain words, such as *of*, will occur rather frequently in almost any English text, while other words, such as *conundrum,* occur infrequently or not at all. The frequency of any specific word may vary widely from one text to another.

Whenever you arrange the various words occurring in a particular text in the order of their frequency of occurrence—first the word that occurred most often in that text, then the word that occurred next most often, and so on—the regularity depicted in Figure 1 will reappear. The twentieth word on your list will occur about one-half as often as the tenth word.[1] If you enjoy this kind of numerology, you will find equally startling regularities at the other end of the distribution among the rare words.

About one-half of the total number of *different* words in the text occur exactly once each, about one-sixth occur exactly twice each, and about one-twelfth occur three times each (see Table 1). The ratio $1/[n(n+1)]$ gives the fraction of all the distinct words in the text that occur exactly n times each. This regularity in frequency of occurrence of the rare words is, of course, the same rank-size law we have been observing at the other end of the distribution, for the rank of a word is simply the cumulated number of different words that have occurred as frequently as it has, or more frequently. Suppose then, as the rank-size rule requires, that $K/(n+1)$ words occur $n+1$ or more times each, and K/n words occur n or more times each. Then the number of words occurring exactly n times will be $K/n - K/(n+1) = K/[n(n+1)]$.

The rank-size law, often called *Zipf's law* in honor of an American linguist who wrote a great deal about it, holds for just about all of the texts whose

---

[1] In most cases, the first two or three frequencies are substantially lower than the rule predicts, as in Figure 1.

TABLE 1. The Numbers of Rarely Occurring Words in James Joyce's *Ulysses*

NUMBER OF OCCURRENCES (n)	NUMBER OF WORDS	
	Actual	Predicted*
1	16,432	14,949
2	4,776	4,983
3	2,194	2,491
4	1,285	1,495
5	906	997
6	637	712
7	483	534
8	371	415
9	298	332
10	222	272

* Predicted number $= K/[n(n+1)]$; $K = 29,899$, the total number of *different* words in *Ulysses*.

vocabularies have been counted, in a great range of languages, not excluding American Indian languages. Though it holds for *Ulysses*, it fails for James Joyce's *Finnegan's Wake*, and it fails for Chinese texts if individual Chinese ideograms are counted as words—which is as good an argument against the latter identification as it is against Zipf's law.[1]

Why does this regularity hold? Why should the balance between frequent and rare words be exactly the same in the daily newspaper as in James Joyce's *Ulysses*, the same in German books as in English books, or the same in most (not all) schizophrenic speech as in normal speech?

Several answers have been proposed, one of which is typical of the explanations that are provided by probability theory. Probability theory often explains the way things are arranged on average by conceding its inability to explain them in exact detail. To explain the laws of gasses it avoids tracing the path of each molecule.

To explain the word distribution, we make some assumptions that might be thought outrageous if applied in detail, but that might be plausible if only applied in the aggregate. We assume that a writer generates a text

[1] Most linguists would consider that more often pairs of ideograms than individual ideograms serve as the basic lexical units corresponding to words in Chinese. The entries in Chinese and Japanese dictionaries are mostly such pairs. It is less easy to say why *Finnegan's Wake* doesn't fit the rule, but the freedom Joyce exercises in creating all sorts of word fragments and variants of dictionary words undoubtedly has something to do with the matter.

by drawing from the whole vast store of his memory, and by drawing from the even vaster store of the literature of his language. The former of these processes we might call *association,* the latter *imitation.* Specifically, we assume that the chance of any given word being chosen *next* is proportional to the number of times the word has previously been stored away—in memory or in the literature. Remember, these assumptions are intended to apply only in the large. To accept them, we need not believe that Shakespeare wrote sonnets by spinning a roulette wheel any more than we believe the individual molecules of a gas chart their courses by shaking dice.

If we accept the assumptions, then it becomes a straightforward mathematical matter but one beyond the scope of this paper to derive the probability distribution they imply. The derivation yields what is known as the *Yule distribution.* In the upper range, among frequently used words, the Yule distribution agrees with the rank-size law of Zipf; in the lower range, among rarely used words, it gives precisely the observed fractions $1/[n(n+1)]$.

Now we see why the *same* distribution can fit texts of diverse kinds drawn from the literatures of many languages. The same distribution can fit because it does not depend on any very specific properties of the process that generated the text. It only depends on the generator being, in a probabilistic sense, an associative and imitative process. We might even suspect that substantial departures from exact proportionality in association and imitation would not greatly change the character of the distribution. To the extent that the consequences of changing the assumptions have been explored, mathematically and by computer simulation, the distribution has indeed proved robust. We can give Shakespeare and Joyce a great deal of latitude in the way they write without altering visibly the gross size-rank relation of their vocabularies, but as *Finnegan's Wake* shows, we can't give them infinite latitude.

MEGALOPOLIS AND METROPOLIS

Having stripped away some of the mystery of the vocabularies of literary texts, we are perhaps prepared to tackle the corresponding regularity in American city sizes. We have seen (Figure 1) that the city populations obey the same rank-size law, to a quite good approximation.[1] If two cities have ranks j and k, respectively, in the list, their populations ratio will approximate k/j.

The regularity is not just a happenstance of the 1960 Census. It holds quite well for all the Censuses back to 1780. It does *not* hold, however, for cities in arbitrarily defined geographical regions of the world, that are

---

[1] We can take either the populations of cities as defined by their corporate boundaries, or populations within metropolitan areas as defined by the U.S. Census. The regularity shows up about as well in either case—perhaps it is a little more satisfactory if we use metropolitan areas.

not relatively self-contained economic units. It does not hold, for example, for Austria, or for individual Central American countries, or for Australia. Nor does it hold if we put the cities of the whole world together (see the uppermost curve in Figure 1). In that case, the distribution is still relatively smooth and regular, but population does not drop off with rank as fast as Zipf's law demands. The distribution is flatter, and the largest metropolises are "too small," though, I hasten to add, this phrase should not be interpreted normatively.

In the case of city sizes, then, we must be prepared to explain *two* things: why Zipf's law has held for nearly two centuries for the cities of the U.S., and why it doesn't hold for many other aggregates of cities. Let's start with the former question and ask what the analogues might be to the association and imitation processes that explained the word distributions. More precisely, let's ask what processes would lead cities to grow at rates proportional, on average, to the sizes already achieved (sometimes called *Gibrat's principle*); for that is the main assumption the mathematical derivation requires.

Cities grow by the net balance of births over deaths, and they grow by the net balance of inward over outward migration. With respect to births and deaths, we need assume only that, on average, birth and death rates are uncorrelated with city size. With respect to migration, we assume that migration outward is proportional, on average, to city size (i.e., that per capita *rates* are independent of size), and migration inward (from rural areas, from other cities, or from abroad) is also proportional to city size. The last assumption means that the cities in a given size group form a "target" for migration, which is larger, in total, as the total population already living in the cities of that group is larger. I leave to the reader the reasons why this might be a plausible assumption, at least as an approximation.

If we make these assumptions, we are again led by the mathematics of the matter to the rank-size law of Zipf. But now it is instructive to ask: under what circumstances would we expect a collection of cities to fit the assumptions? The answer is that the cities should form a "natural" region within which there is high and free mobility of population and industry, and which is not an arbitrary slice of a still larger region. The U.S. fits these requirements quite well, while an area playing a specialized role in a larger economic entity might not fit at all (for example, Austria after dissolution of the Empire, or a country specializing in agricultural exports and having a single large seaport).

If we put together a large number of distributions, each separately obeying the rank-size law, we get a new distribution of the same shape, simply displaced upward on the graph, but with the top few omitted. We would expect the totality of the world's cities to fit the rank-size distribution, except for a deficiency of extremely large metropolises at the very top, and so it does. If we take the published figures at their face value (the definitional problems are severe, and the census counts of varying accuracy), there are somewhat more than 50 urban aggregations in the world having over 2 million people each. Zipf's law would then call for a New York or a Tokyo of 100 million people, instead of the mere 16 million who now inhabit each of those cities. However, the deficiency of cities at the very top (mostly the top ten) is soon largely made up by the numerous cities over 5 million population each. Already, the tenth city on the list, Chicago, has a population of 7 million, only one-third fewer than the number demanded by Zipf's law.

The sizes of cities are of obvious importance to the people who live in them, but it is not obvious what practical conclusions we are to draw from the actual size distribution. One *possible* conclusion is that the distribution isn't going to be easy to change without strong governmental or economic controls over places of residence and work. Or to put the matter more palatably—because we generally wish to avoid such controls—the mathematical analysis that discloses the forces governing the phenomena teaches us that any attempt to alter the phenomena requires us to deal with those forces with sophistication and intelligence.

BIG AND LITTLE BUSINESS

Economists have generally been more interested in the sizes of business firms than they have in the sizes of cities. Concentration of industry in the hands of a few large firms is generally thought to be inimical to competition and is gnerally also supposed to have proceeded at a rapid rate in the U.S. during the past half century.

It has long been known that business firms in the U.S., England, and other countries have size distributions that resemble Zipf's rank-size law, except that size decreases less rapidly with rank than in the situations described previously (that is, the ratio of the largest firm to the tenth largest is generally less than ten to one[1]). The slower the decrease in size with increase in rank, the less concentrated is business in the largest firms.

Economists have been puzzled by the fact that the rate of decrease in size with rank, which is one way of measuring industrial concentration, appears to be about the same for large American manufacturing firms at the present

---

[1] Let m and n be the ranks of two members of a rank-size distribution, and let S_m and S_n be their respective sizes. Then the rank-size law, in this generalized form requires $S_m/S_n = (n/m)^k$, where the exponent k is a proper fraction. When k approaches unity as a limit, we get the special case of the Zipf distribution. The general distribution is usually called by economists the *Pareto distribution*. If we graph the logarithmic distribution, taking logarithms of both ranks and sizes, we again obtain a straight line with a slope equal to the fraction k. The steeper this straight line (the larger k), the larger are the first-ranking firms compared with the firms further down the list (i.e., the larger is k, the greater the concentration).

time as it was 25 years ago or even at the turn of the century. Even during periods of frequent mergers, the degree of industrial concentration, as measured by the rank-size relation, has changed slowly or not at all.

From our previous analyses, we should be ready to solve the puzzle. Indeed, it can be shown mathematically that under appropriate assumptions about the firms that disappear by merger, and those that grow by merger, mergers will have no effect on concentration. Moreover, the assumptions required for this mathematical derivation fit the American statistical data on mergers quite well. In analogy to the processes for words and cities, we can guess what those assumptions—and the data that support them—are like:

(1) The probability of a firm "dying" by merger should be approximately independent of its size

(2) The average assets acquired by surviving firms through mergers should be roughly proportional to the size they have already attained.

And these are indeed true.

Thus, a line of scientific inquiry that began with a linguistic puzzle over word frequencies leads to an explanation of a paradox about industrial concentration in the U.S. That explanation opens new lines of research for understanding business growth and arriving at public policy for the maintenance of business competition.

Our fascination with rank-size distributions need not stop with the three examples we have examined here. We may expect the Zipf distribution to show up in other places as well, and each new occurrence challenges us to formulate plausible (and testable) assumptions from which the rank-size law can be derived and the occurrence explained. We will leave a final example as an exercise for the reader. List the authors who have contributed to a scientific journal over 20 years, or whose names have appeared in a comprehensive bibliography, such as *Chemical Abstracts*. Note the number of appearances for each author, and rank the authors by that number. Then about one-half of all the authors will have appeared exactly once, one-sixth will have appeared twice, and so on; the data will not stray far from the Yule distribution. What are the ways of authors that can provide a naturalistic explanation for that fact?

PROBLEMS

1.   From Figure 1, what roughly is the population of Philadelphia, the fifth largest city in the U. S.?

2.   Consider Table 1.
     (a) 483 distinct words appear 7 times in the text of *Ulysses*. How was the predicted value of 534 computed?
     (b) Suppose the predicted number of words occurring n times is 164. Approximately what is n? [Hint: You will have to use the quadratic formula from high school algebra.]

3.   What does the author mean by *association*? By *imitation*?

4.   How can the same distribution fit the population of U.S. cities and the frequency of words in a text?

5.   What is *Gibrat's principle*? How does it relate to Zipf's law? To the sizes of U.S. cities?

6.   State in words the mathematical assumptions which lead to the Yule distribution, first in the case of literary texts, then in the case of city sizes.

7.   Can you think of a reason why the largest few cities in the U. S. might not satisfy the rank-size law?

8.   (a) How are the Zipf and Pareto distributions related?
     (b) In a logarithmic graph of the Pareto distribution, what is k?

9.   Suppose IBM's and Xerox's sales rank first and fourth respectively, Xerox's sales are $1 billion, and business firm sales obey a Pareto distribution with k = 1/2. What would IBM's sales be?

10.   If R_n is the rank of a thing of size n, state Zipf's law.

11.   Assume the fourth largest city has a population of 10 million. What ranking would you expect a city of 2½ million to hold, according to Zipf's law?

12.   The author states that the number of different words occurring exactly n times in a given text equals K/[n(n+1)]. What is K? (Hint: see Table 1.)

13.   Refer to the exercise outlined in the final paragraph. What two assumptions would you postulate for this distribution? [Hint: Use the assumptions for mergers or city sizes as a close guide.]

1 ENGINE # 421
2 BARS # 327
3 SHINGLES # 478 A
4 GLASSWARE #58C FRAGILE

HOW ACCOUNTANTS SAVE MONEY
BY SAMPLING

John Neter   *University of Minnesota*

ACCOUNTANCY AND statistics are regarded by many people as two of the dullest subjects on earth. The essays in this volume, it is hoped, will change people's views about statistics. This essay deals with important uses of statistics in accounting practice, and it may also reveal some interesting facets of accounting.

All of us, after all, want to use our money efficiently and effectively. We shall see how the use of statistical sampling in accounting saves money for railroads and airlines as they face problems of dividing revenues among several carriers. Similar statistical sampling methods are used in other areas of accounting and auditing work. Indeed they are used in many fields of business, government, and science.

Accountants and auditors traditionally have insisted on accuracy in the accounting records of firms and other organizations. This insistence has led them to do much work on a complete, 100% basis. For instance, an auditor may want to check the value of the inventory that a firm has on hand. To do this, he may examine the entire inventory; that is, he may actually count how many units of each type of inventory item are on hand, determine the value of each kind of unit, and thus finally obtain the total value of the inventory.

As another instance, an auditor may want to know the proportion of accounts receivable that have been owed for 60 or more days. This information may be needed to verify a reserve for bad debts. Accounts that have not been paid within 60 days are more susceptible to bad debt losses than accounts that have been open a shorter time. In order to establish the proportion of accounts receivable that are 60 days old or older, the auditor may examine every single account receivable held by the firm and determine for each the amount of money owed for 60 or more days.

Is it necessary to conduct these 100% examinations of inventory, of accounts receivable, or of similar collections, in order to obtain the figures that the accountant needs? More specifically, could a sample adequately provide the information needed by the accountant without all of the tedious work necessary for a complete, 100% examination? Let us focus on the inventory items.

In statistical terminology, the group of inventory items for which the total value is to be ascertained is called the *population* of interest. A *sample* selected from such a population consists of some, but not all, of the items in the population. A sample is selected to find out about characteristics of the population without looking at every element of the population.

The cost of examining a relatively small sample of inventory items is usually less than the cost of a complete examination because the sample requires an examination of fewer items. But are the results based on the relatively small sample almost as good as those from a complete examination?

Experience with sampling in many areas has shown that relatively small samples frequently provide results that are almost as good as those obtained from a complete examination, while at the same time the sample results cost considerably less. Indeed, sometimes the sampling results are even better than those from a 100% examination. That statement may seem startling, but consider the task of taking an inventory in a large company. Many persons are required for the task. Because of the size of the undertaking, it may be hard to give thorough training to these persons, and the quality check on the work may have to be limited. On the other hand, a small sample of the inventory items would require fewer persons, and therefore they could be trained better. Furthermore, the quality control program for the inventory could be more rigorous when a smaller number of persons are involved. The net effect might well be that the sample results are more accurate than the 100% enumeration! That is, the gains in accuracy from better training and quality control with a small sample may more than balance the sampling error introduced by selecting only a sample of inventory items instead of all of them. Of course, the sampling must be done intelligently and properly. The study of sampling is an important part of statistics.

THE CHESAPEAKE AND OHIO FREIGHT STUDY

Statements that relatively small samples can provide results almost as good as those from a complete examination, or indeed sometimes even better, are often not convincing by themselves. Statisticians have therefore often found it helpful to conduct studies that compare the results of a sample with those of a complete enumeration. Such a study was made by the Chesapeake and Ohio Railroad Company in determining the amount of revenue due them on interline, less-than-carload, freight shipments. If a shipment travels over several railroads, the total freight charge is divided among them. The computations necessary to determine each railroad's revenue are cumbersome and expensive. Hence, the Chesapeake and Ohio experimented to determine if the division of total revenue among several railroads could be made accurately on the basis of a sample and at a substantial saving in clerical expense.

In one of these experiments, they studied the division of revenue for all less-than-carload freight shipments traveling over the Pere Marquette district of the Chesapeake and Ohio and another railroad (to be called A for confidentiality), during a six-month period. The waybills of these shipments constituted the population under examination. A waybill, a document issued with every shipment of freight, gives details about the goods, route, and charges. From it, the amounts due each railroad can be computed. The total number of waybills in the population was known, as well as the total freight revenue accounted for by the population of waybills. The problem was to determine how much of this total revenue belonged to the Chesapeake and Ohio.

For the six-month period under study, there were nearly 23,000 waybills in the population. Since the amounts of the freight charges on these waybills vary greatly (some freight charges were as low as $2.00, others as high as $200), it was decided to use a sampling procedure which is called *stratified sampling*. With this procedure, the waybills in the population are first divided into relatively homogeneous subgroups called strata. The subgroups in this instance were set up according to the amount of the total freight charge, since the amount due the Chesapeake and Ohio on a waybill tended to be related to the total amount of the waybill. That is, the larger the total amount of a waybill, the larger tended to be the amount due the Chesapeake and Ohio on that waybill. Specifically, the strata were as follows:

Stratum	Waybills with Charges Between
1	$ 0 and $ 5.00
2	$ 5.01 and $10.00
3	$10.01 and $20.00
4	$20.01 and $40.00
5	$40.01 and over

Note that each stratum contains waybills with total freight charges of roughly the same order of magnitude. Because of the general tendency by which the amount due the Chesapeake and Ohio varied with the total freight charge on a waybill, each stratum is relatively more homogeneous with respect to the amount of freight charges due the Chesapeake and Ohio. At the same time, the strata differ substantially from one another.

Statistical theory then was used to decide how large a sample from each stratum must be selected so that the amount of the revenue due the Chesapeake and Ohio could be estimated with required precision from as small a sample as possible. One piece of information needed for this determination is the number of waybills in each stratum. The sampling rates decided on for the strata were:

Stratum	Proportion to Be Sampled
1	1%
2	10%
3	20%
4	50%
5	100%

Note that this theory led to larger sampling rates in the strata containing wider ranges of freight charges and smaller sampling rates in the strata containing narrow ranges of freight charges. To understand this, consider stratum 1, containing waybills with charges between $0 and $5.00. Here the variation between the waybill amounts is small, and therefore a small sample will provide adequate information about the amounts of all of the waybills in that stratum. On the other hand, stratum 4, containing waybills with charges between $20.01 and $40.00, has much greater variation. A larger sample is therefore required in this stratum to obtain adequate information about the amounts of all waybills in that stratum. In an unreal extreme situation with all the waybills in a stratum having the same amount due the Chesapeake and Ohio, a sample of just one waybill would provide all the information about the waybill amounts in that stratum.

Once the sample sizes were determined, the next problem was to select the samples from each stratum. For a statistician to be able to evaluate the precision of the sample results, that is, how close the sample results are likely to be to the relevant population characteristic, the sample must be selected according to a known probability mechanism. Various methods of probability sampling are available. One is called *simple random sampling.* This type of sample may be directly selected by use of a table of random numbers, a portion of which is illustrated in Table 1. How might Table 1 be used to select a simple random sample from each of the strata? Consider stratum 1 and suppose it contains 9000 waybills, which we label with four-digit numbers from 0001 to 9000. We want to obtain four-digit numbers from the table; we might start in the upper left-hand corner, using columns 1 through 4. The first number obtained is 1328. Our first sample waybill is then the one numbered 1328. Our second sample waybill would be 2122. The next number from the table of random digits is 9905, but there are only 9000 waybills in the stratum, so we pass over this number and go on to the next one, which is 0019. This process would be continued until the required sample of 90 (1% of 9000) has been obtained. The digits in the table of random digits are generated so that all numbers (four-digit numbers in our case) are equally likely.

Another method of selecting waybills from each stratum is called *serial number sampling,* and this was the method actually used by the Chesapeake and Ohio Railroad. In this procedure, the sample within each stratum is selected according to certain digits in the serial number of the waybill. In this particular case, the last two digits in the serial number of the waybill were used. To explain how these last two digits are used to select the sample, consider stratum 1, with its 1% sample. The number of possible pairs of digits appearing in the last two places of the serial number (00, 01, 02, . . . , 99) is 100. If one of these pairs is chosen from a table of random digits and all waybills with these last two digits in their serial number selected for the sample, it will be found that about 1% of the stratum is included in the sample. For stratum 1, the random number turned out to be 02. Therefore, all waybills with freight charges of $5 or less whose last two serial number digits are 02 were selected for the sample. The serial number digits used for the other strata, as well as the sampling rates, were as follows:

Stratum	Proportion to Be Sampled	Waybills with Numbers Ending in:
1	1%	02
2	10%	2
3	20%	2 or 4
4	50%	00 to 49
5	100%	00 to 99

TABLE 1. Portion of a Table of Random Digits

LINE	(1)–(5)	(6)–(10)	(11)–(15)	(16)–(20)	(21)–(25)	(26)–(30)	(31)–(35)
101	13284	16834	74151	92027	24670	36665	00770
102	21224	00370	30420	03883	94648	89428	41583
103	99052	47887	81085	64933	66279	80432	65793
104	00199	50993	98603	38452	87890	94624	69721
105	60578	06483	28733	37867	07936	98710	98539
106	91240	18312	17441	01929	18163	69201	31211
107	97458	14229	12063	59611	32249	90466	33216
108	35249	38646	34475	72417	60514	69257	12489
109	38980	46600	11759	11900	46743	27860	77940
110	10750	52745	38749	87365	58959	53731	89295
111	36247	27850	73958	20673	37800	63835	71051
112	70994	66986	99744	72438	01174	42159	11392
113	99638	94702	11463	18148	81386	80431	90628
114	72055	15774	43857	99805	10419	76939	25993
115	24038	65541	85788	55835	38835	59399	13790
116	74976	14631	35908	28221	39470	91548	12854
117	35553	71628	70189	26436	63407	91178	90348
118	35676	12797	51434	82976	42010	26344	92920
119	74815	67523	72985	23183	02446	63594	98924
120	45246	88048	65173	50989	91060	89894	36036

Source: *Table of 105,000 Random Decimal Digits.* Interstate Commerce Commission, Bureau of Transport Economics and Statistics, May 1949.

Since the serial numbers appear prominently on the waybills, this procedure is a simple one for selecting the sample. Furthermore, in this case, experience indicates that it provides essentially the equivalent of a simple random sample from each stratum.

Altogether, 2072 waybills out of 22,984 in the population (9%) were chosen according to this procedure. For each waybill in the sample, the amount of freight revenue due the Chesapeake and Ohio was calculated. For each stratum, the total amount due for the population of waybills was then estimated, and these estimates were added to obtain an estimate of the total amount of freight revenue due the Chesapeake and Ohio on the almost 23,000 waybills in the population. Because this was an experiment, a complete examination of the population was also made, so that the sample result could be compared with the result obtained from an analysis of all waybills in the population. The findings were:

Total amount due Chesapeake and Ohio on basis of complete
  examination of population $64,651
Total amount due Chesapeake and Ohio on basis of sample   64,568
Difference $ 83

Thus, a sample of only about 9% of the waybills provided an estimate of the total revenue due the Chesapeake and Ohio within $83 of the figure obtained from a complete examination of all waybills. Because the sample cost no more than $1000, while the complete examination cost about $5000, the advantages of sampling are apparent. It just does not make sense to spend $4000 to catch an error of $83. Furthermore, although the error in this instance was against the Chesapeake and Ohio, the next time it may be against the other railroad, so that the long run cumulative error is relatively even smaller.

OTHER RAILROAD AND AIRLINE SAMPLING STUDIES

The Chesapeake and Ohio conducted the same type of test for interline passenger receipts. They studied tickets sold during a five-month period to commercial passengers traveling only on the Chesapeake district of the Chesapeake and Ohio and on two other railroads, A and B. The findings are shown in Table 2. Again, these results dramatically demonstrate the ability of relatively small samples to provide precise estimates of the total revenue due the Chesapeake and Ohio.

Airlines also have used statistical sampling to estimate their share of the revenue on tickets for passengers traveling on two or more airlines. Three airlines tested statistical sampling during a four-month period. In that time,

TABLE 2.   Results of Passenger Ticket Study

	100% EXAMINATION	5% SAMPLE	DIFFERENCE	
			Dollars	Percent
Railroad A				
Total number of tickets	14,109			
Total revenue	$325,600			
Chesapeake and Ohio portion of total revenue	$212,164	$212,063	−$101	−0.05%
Railroad B				
Total number of tickets	7,652			
Total revenue	$128,503			
Chesapeake and Ohio portion of total revenue	$ 79,710	$ 80,057	+$347	+0.44%

the degree of error in the sample estimate based on relatively small samples did not exceed 0.07% (that is, $700 in $1,000,000) for any of the three airlines. On the basis of this experiment, wider use of statistical sampling in settling interline accounts has been made. At one point in time, the sample consisted of about 12% of the interline tickets and the cumulative sampling error was running at less than 0.1%. The clerical savings were estimated to be near $75,000 annually for some of the larger carriers and more than $500,000 for the industry.

Statistical sampling in accounting and auditing has also been used to estimate the value of inventory on hand, the proportion of accounts receivable balances that are 60 days old or older, and the proportion of accounts receivable balances that are acknowledged as correct by the customer. In each instance, it has been demonstrated that a relatively small sample, carefully drawn and examined, can furnish results that are of high quality and at a much lower cost than with a complete examination.

To summarize, statistical sampling consists of the selection of a number of items from a population, with the selection done in such a way that every possible sample from the population has a known probability of being chosen. Frequently, a statistical sample can provide reliable information at much lower cost than a complete examination. Also, a statistical sample often can provide more timely data than a complete enumeration of the population because fewer data have to be collected and smaller amounts of data need to be processed. Finally, a statistical sample can sometimes provide more accurate information than a complete enumeration when quality control over the data collection can be carried on more effectively on a small scale.

PROBLEMS

1. Explain the difference between simple random sampling and serial random sampling.

2. Suppose a university administrator is considering ordering some new desks for classrooms. He needs to find out how many desks already in use need to be replaced.

   (a) Should he consider using sampling methods in this situation? What are the arguments for sampling? Against?

   (b) If he did use sampling methods, what would the *population* be?

3. Why was stratified sampling used in the C & O freight study?

4. Refer to Table 2. Add the Railroad A and Railroad B ticket revenues, and find the difference in percent between a five percent sample and a one hundred percent examination.

5. In the C & O freight study, how large a percentage of the total amount due C & O was the result of error due to sampling?

6. An army psychologist wants to take a sample of 1000 enlisted men to find out their attitudes towards the "new Army." He obtains a list of 10,000 enlisted men arranged by squads; each squad has ten men, with a sergeant heading the list, then a corporal, followed by eight privates.

   (a) Would you recommend that the psychologist use serial number sampling (using the digits 0–9) to choose a sample of 1000 from this list of 10,000 men? Why?

   (b) If the psychologist used serial number sampling, what would be the chance of getting only sergeants in his sample? What if he used simple random sampling?

   (c) Answer the questions in (a) and (b) if the psychologist used a list which placed the 100 enlisted men in alphabetical order.

7. Suppose C & O and railroad A sampled tickets to determine their share of revenue from interline passenger receipts every month for a year. For how many months would you expect the sampling error to favor C & O?

8. Use Table 1 to draw a random sample of twenty-five two-digit numbers. How many are even? How many have both digits even? Do the same for a random sample of 100 two-digit numbers. Compare your answers to those obtained by the other students in your class. What conclusions can you draw?

REFERENCES

"Can Scientific Sampling Techniques Be Used in Railroad Accounting?" *Railway Age,* June 9, 1952, pp. 61–64.

R. M. Cyert and H. Justin Davidson. 1962. *Statistical Sampling for Accounting Information.* Englewood Cliffs, N. J.: Prentice-Hall.

W. Edwards Deming. 1960. *Sample Design in Business Research.* New York: Wiley.

Henry P. Hill, Joseph L. Roth, and Herbert Arkin. 1962. *Sampling in Auditing.* New York: Ronald.

Elbert T. Magruder. 1955. *Some Sampling Applications in the Chesapeake and Potomac Telephone Companies.* Chesapeake and Potomac Telephone Companies.

John B. O'Hara and Richard C. Clelland. 1964. *Effective Use of Statistics in Accounting and Business.* New York: Holt.

Morris J. Slonim. 1960. *Sampling in a Nutshell.* New York: Simon and Schuster.

Robert M. Trueblood and Richard M. Cyert. 1957. *Sampling Techniques in Accounting.* Englewood Cliffs, N. J.: Prentice-Hall.

Lawrence L. Vance and John Neter. 1956. *Statistical Sampling for Auditors and Accountants.* New York: Wiley.

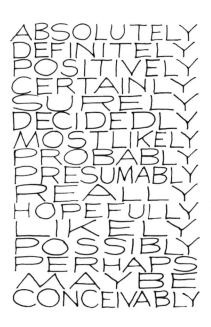

ABSOLUTELY
DEFINITELY
POSITIVELY
CERTAINLY
SURELY
DECIDEDLY
MOSTLIKELY
PROBABLY
PRESUMABLY
REALLY
HOPEFULLY
LIKELY
POSSIBLY
PERHAPS
MAYBE
CONCEIVABLY

THE USE OF SUBJECTIVE PROBABILITY METHODS IN ESTIMATING DEMAND

Hanns Schwarz   *Vice President, Daniel Yankelovich, Inc.*

THE CONTRIBUTIONS of statistics to sampling and analysis of data in marketing research are widespread and liberally documented. For example, we generate facts about the behavior and attitudes of very large populations from studies of relatively small (and inexpensively gathered) samples of respondents. The methods for selecting such small groups to represent large populations come from statistical sampling theory. Such sample surveys are now common practice and indispensable in research. Methods of analyzing data emanating from surveys (as well as from other sources) are also research tools provided by statistics.

Moreover, new statistical applications are being developed continually. Some of these relate neither to survey sampling nor to analysis of data, but to the basic questioning process itself—the essential core of marketing research. In 1971, these applications are not yet in widespread use nor are they well

documented. This paper describes one technique which has been used successfully in recent marketing research: the use of subjective probability methods
in predicting demand.

THE NEED TO ESTIMATE DEMAND BY SURVEY

One important function of commercial research is to make estimates of consumer demand for a product or service. On occasion, such estimates are
required for existing products, although forecasting methods based on past
sales data are then usually entirely adequate. When new or modified products
or services are involved, however, other techniques are called for. In some
cases, test markets are used and sales results in these few markets become
the basis for national estimates. Experimental variations of market testing
using simulated retail outlets are used in instances in which information is
required prior to a commitment to large-scale production, in which a product
is not far enough along in production to supply one or more entire test markets,
in which secrecy is considered desirable, in which there is not enough time
for a test market, or in which the test market method has not proved to
be particularly useful in the past.

In many instances neither a test market nor an experimental simulation
can be used to predict demand, notably in the case of large, durable products
such as cars or major appliances, new services not as yet adapted to the
consumer market, or new products only in the concept stage that can neither
be sold nor shown except, perhaps, in the form of pictures. In such instances,
a research survey of the likelihood of purchase is needed.

THE TROUBLE WITH ESTIMATING DEMAND BY SURVEY

Studies of the likelihood of purchase are probably as old as research itself,
and over the years, they have been tried with many variations and with various
degrees of technical expertise. But the issue almost always boils down to
a question such as "How likely are you to buy_____?" All too often, results
of such studies have been impressively wrong, usually ending in estimates
of buying intention substantially higher than subsequent levels of actual purchases. In certain cases—notably in attempts to assess interest in a new type
of product that does not yet exist—there are instances in which 10% or
less of those who said they would buy actually did so when the product
reached the market. Of course, the same problem occurs in the case of existing products, too. Research literature abounds with examples of gross discrepancies between stated purchase intentions and subsequent purchasing behavior. A recent example from our own work concerns automobile buying.
Of a total of 72 persons who reported that they planned to buy a car within

the next six months, only 33 (46%) actually had done so when they were reinterviewed after six months.

The reasons for such overstatement by survey respondents have been explored and discussed many times. The desire of survey respondents to be agreeable—to say "Yes"—particularly because it costs them nothing to do so, has often been cited. Moreover, there is an ego-enhancing aspect to reporting that the purchase of a new car, for example, is imminent. It has been generally agreed that this overstatement phenomenon is not due to conscious falsification but to a failure to consider seriously all of the factors which will play a part when the time of actual purchase comes. In the absence of such serious consideration, it is easiest to say "Yes."

The technique discussed in this paper is based on the experience that most survey respondents are not capable of giving considered answers to direct questions on purchase likelihood. The technique is based on a thorough study and understanding of the particular purchasing process involved, the identification of key factors which are likely to affect this purchase and a quantification of how and to what degree these factors are operating for each survey respondent to affect his likelihood of actual purchase. The method may be termed "a systematic, subjective computation of purchase probability," and it entails the use of subjective probabilities that differ somewhat from classical statistical probabilities. The latter derive either from the full understanding of the mechanics of a process (e.g., dice shooting, coin tossing) or from extensive empirical observation so that they are precise or very nearly precise probabilities. Subjective probabilities on the other hand are not nearly so precise. They are essentially educated guesses derived from judgment, experience, and so on, and generally are used under conditions of uncertainty which very frequently characterize business decisions.

THE USE OF PROBABILITIES IN ESTIMATING DEMAND

In order to demonstrate most clearly how the procedure works, a hypothetical study involving an estimate of consumer demand for a new, atomic-powered cabin cruiser is used for illustrative purposes. While this is a greatly simplified example, it will serve to show how this statistical technique could be used in a variety of subject areas. The essential details of this hypothetical study are that it consists of personal interviews with 200 men who report that they expect to buy a cabin cruiser within the next year. This sample of men is selected through telephone screening to be representative of those "in the market" for a cabin cruiser. These men are shown pictures of the new cruiser as well as a detailed written description and are questioned as to their likelihood of buying it. In addition, answers are solicited to questions relating to the various factors that will be used later to assess overall purchase probability.

TABLE 1.   Likelihood of Purchasing Cabin Cruiser
as Expressed by Respondents

	NUMBER	PERCENT
Total Respondents	**200**	**100**
Definitely intend to buy	10	5
More likely than not to buy	70	35
Not likely to buy	120	60

The results of the key purchase likelihood question are given in Table 1. Forty percent of the respondents have expressed at least a moderate level of interest in buying the cabin cruiser. Yet it is unreasonable to assume that all of these, in fact, will purchase one. In other words, not every person who has reacted positively has a probability of 1.0 of purchasing it. At this point, the process of making a more realistic assessment of the purchase probability of each respondent starts.

To begin with, the assumption is made that those who have expressed a definite purchase intention are more likely to become actual purchasers than those who have merely admitted to the likelihood of purchase. Therefore, the latter group will be "weighted down" to reflect this lower purchase probability. There is no sure way to determine precisely what weight to apply in the absence of any hard data comparing expressed intention and subsequent purchase of cabin cruisers. Such data have been compiled for other products at other times, however, and can at least serve as a rough guide. For example, in the automobile study cited earlier, the probability of purchase for those who claimed to be planning a purchase was 0.46. Another guide comes from the experience of the researcher and his associates whom he may consult in order to arrive at a weighting scheme based on a "jury of informed opinion." In this hypothetical instance, the consensus is that respondents who have said they are "likely to buy" will receive a weight of 0.4, so that at this intermediate stage, the estimate of demand would be

$$\begin{aligned} &10 \text{ "definitely intend to buy"} \times \text{weight } 1.0 = 10 \\ \text{plus } &(70 \text{ "likely to buy"} \quad\quad \times \text{weight}) \, 0.4 = 28 \\ \text{divided by } &200 \quad\quad\quad\quad\quad\quad\quad\quad\quad\quad = 19\% \text{ purchasing.} \end{aligned}$$

The remainder of the process entails similar weighting of respondents based on various purchase-related factors, thereby further modifying purchase probability. We could argue that further modification may be unnecessary as respondents have considered these purchase factors in answering to the buying intention question. However, there is considerable opinion (Juster 1966) to the effect that responses of this type are, on the whole, not considered ones and that modifying factors do need to be taken into account.

TABLE 2. Weighting the Likelihood that Respondents Will Buy Cabin Cruisers by Time of Intended Purchase

EXPECT TO BUY	WEIGHT
Within 3 months	1.0
In 3 to 6 months	0.8
In 6 to 12 months	0.6

There is likely to be a considerable number of such factors, but for the sake of simplicity, only four will be considered here. Two types of factors exist, the first bearing upon the likelihood of respondent's purchase of *any* cruiser at all within the next year despite his declaration that he will (because this was a requirement for eligibility in this study) and the second bearing upon the likelihood that he will select the particular cruiser being studied.

Of those factors bearing on the purchase of any cruiser, the first concerns just how soon the purchase is expected to be made—within three months, in three to six months or in six to twelve months. Experience indicates that the closer the intended purchase, the more likely that the purchase, in fact, will be made; in other words, those who say they expect to buy within three months are more likely to make actual purchases than those who expect to buy in three to six months, who, in turn, are more likely to become actual purchasers than the six-to-twelve month group. After another round of discussions among several knowledgeable researchers, the weighting scheme shown in Table 2 is developed. In the automotive study cited previously, the data clearly showed that the discrepancy between stated intention and actual purchase was substantially less when the intention was to buy within three months of the time of interview than when the projected purchase data was four or more months away.

The second purchase-related factor deals with the attitudes of the wives of married respondents toward the impending purchase. In this hypothetical instance, the three types of responses coded in the study are shown in Table 3, along with the weights that again are decided by consensus among a group of researchers.

TABLE 3. Weighting the Likelihood that Respondents Will Buy Cabin Cruisers by Wives' Attitudes

RESPONSE	WEIGHT
Wife has no voice in decision	1.0
Wife has a voice and wants to buy a cruiser	1.0
Wife has a voice and prefers alternative purchase	0.6

TABLE 4.   Weighting the Likelihood that Respondents Will
Buy Cabin Cruisers by Concern over Atomic Power

RESPONSE	WEIGHT
Unconcerned over use of atomic power	1.0
Evidences concern over atomic power	0.4

The third purchase factor bears on the likelihood of purchasing the particular cruiser being studied and deals with the degree of apprehension on the respondent's part about the fact that this is an atomic powered cruiser and might therefore be dangerous in the event of improper shielding of the power source.   The weights agreed upon after discussion are shown in Table 4.

The final purchase factor that also bears on the likelihood of selection of the particular cruiser studied concerns the degree to which it fits in with individual price requirements.   Those who are planning to spend either more or less than the expected retail price of the cabin cruiser are weighted down as in Table 5.

Using the various weights which have been described, each respondent's probability of purchase is assessed.   To illustrate the process:

(1) Respondent A, after seeing a picture of the cabin cruiser and reading a description, reports that he "definitely intends to buy": weight = 1.0.

(2) He claims he will make the purchase within three to six months: weight = 0.8.

(3) He is married but reports that his wife will have no voice in the purchase decision: weight = 1.0.

(4) He reports no apprehension over the use of atomic energy as the power source for this cruiser: weight = 1.0.

(5) The amount he is planning to spend in his upcoming purchase is somewhat less than the expected price of the cruiser: weight = 0.3.

TABLE 5.   Weighting   the   Likelihood
that Respondents Will Buy Cabin
Cruisers by Consistency with Price
Requirements

RESPONSE	WEIGHT
Price in line with plans	1.0
Price higher than planned	0.3
Price lower than planned	0.8

These weights provide the basis for estimating overall purchase probability. Each weight itself is a limited sort of purchase probability—what is called a *marginal probability*. To illustrate how several marginal probabilities are combined to form an overall probability, consider the following illustration.

> The probability of obtaining "heads" in a simple coin toss is 0.5, a marginal probability. The probability of tossing two consecutive heads is 0.25, the product of the marginal probabilities for tossing "heads" on the first toss and for tossing "heads" on the second toss $(0.5) \times (0.5) = 0.25$.

Thus, respondent A's overall purchase probability is simply the product of the various weights (or marginal probabilities): $(1.0) \times (0.8) \times (1.0) \times (1.0) \times (0.3) = 0.24$. The sum of these purchase probabilities across all respondents represents the estimate of demand for the new cabin cruiser. In all, 80 respondents had originally reported that they "definitely intend to buy" or are "likely to buy." If one were to take these responses at face value, he might conclude that as many as 40% (80 of 200) of those in the market may buy this cruiser, though one would be hard pressed to find a modern researcher who would be willing to make so liberal an interpretation of these data. Implicit in this type of estimate is that each respondent who reacts in a positive way is treated as having a purchase probability of 1.0. A more stringent estimating method reduces this purchase probability to 0.4 for those who are only "likely to buy," but leaves the probability at 1.0 for those who report that they will "definitely buy." This, in fact, is how the intermediate estimate of 19% cited earlier was obtained.

Using the system of weighting for various purchase factors brings the estimate lower still; for no one receives a purchase probability of 1.0 without specifically "earning it." In other words, even respondents who will "definitely buy" must give all the "right answers" with respect to the four purchase factors in this illustration before they receive a purchase probability of 1.0. Thus, it would not be at all unlikely that the sum of probabilities for the 80 respondents who reacted positively to the cruiser might be in the area of 5.0 leading to an estimate of demand of 2.5% (5/200), a far cry from 40% and 19%.

In using this subjective probability technique, we must observe one caution. We must overcome the temptation to include an overlong list of factors because if one keeps on multiplying probabilities for too long, the results will come dangerously close to 0. Thus, the factors chosen should be only those that are important purchase influences. In any given case, there are many marginal factors which have only small effects. The analyst must exercise good judgment in limiting the list to the key factors.

Estimates developed through this technique have been accurate. In the study of automobile-buying intentions, for example, the data were highly predictive of actual consumer demand six months later. Although the rule of confidentiality of client data precludes disclosure of specific figures, predicted total sales of the make of automobile as well as various component models were very close to subsequent actual demand.

A PRACTICAL RATIONALE

The technique that has been described here is very unlike the exacting and precise statistical procedures used in sampling and data analysis, described elsewhere in this volume. The development of the weighting schemes appears to be a rather subjective process that is seemingly defenseless against the question: "Why did you choose to use a weight of 0.8 and how can you prove it is better than a weight of 0.6?" It is *not* possible to prove in advance that 0.8 is a better weight than 0.6. However, this lack of certainty should not be permitted to create a "hang-up" to eliminate the use of weights because the means of arriving at them is less than precise. It seems quite clear that the use of weights that are arrived at intelligently is better than having to assume that the person who says "Yes, I will buy," in fact, will do so. In other words, it makes sense to assume that, all other things being equal, the man who is concerned over radiation is less likely to buy an atomic-powered cruiser than one who is not. The method described here makes use of this knowledge on the theory that it would be less accurate to ignore it.

In other respects, too, the method described here is less precise than classical statistical theory would dictate; for example, the assumption of the independence of the various weights is made. Despite some of these loose ends in the theoretical framework of the technique, however, we see only one practical limitation on its usefulness: the skill of the practitioner. Successful application depends on the ability to understand the essential workings of the particular purchasing process being studied and the pinpointing of those key factors that will affect purchasing behavior. As we gain more experience, it is not unlikely that some standardization of weights according to product class may be achieved (e.g., for cars, major appliances, minor appliances) thereby increasing the general usefulness of the method.

On a pragmatic basis, the technique has proved to be a valuable one. It gives good, realistic results and appears to offer a viable solution to the problem of determining purchase likelihood via questionnaire.

PROBLEMS

1. Respondent B after reading about the cruiser and seeing a picture of it is interviewed by the research group. He definitely plans to buy within three months. His wife, however, has a voice in the decision and prefers another type of cruiser. He was planning to spend more than the price of the cruiser. The use of atomic power for propulsion does not bother him.

Use tables two through five to find his marginal probability.

2. Who decides what weights to assign to the purchase factors?

3. Give one reason in favor of subjective probability methods and one opposed to them.

4. Each respondent could be asked to select a weight for himself. For example, a respondent giving a weight of 0.4 would believe that he has a forty percent chance of making a purchase. Comment on this possible method.

5. At the beginning of the school year, a high school guidance counsellor wants to find out how many students in the senior class will go to college next year. 1000 say they plan to go to college, but experience tells him this figure is too high. He decides that two factors are important in shaping the students' final decision: grades and professional plans. He devises a questionnaire and assigns weights as follows.

Group	Grade Point Average	Weight	Group	Professional Plans	Weight
I	3.5–4.0	1	A	"Plan to pursue a definite profession which interests me"	1
II	3.0–3.5	.8	B	"Not sure—will decide by taking courses which interest me"	.8
III	2.0–3.0	.5	C	"Not sure—want to make money somehow"	.4

He submits the questionnaire to the students who say they plan to go to college. He obtains the following result:

PLANS \ GPA	I	II	III
A	150	100	50
B	75	150	175
C	25	100	175

How many seniors does the guidance counsellor estimate will attend college next year?

6. A clothes designer is bringing out a new line of "leisure clothes." He estimates that, of those who claim a "definite interest" in buying these clothes, only 50% of those disagreeing with the statement "the clothes he wears always express a man's *real* personality" will actually purchase; while only 60% of those who say they must wear conventional business suits with ties for work will buy his clothes. The designer figures that 50% × 60% = 30% of those who declare a definite interest in buying, disagree with the statement about clothes and personality, and must wear conventional clothes to work will actually buy his clothes. Is the designer's reasoning correct? Why?

REFERENCE

F. Thomas Juster. 1966. *Consumer Buying Intentions and Purchase Probability.* National Bureau of Economic Research.

PRELIMINARY EVALUATION OF A
NEW FOOD PRODUCT

Elisabeth Street   *General Foods Corporation*

Mavis B. Carroll   *General Foods Corporation*

MANY AMERICANS like to have at hand an easy-to-prepare, nutritious, on-the-run meal. Our company has been developing such a product, called H.

Development of such a product calls for a thorough evaluation. In this essay, we limit ourselves to two aspects of the evaluation, the protein content of H and its tastiness. We shall show how statistics was central in answering our questions.

The palatability of a product can be determined by having people taste the prepared product and evaluate its acceptability both overall and by specific attributes. The nutritive quality of a food product can be determined by feeding it to animals whose metabolic processes are very similar to ours. Both types of tests should be designed, that is, planned in detail in advance. This helps avoid a consistent error in one direction (bias) and ensures that the proper number of people and animals are included in each study to answer with a fair degree of confidence the questions being posed. Some of the steps in the design and analysis of such tests will be described here.

PROTEIN EVALUATION

The protein content of H, in one sense, was satisfactory, as calculations based on the constituents of H showed. We were not certain, however, about the efficiency of the protein in H under conditions of actual use, and in addition, we wanted to compare two forms of H, one solid and the other in liquid form.

A rat-feeding study can give practical support to the high protein claim and a way of comparing the two variants of H because the rats' responses would be affected by any interaction of the ingredients in the formula or by a shortage of an essential amino acid such that the protein would be less efficient. Neither of these conditions would be indicated by the paper calculation.

Previous experience had shown that 28-day feeding of 10 to 15 rats on a diet gives a fairly reliable estimate of the diet's protein efficiency. For this feeding study, as for all such studies, male rats, newly weaned, were used. Male rats grow faster than female rats, and while weanlings, they are in a period of maximum growth rate. Adult rats are not used, for their weights are stabilized, and it is the animals' weight gain that is of primary interest.

Besides comparing the two forms of H, the experiment also compared each with a casein control diet, which served two purposes. First, it is a standard diet to which many experimental diets are compared; second, because we have had much experience with the casein diet, it would provide a check on whether something was amiss with the batch of rats or with some other aspect of the study. Things sometimes go wrong in mysterious ways, and it is important to have some sort of check.

At the start of the experiment, the 30 animals used varied in weight from 50 to 63 grams. They were arranged in ascending order of weight, and from the three lightest ones, one was assigned to each of the three diets in the assay. Such a trio of approximately equal weight animals is called a *block*. The assignment of rats to diets within this block of three was by chance or at random; that is, random numbers determined which rat went on each diet. The second block of the next three lightest rats was assigned one to each diet in the same way. This process of randomized block assignment continued until all 30 rats set aside for this study were distributed among the three diets, 10 per diet.

Using blocks of rats of comparable body weight ensures that each diet has its proper share of light and heavy rats. Randomization helps balance other factors which may influence the outcome of the study. For example, the shipment of 30 rats probably included many who were littermates: they were all newly weaned male rats and so were all born at approximately the same time. The probability is high that the randomization spread the rats from one litter among the three diets. Of course, to be sure that a litter is equally divided among the three diets it would be necessary to identify the rats by litter, reduce a litter size to a multiple of three by random withdrawal of animals, and randomly assign the remaining rats equally to the three diets. This would be done if some inheritable trait might importantly affect the results of a study.

The rats were assigned to cages by random numbers because prior studies had shown that rats at the top of a rack of cages gain more weight than those at the bottom. This randomization guarded against the possibility that most of the rats on one diet were put in top cages while rats on another diet went into bottom cages.

During the H feeding study each rat was permitted to eat as much as he wished. His food consumption was measured, and his weight was checked weekly. The final weight gain was simply the difference between his final, 28-day body weight and his initial weight. Intermediate weights are used to check on any untoward event such as respiratory infection, since, in the rat, body weight is a sensitive indicator of general well-being. The total food consumed times the proportion that is protein—a value obtained by chemical analysis of samples from each diet—gives a rat's protein intake. For the H study, the results for these two measures are shown in Figure 1. Each point on the graph represents one rat. All three diets were made up to have about the same proportion of protein, 9%. The chemical analysis showed them to be close to that: 9.875, 9.875, and 9.50%. Therefore, the higher intake of protein for rats fed the H diets means they ate more food which may indicate that these diets were more palatable to the rats than the casein diet. However, experienced nutritionists claim that rats generally will eat heartily any balanced high protein food irrespective of texture and flavor. All that is known from the data is that the intake of H was substantially above the intake of casein. The other most obvious fact about the points in Figure 1 is that they tend to fall close to a slanted straight line drawn through the middle of them. However, closer inspection of Figure 1 with ruler in hand indicates that a freehand straight line through liquid H points lies slightly above a line through solid H points and has a steeper slope. A line through the casein points alone would lie below either of the H lines, if it were extended to the higher intakes. Thus some doubt was cast on the initial conclusion that all points fell on one line and thus on the hypothesis that any differences between the dietary weight gains could

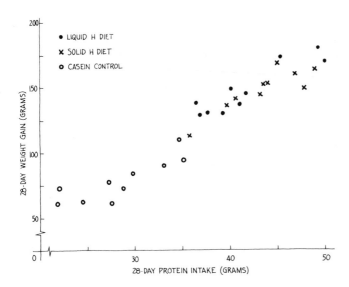

FIGURE 1

Relationship of 28-day protein intake and weight gain in young male rats

be attributed entirely to differences in protein intake; that is, we began to think that the difference in weight gain was not due simply to the difference in protein intake.

Using a computer program, we fitted a straight line to the points for each diet by the method of least squares; that is, the line was selected which minimized the sum of the squared vertical distances of the points from the line. These fitted lines, called *estimated regression lines,* are described by equations of the form $Y = \bar{Y} + b(X - \bar{X})$, where Y is the predicted weight gain and X the protein intake. $\bar{X}$ *and* $\bar{Y}$ are the averages of the X's and Y's separately for each set of data, and each b represents the slope of that straight line, that is, the predicted increment in weight gain per unit of protein intake.

The three regression equations were:

Liquid H: $Y = 151 + 3.72(X - 41.7)$
Solid H: $Y = 150 + 3.66(X - 43.3)$
Casein: $Y = 79 + 2.91(X - 28.4).$

Inspection of the three lines—they are graphed in Figure 2—shows that the casein line is decidedly lower and slopes less than the two H lines. The two H lines were then compared, and we found that they did not differ more than one would expect by chance. Thus our suspicion was confirmed that the greater gains of the H fed rats relative to the gains for rats on

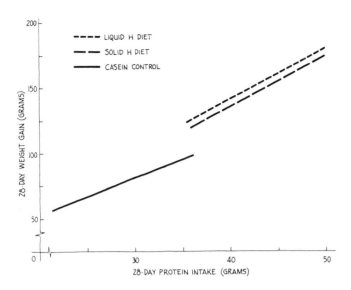

FIGURE 2

Estimated regression of 28-day weight gain on protein intake for young male rats

casein were not due simply to greater protein intake, because the slope of the H lines is higher.

To summarize the results of the protein evaluation: differences in protein intake between the casein fed rats and H fed rats did not account for all the differences in weight gain. For a given increase in protein intake the H diets resulted in a greater increase in weight gain than did the casein diet. There was little difference between solid and liquid H in terms of expected weight gain. If the addition of solidifying agents is detrimental it was not evident in this assay.

PALATABILITY EVALUATION

While the study to determine and compare the nutritive values of the two H variants was being run, a preliminary consumer test to determine the palatability of the two variants also was run. The aim was to compare the two H products and to see if they were as acceptable overall as a competitive product already on the market, designated C.

Previous experience had shown that having 50 people taste and evaluate a product under controlled conditions is adequate to reveal a major problem in acceptability, if one exists. Controlled conditions were obtained by bringing individuals into a central testing location and having trained personnel prepare, serve, and interview the 150 individuals required—50 to taste C and

PLEASE CHECK THE BOX UNDER THE PICTURE WHICH EXPRESSES HOW YOU FEEL
TOWARD THE PRODUCT WHICH YOU HAVE JUST TASTED

FIGURE 3

Pictorial taste-test ballot (the scores +3 to −3 were assigned the figures from left to right)

50 to taste each of the two variants of H. Because men and women might respond differently, the test specified that each product would be tasted by 25 males and 25 females. The tasters, who were paid for their help, were recruited from local churches or club groups. As they were recruited, individuals were assigned randomly to taste each of the three products until 25 of each sex had tasted and evaluated each of the three products.

After tasting the product, each person was asked to mark a ballot rating overall acceptability. Figure 3 shows the pictorial scale used to measure acceptability. Individuals seem to find it easier to express their feelings about a product using a pictorial scale rather than a scale using words such as "excellent," "very good," "good," "average," "poor," "very poor," and "terrible."

Each of the pictures is assigned a number (or score) in the sequence +3, +2, +1, 0, −1, −2, −3, with +3 being most acceptable.

For each of the three products the total number of votes for each of the pictures was tallied separately for males and females. The results, called frequency distributions, are shown in Table 1 with the pictures being replaced by the assigned number or score, and M being for males and F for females.

From the frequency distributions it is apparent that all those tasting the same product did not agree on its acceptability. It is difficult to look at the six distributions and decide whether any one group of 25 is scoring what they tasted as more or less acceptable than each of the other five groups scored what they tasted. To simplify the comparison among the six groups of 25 people an average score is obtained by multiplying each score by its frequency, summing the results, and dividing by 25.

We note that the differences in average scores are not large when male tasters are compared to female tasters. We note too that the averages for

TABLE 1. Frequency Distribution of Scores

| | PRODUCT TASTED | | | | | |
| | C | | Liquid H | | Solid H | |
SCORE	M	F	M	F	M	F
+3	1	0	4	3	2	3
+2	2	3	6	7	6	5
+1	7	8	9	7	10	9
0	8	9	5	6	6	7
−1	5	3	0	2	1	0
−2	2	1	1	0	0	1
−3	0	1	0	0	0	0
Total tasters	25	25	25	25	25	25
Average score	0.20	0.24	1.24	1.12	1.08	1.04
Variance	1.50	1.43	1.44	1.36	0.993	1.2

liquid H and solid H don't differ much, but all four of the H averages are well above the two C averages. The question to be answered is whether these differences in average scores are larger than can be expected by chance, considering the way people vary when they rate the same product. What is needed is a yardstick to permit us to say what difference between any two averages, each based on 25 tasters, is larger than can be expected by chance alone.

To arrive at such a yardstick we must first measure how variable people within each of the six groups are. This measure, called *variance,* is obtained by taking the difference of each person's score from the average, squaring the differences, summing the squares, and dividing by one less than the number of people. For each of the six groups the variance was computed; it is shown at the bottom of the table of frequency distributions. The variances for all six groups in this study were then averaged to give 1.32, a good measure of variability among males or females tasting the same product.

If individuals vary, so will the averages based on individuals. How much the averages vary will depend on the number of tasters: the larger the number the more representative and less variable the average will be. Because of this variability we can never be absolutely sure that two averages are different, but we must take some risk in drawing conclusions. So the best we can do is state the risk in setting up our yardstick. We chose to take a "1-in-20" chance of being wrong when we calculated the amount by which two averages had to differ in order for us to conclude they were different. Using the variance of the average of 25 scores and taking into account the risk gives 0.64 as our yardstick for comparing any two of the average scores.

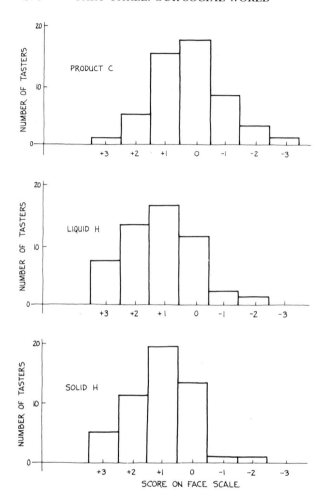

FIGURE 4

Combined male and female dis-
tributions of face-scale scores
by product tasted, 50 tasters
per product

This tells us that there is no evidence that males and females who rate the same product rate them differently because we have for the differences in averages the following:

C: $0.24 - 0.20 = 0.04$
Liquid H: $1.24 - 1.12 = 0.12$
Solid H: $1.08 - 1.04 = 0.04.$

We can now combine the distributions for males and females. The resulting distributions are shown in Figure 4. We note that the distribution for product C is shifted to the right, the less acceptable scores, as compared to the H results. There is considerable overlapping of the distributions representing the variation, so even on a pictorial basis we have problems interpreting the relative acceptability of the test products. We must resort to the average scores, which for the 50 individuals who tasted the same product are the following:

$$C = 0.22 \qquad \text{Liquid H} = 1.18 \qquad \text{Solid H} = 1.06.$$

Our yardstick for judging differences decreases, because these new average scores are based on 50 people. Our new yardstick, taking a 1-in-20 chance of being wrong, is 0.45.

It is apparent the difference between the acceptability of the two H products is not large enough to say one is more acceptable than the other. There is very little risk, however, in concluding that both H products are more acceptable than C, the competitive product.

Thus, through testing and with the application of statistical methodology these new products were shown to be palatable and to live up to the concept of a high-protein food. Two of the early criteria in the long process of introducing a new food product have been met.

PROBLEMS

1. For the protein evaluation, why were rats chosen as the experimental animals?

2. (a) Explain the idea of blocking.

   (b) Suppose the rats had not been assigned via blocking but instead were assigned in the following way: For each of the thirty rats a six-sided die is rolled. A 1 or a 2 means the casein diet for that rat, 3 or 4 liquid H, and 5 or 6 the solid H diet. Under this scheme how many rats would be assigned to each diet? Would it be possible that all rats are assigned to the casein diet?

3. In the protein evaluation experiment, what was the least protein intake for any rat on the casein diet? The most? Answer the same questions for the liquid H diet and the solid H diet.

4. What was the median weight gain of the 30 rats in the protein evaluation experiment?

5. Using the appropriate equation derived from the protein evaluation experiment, how much weight would you predict a rat would gain in 28 days if he consumed 45 grams of protein on the casein diet? State the assumptions that led you to this prediction.

6. Suppose you were setting up the protein evaluation experiment. Explain how you would use the random number table on page 88 to
   (a) assign rats in each block to the three different diets;
   (b) assign rats to the top or bottom cages.

7. (a) The yardstick for comparing average scores between males and females tasting a food in the palatibility evaluation was .64; for comparing average scores between two different food products, it was .45. Why are these yardsticks different?

   (b) Do you think you could ever achieve a yardstick smaller than .1? If your answer is yes, how might you do this?

8. Calculate the variance of the following scores for a product B.

SCORE	FREQUENCY OF SCORE
+3	2
+2	7
+1	7
0	7
−1	2
−2	0
−3	0

TOTAL TASTERS 25
AVERAGE SCORE 1

MAKING THINGS RIGHT

W. Edwards Deming   *Consultant in Statistical Surveys*

WHAT IS THE STATISTICAL CONTROL OF QUALITY?

The statistical control of quality is the use of statistical methods in all stages of production—in design of product, in tests of product in the laboratory, in tests in service, for specifications and tests of incoming materials and assemblies, and for achieving economies in production, maintenance, and replacement of machinery and equipment, economies in inventory of parts for repairs of machinery, even economies in inventory to meet predicted demand.

Inspection is a very important function in production. The effects of instruments, machines, and human observations jointly create figures that must be transcribed onto forms constructed for the purpose. Faults recorded in inspection may be inherent to the product, or they may be caused by faulty instruments or gauges, or even by poor measuring practice.

We must be content in this article to limit ourselves to a few simple examples of statistical control of quality drawn from the production line. In the first two examples the aim will be to detect the existence of special causes of trouble, for the operator to correct. In the third example the aim will be to measure the effects of common (environmental) causes of trouble, for management to correct. In the real world, we are always working on both kinds of causes. We hope the reader will see in the examples the distinction between special causes and common causes and how they affect the variability of the process or lead to other kinds of trouble.

EXAMPLE 1: FUDGING THE DATA

Figure 1 shows the distribution of diameters in centimeters, these being the results of the inspection of 500 steel rods. Such a graphic representation of a distribution is called a histogram. The lower specification limit (abbreviated **LSL**) of the diameter of these rods was 1 centimeter. Rods smaller than 1 cm. would be too loose in their bearings, and such rods would be thrown out (rejected) in a later operation, when they must be fitted to a hole. Rejection means loss of all the labor that was expended on the rod up to this point, as well as loss of material and of overhead expense.

The horizontal axis in Figure 1 shows the centers of intervals of measurements; for example, 0.998 stands for rods that measured between 0.9975 and 0.9985 cm. The vertical axis is labeled to show the number of rods that fell into an interval of 0.001 cm. on the horizontal axis. For example, about 30 rods were in the interval centered at 0.998. It appears from the distribution that $10 + 30 + 0 = 40$ rods failed because they were too small.

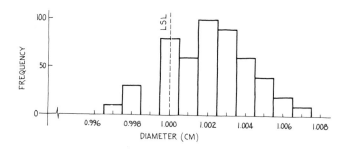

FIGURE 1

Distribution of measurements on the inside diameters of 500 steel rods. The chart detected the existence of a special cause of variation, a fault in recording results of inspection

A distribution is one of the most important statistical tools, when used with skill, yet it is extremely simple to construct and to understand.

Figure 1 is trying to tell us something. The peak at just 1 cm. with a gap at 0.999 seems strange. It looks as if the inspectors were passing parts that were barely below the lower specification, recording them in the interval centered at 1.000. When the inspectors were asked about this possibility, they readily admitted that they were passing parts that were barely defective. They were unaware of the importance of their job, and unaware of the trouble that an undersized diameter would cause later on.

This simple chart thus detected a special cause of trouble. The inspectors themselves could correct the fault. When the inspectors in the future recorded their results more faithfully, the gap at 0.999 filled up. The number of defective rods turned out to be much bigger, 105 in the next 500, instead of the false figure of $10 + 30 + 0 = 40$ in Figure 1.

The results of inspection, when corrected, led to recognition of a fundamental fault in production; the setting of the machine was wrong. It was producing an inordinate number of rods of diameter below the lower specification limit. When the setting was corrected and the inspection carried out properly, most of the trouble disappeared.

The upper specification limit had its problems also, but they were not so serious. A rod that is too large in diameter can be tooled off to fit. This is not the economic way to achieve the right dimension, but it is cheaper than to lose all the labor expended up to that time on the rod. The next problem was accordingly to increase uniformity and work on the correct centering of the average diameter, to reduce the number of defectives with wrong diameters.

EXAMPLE 2: DETECTING A TREND

The second example deals with a test of coil springs one after another as they come off the production line. These springs are used in cameras of a certain type. According to the specifications, the spring should lengthen by 0.001 cm. for each gram of pull. These springs are relatively expensive, and are supposedly made to exacting requirements. The length of any horizontal bar in the histogram at the right in Figure 2 shows how many springs the inspectors recorded with the elongation shown. We have turned it sidewise for convenience. This histogram represents measurements on 50 springs manufactured in succession. It will be noted that the distribution is symmetrical and is centered close to the specification; furthermore all 50 springs were within the upper and lower specification limits. One might be tempted to conclude from this histogram alone that the production of this spring presents no problems. However, another simple but powerful statistical tool, called a *run chart,* indicates trouble, as we now explain.

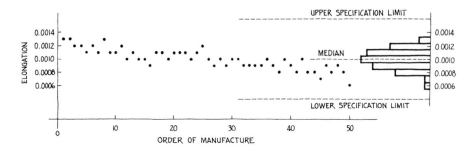

FIGURE 2

Run chart for 50 springs tested in order of manufacture. The chart shows a definite trend downward and thus reveals the existence of a special cause of variation, which it is important to correct. The frequency distribution alone could not detect this trouble

A run chart is merely a running record of the results of inspection. The horizontal scale shows the order of the item as produced, and the vertical axis shows the measurement for that item. In Figure 2, the elongations of the 50 successive springs are plotted on the vertical scale. A run chart has several simple uses. For example,

(1) A run of six or seven consecutive points lying all above or all below the median—the middle point in height—signifies with near certainty the existence of a special cause of variation, usually a trend.

(2) A run of six or seven points successively progressing upward or successively progressing downward has the same significance.

In no instance in Figure 2 is there a run up or a run down of length 6. It so happens that the median of the 50 points falls midway between the upper and lower specification limits. This would be good, but we note that the opening burst of points at the left of the figure has 10 points in succession above the median. Fifteen out of 18 points after point 29 fall below the median. These observations give a statistical foundation for the conclusion that, although the points vary up and down, there is a general drift downward. You may feel that your eye was good enough to detect this trend without knowing from theory that a run must have six or seven points above or below the median to detect with near certainty the existence of trouble, and in this example, you would be correct, but in more complicated examples such trends are often not detectable by eye.

Knowledge of what lengths of runs are required to indicate trouble is also valuable but secondary in problems of production. Indeed, it is an important statistical point that some of the most powerful statistical techniques are simple, as in our examples here. It was their widespread use, which began about 1942, that laid the foundation for the statistical control of quality, which of course has since grown into all phases of management. This movement led to the organization years ago of the American Society for Quality Control, over 23,000 strong in 1970.

In our camera-spring example, either the production process is in trouble or the apparatus used for testing is giving false readings. Correction is vital, whatever be the source of the trend. If it is the tension of the spring that is drifting downward (and not the testing apparatus), defective springs will be produced in the immediate future. If the source of the trend is faulty testing, then the tests are misleading, and may have been giving faulty reports on all the springs produced recently.

In this particular case, the trouble lay in a thermocouple that permitted the temperature to drift during the annealing of the springs. The process was headed for trouble. The simple run chart detected the trend before trouble occurred. The operator himself, seeing the trend, was able to head off trouble.

The reader may note that the histogram and the run chart in Figure 2 were plotted from the same data, yet they tell different stories. The histogram by itself gives no indication of anything wrong; it could have indicated unsatisfactory positioning. The run chart, however, leads us to suspect the existence of something wrong, a trend that, unless corrected, would soon lead to the production of defective springs.

It is interesting to note that if the points in Figure 2 had been plotted in random order instead of one after another in the order of production (1, 2, 3, and onward to 50), the run chart would have lost its power to detect a trend. Statisticians are thus not only concerned with figures, but with the relevant figures. In this instance, the order of production was relevant—very relevant—and was used to make the run chart. The histograms in Figures 1 and 2, on the other hand, do not make use of the order of production. They would remain unchanged, regardless of order: they depend only on the numbers recorded as results of inspection. The histogram in Figure 1 nevertheless did its work; it told us that something was wrong (namely, in the inspection itself). A run chart in connection with Figure 1 would not have added any relevant information. The histogram in Figure 2, however, was helpless to detect the existence of anything wrong. Judging by it alone, without the run chart, we could not have detected impending trouble.

EXAMPLE 3: MEASUREMENT OF COMMON
(ENVIRONMENTAL) CAUSES

The first two examples dealt with special causes, specific to a designated worker or to a machine or to a specific group of workers. Statistical techniques point to specific sources of trouble when the process is nonrandom. The same statistical methods also tell the worker to leave things alone, to avoid overadjusting when attempts at adjustments would be ineffective or cause even greater variation than now exists.

There is another kind of problem that faces the management of any concern. No matter how skilled the workers, and no matter how conscientious, there will be at least a bedrock minimum amount of trouble in production owing to common or environmental causes. All the workers in a section work under certain conditions fixed by the management, or one might say, by the environment, which only the management can alter. For example, all the workers use the same type of machine or instrument. They are all doing about the same thing, and are using the same raw materials (which might be semifinished assemblies). They must put up with the same amount of noise and smoke.

It used to be supposed by management that all troubles came from the workers: that if the workers would only carry out with care the prescribed motions (soldering a joint, placing a part, turning a screw), then the product would be right with no defectives. This kind of reasoning does not solve the problem. Alert management can look into the problem with "infrared vision," supplied by statistical techniques.

An example was a small factory that made men's shoes. The machinery that sews soles is expensive. Time that an operator spends rethreading the machine and adjusting the tension after a break in the thread is time lost. Minutes lost may add up to hours and even days in the course of a month. There is not only the loss of rent paid for the machine and wages of the operator, but loss of labor and materials, nonproductivity of floor space, light, and increases in general overhead expenses. In this factory, about 10% of the working time was being spent rethreading the machines and adjusting the tension. Management was rightly worried. The trouble became obvious with a bit of statistical thinking. Observations on the proportion of time lost by the individual workers provided data for a chart similar to Figure 3. This figure showed that all the operators were losing about the same amount of time rethreading their machines. In fact, the time lost per day per man showed a pattern of randomness. This uniformity across operators could only point to the environment. What was the trouble?

The trouble turned out to be the thread. The management of the company was trying to save money by buying second-grade thread that cost 10¢ per spool less than first-grade thread. Penny wise and pound foolish! The

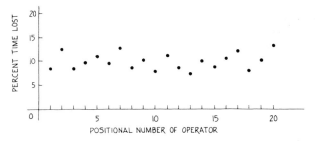

FIGURE 3
Time lost by each of 20 oper-
ators

savings on thread were being wiped out and overwhelmed many times over by troubles caused by poor thread.

A change to thread of first grade eliminated 90% of the time lost in rethreading the machines, with savings many times the added cost of better thread.

What is the distinction between this example and examples 1 and 2? In examples 1 and 2, the workers themselves could make the necessary changes, and they did. In this example, the operators were helpless. They could not put in an order for thread of first grade and scrap the bad thread. Their jobs were rigid—work with the materials and machines supplied by the management. They all worked with the same bad thread; that is, they all worked under a common cause of trouble. Management is responsible for common (environmental) causes; therefore only management could change the thread.

But how is management to know that there are common causes of trouble? The answer is simple: common causes are always present. Management needs a better answer, however; management needs a graphical or numerical measure of the magnitude of the trouble wrought by common causes. Without statistical techniques, management can have no accurate idea about the magnitude of the trouble being caused by conditions that only management can change.

Charts such as Figure 3 tell the management that there is a problem, that the time lost on rethreading will not go below 10% until management makes some fundamental change. The change in thread in example 3 was a fundamental change. What to change is not always as easy to perceive as it was in this example, however. Sometimes a series of experiments is required to discover the main causes of trouble. Statistically designed experiments have led to the identification of common causes such as raw materials not suited to the requirements, poor instruction and poor supervision (almost synonymous with unfortunate working relationships between foremen and production work-ers), and vibration.

Shift of management's emphasis from quantity to quality is one common environmental cause of trouble. The production workers continue to work with emphasis on quantity, not quality. Discussion of methods by which management may direct the shift from quantity to quality, however important, is beyond the scope of this essay.

PROBLEMS

1. In Example 1, why is having too large a diameter rod less serious than having too small a diameter?

2. While conducting a large-scale community health audit, a medical re-searcher developed a fast method of measuring blood pressure. A statisti-cian claimed to have discovered a flaw in the method just by looking at the following histogram of 1000 blood pressures (as measured by the new method).

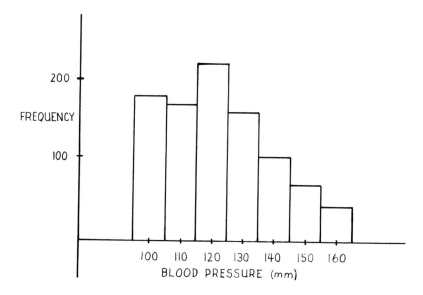

What seems to be wrong with the method?

3. Suppose the thermocouple problem had not existed for the camera spring example. Do you think the histogram of 50 spring pulls would be as spread out, more spread out or less spread out than the one in Figure 2?

4. In automobile manufacture, it has often been found that cars produced on Mondays and Fridays are more frequently defective than those on Tuesdays, Wednesdays and Thursdays. Why do you think this is so? If you were part of the management, how might you attempt to correct this?

5. When is it desirable to use a run chart for a manufacturing process? Why? Justify your answer.

6. Referring to Figure 3, which worker lost the most time rethreading? How much time? Which lost the least? How much time?

7. In a paper-cup factory, the cups are conveyed along a belt with four operating positions: the first operator cuts and glues the cups, the second coats them with plastic, the third counts and packages them, and the fourth places the packages in large cartons and seals the cartons. Management is concerned that too many cups are ruined during production. The quality control department studied the site at which cups were ruined during a shift (there were 8 belts in operation).

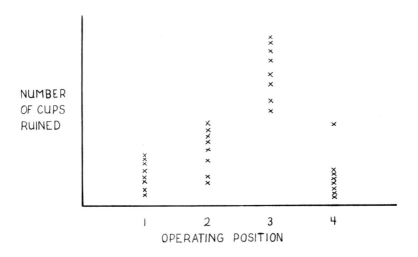

Is this a problem which requires a solution at the level of management?

8. In the 7 to 4 shift at the paper-cup factory, the operators of the plastic-coating machines get a break from 9–9:30 and a lunch break from 1–1:30. During these times their machines are idle. Management is worried because too many cups are being produced with an unacceptable amount of plastic coating, some with too much and some with too little. The quality control department prepared charts for each machine, all of which had the following appearance (each dot stands for a batch of cups).

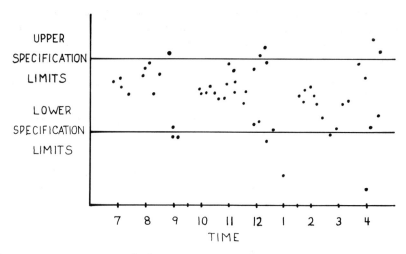

One manager suggested that the plant might save money by giving the operators two extra coffee breaks. Why?

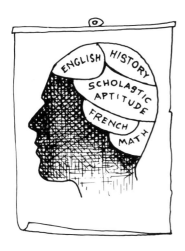

CALIBRATING COLLEGE BOARD SCORES*

William H. Angoff   *Educational Testing Service*

MOST DIRECTORS of testing programs are reluctant to use the same form (edition) at different administrations of a mental test and prefer instead to introduce different forms at different times. Their reluctance is understandable. Continued reuse of the same form encourages the collection of files of test questions and makes it possible for some students to acquaint themselves in advance with the questions that will later appear on the test. This practice not only invalidates their performance on the test, but it clearly works to the unfair disadvantage of those students who have not had access to the items. There are other problems too. For example, the measurement of growth, practice, fatigue, and so on, which require two or more administrations of a test, is often

* The present article draws heavily upon "How We Calibrate College Board Scores," which the author published in *The College Board Review,* No. 68, (Summer) 1968. We express our appreciation to the College Entrance Examinations Board for their permission to publish this revision.

rendered infeasible because the second measurement is contaminated by the student's recollection of the questions he was exposed to at the time of the first testing. Although many such problems can be circumvented by the practice of using different test forms at different administrations, this practice brings with it other problems that also require solution. In any testing program which makes use of a number of different forms of the same mental test there will inevitably be variations in difficulty from form to form. Therefore, if the scores of individuals who take the different forms are to be compared with one another for the evaluation of their relative abilities, it is necessary in the interests of equity to calibrate, or "equate," the scores on the different forms.

The process of equating is a statistical one which in our testing programs at Educational Testing Service (ETS) ultimately yields an equation for converting raw scores to scaled scores. Thus, except for random error, one can assume that a student who has earned a scaled score of, say, 563 on a particular test form would have earned that same scaled score whether he had taken a more difficult or less difficult form of the test than the one he actually took. That is, the score of 563 that is reported to him is *his* score and represents *his* level of performance (and, by inference, his level of ability) at the time that he took the test.

More generally speaking, the score of 563 represents a particular level of ability—let us say, verbal ability as measured by the verbal section of the Scholastic Aptitude Test (SAT)—in the same sense that the measurement 62°F represents a measurement (much more precise, of course) of temperature. It is a measurement which is independent of the type of Fahrenheit thermometer used, the time of year in which the measurement was taken, or the temperatures of other places in the world at the time. Similarly, the score of 563 is taken to represent the same level of ability, whoever earns it, in the same sense that 62°F represents the same degree of temperature, whether the object measured is air, water, coffee, or a martini.

THE MEANING OF SCORES

The questions are sometimes asked "What does the score signify? Is it high or is it low? How far is it above the average?" Once again, the temperature analogy is appropriate because the same questions may be asked with respect to the measurement of 62°F: "What does 62° mean? Is it warm or is it cool? How close is it to the average?" It needs little elaboration to say that these questions cannot be answered as they are stated. Sixty-two degrees represents a high temperature when compared with mean temperatures in New York City in January; it represents a low temperature when compared with mean temperatures in New York in July. At any given time it is warm compared with temperatures at the poles, but cool compared with most temperatures at the equator. It is cool for acceptable morning coffee and, most experts will agree, intolerably warm for a martini. Similarly, 563 is high or low, promising or disappointing, acceptable or unacceptable, depending on the choice of the particular reference group and on the standards set in the educational endeavor to which the student aspires.

The question that *should* be asked in interpreting scores is "How does this student compare with other students who are competing with him?" or, more fundamentally, "Is it possible to compare this student with other students, even though they may not all have been measured with precisely the same form of the test?"

Viewed slightly differently, the purpose of equating is to maintain a constant scale over time in the face of changing test forms and different kinds of students. Only if this purpose is achieved will it be possible to compare students tested today with students tested five years ago, to plot trends, and to draw conclusions regarding, for example, the effects of practice, the effects of growth, the effects of changing curricula, or the effects of changes in the composition of the student group over the course of time.

CONSTANCY OF SAT SCALED SCORES

One popular misconception is that SAT scaled scores, which are expressed on a 200–800 scale, are reported "on a curve," separately defined and separately determined at each administration of the test. This type of scaling is intentionally *not* carried out. For if it were, it would be impossible to compare students tested at different administrations, since the average scaled score for all administrations, by definition, would be the same. Moreover, under such a system a student's scaled score would depend in part on the caliber of the group with which he happened to take the test. It would be to his advantage, therefore, to take the test with a generally lower-scoring group, because he would stand relatively high in comparison with that group, and would consequently receive a higher score than if he were to take the test with a higher-scoring group. The scale that is used in the College Board program (among others) is a constant scale, defined once and *only* once, and perpetuated in that form from that time on.

To do this we construct for each form of a test an equation by which raw scores on a test form are converted to scaled scores. The methods that are followed in the equating process are so designed as to produce an equation that is *characteristic of the test form itself* and relatively unaffected by the nature of the group of individuals on whom the data were collected to form the basis of the equating process.

The equating of raw scores on two forms of a test requires an assessment of the relative difficulty of those forms. This requirement implies that ideally the same group of individuals should take both forms. But because the data for equating are drawn from operational administrations at which only one form is administered to each student, two separate groups of students must be chosen for analysis, one group taking one of the two forms and the other group taking the second. However, these two groups are likely to be different, with respect to both average ability and dispersion (spread) of ability. Therefore, any evaluation of the relative difficulties of the forms on the basis of a direct examination of the data for these two groups could well be misleading and biased. Such an evaluation could easily result in a conversion equation for Form X that is contaminated by the characteristics of the groups rather than one that is based solely on the characteristics of the test forms. For example, if the group taking Form X is brighter, we might erroneously decide that Form X is easier. Some means, therefore, is needed to adjust for differences between the two groups.

The device used for making these adjustments is a short "equating" test administered to both groups, A and B, at the time that they take the regular operational test. Sometimes the equating test is a separately timed test administered during the course of the testing session; sometimes it is not a separate test at all, but instead, a collection of questions interspersed throughout the operational test. This collection of questions, nevertheless, is treated statistically as though it were a separate test. At the time that the equating of the operational forms (Forms X and Y) is carried out, scores are derived for the two groups on *both* the equating test *and* the operational forms that were administered to them. Appropriate formulas are then applied to the statistics observed for the two groups to yield estimates of the behavior of the two forms as though they had been administered to the same group.

If the equating test is to be used as a basis for comparing the two groups and making adjustments for differences between them, then it must represent precisely the same test for both groups. The restrictions are easy to satisfy when the equating test is separately timed; in most instances a separately timed equating test and its directions need only to be reprinted.

The restrictions are not so easily satisfied, however, when the equating test is a collection of individual questions interspersed throughout the test. Here special care must be taken to avoid differential contextual effects. For example, questions such as those in reading comprehension or data interpretation sections (questions that pertain to a single passage or to a single graph or set of numerical data) should ordinarily be used as a block because there is a real possibility that they will be interdependent. If they are, any change in the composition of the block could well disturb the meaning of the individual questions. Matching questions (e.g., questions that call for matching people with events, works, philosophies) must be taken as a group. Care should also be taken to put the equating questions in one form in about the same relative position as in the other form. It is advisable to avoid using equating questions that appear near the end of the test, where failure to answer the questions may be caused as much by insufficient time to respond to them as by their inherent difficulty.

The conversion equation for a test form provides a description of its overall difficulty and discriminating power because it essentially "locates" or "places" the test form on the scaled-score scale.   For example, if a test form is relatively easy, then the scaled score corresponding to a given raw score will be lower than the corresponding scaled score on a more difficult form.   A score of 57, for example, on an easy test form might correspond to a scaled score of 590.   On a more difficult form the very same raw score of 57 might correspond to a scaled score of 640.   Obviously, the score of 640 can be earned on the easier form only by getting a raw score higher than 57.   This result is intuitively equitable, since the successful completion of 57 difficult questions *merits* a higher scaled score than the successful completion of 57 easy questions.

SCALING ACHIEVEMENT TEST SCORES

The foregoing discussion dealt with the problem of adjusting the scores on alternate and interchangeable forms of a test, so that a person of given ability will earn the same scaled score regardless of the form of the test he happens to have taken.   The type of solution that is given here is appropriate to the problem that we face when, for example, we wish to compare or merge data for individuals or groups of individuals who have taken different forms of the same test.   From a theoretical point of view, at least, this is a relatively simple problem.   The problem is different and far more complex, however, when we wish to compare two individuals (or groups) who have taken entirely different tests, for example, in chemistry and in French.

Of course, from a logical or educational point of view it makes no sense to compare the scores of two individuals who have taken entirely different tests.   But like it or not, such comparisons are inevitable, and logical or not, they are made.   Just as the grade-point average is a composite of marks earned by different students in different combinations of courses and just as it is used for comparing and evaluating the relative accomplishments of different individuals (e.g., for determining pass, fail, and honor status), scores on different Achievement Tests are also compared, merged, averaged, and ranked in many college admissions offices as they must be.   Recognizing as a fact of life that incomparable things *will* be compared, it behooves us to construct a system that, while it cannot be wholly satisfactory, will represent an improvement in the status quo and avoid some of its obvious imperfections.   The problem and its solution are as follows.

In the College Board Admissions Testing Program, all candidates typically take the Scholastic Aptitude Test.   No option is given.   In the Achievement Test series, however, there are options.   Candidates may take one, two, or three of the 16 Achievement Tests in the series; moreover, they may take any one, two, or three they feel inclined or prepared to take.   Therefore, while virtually all candidates take the SAT, the group of candidates taking any

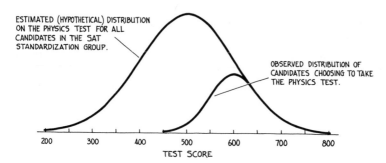

FIGURE 1

Hypothetical distribution of physics test scores for everyone and for students electing to take the physics test. Source: Angoff (1968)

one of the Achievement Tests is a self-selected group, each one different from the next in overall level and dispersion of ability. The problem arises when we wish to compare the scores of the groups of students who take different Achievement Tests.

In order to make these comparisons more equitably, the system of scores reported for the various Achievement Tests (which all use the same 200–800 score scale used for the SAT) is constructed to reflect the level and dispersion of abilities (as measured by the SAT) of the students who characteristically take each of the tests. The statistical process that results in this type of scale construction involves making an estimate of the performance on each Achievement Test of the entire original College Board standardization group for the SAT, assuming that they all had the appropriate instruction in the subject and all took the test. As a result, the scores for a test such as the physics test, which is typically taken by a relatively high-scoring segment of all the candidates who take the SAT, are automatically "placed" relatively high on the scale. Correspondingly, the scores for a test which is typically taken by a low-scoring segment of all the candidates who take the SAT are "placed" relatively low on the scale. A schematic diagram describing the distribution of scores on the physics test for *all* candidates taking the SAT and also for those choosing to take the physics test appears in Figure 1.

PURPOSE OF SCALING

The purpose of this type of scaling is to ensure that a candidate who chooses to compete with more able candidates is not put at a disadvantage, that is, that a candidate who is average in a highly selected group of candidates will earn a higher scaled score than a candidate who is average in a less able group. The intent is to make it impossible for a candidate to "beat the game" by taking advantage of the machinery of the testing program and making a *strategic* choice of the particular Achievement Test (or tests) that will yield the highest score(s) for him.   Moreover, since the scores on the various tests are scaled in accordance with the abilities of the candidates who typically choose to take them, it is possible for college admissions officers to take an average of the scores offered by each candidate with relative confidence that the average represents an equitable basis for comparing students who have taken different combinations of tests.

Given, then, that there are wide variations in the ability levels (and dispersions) of the groups of candidates taking the various tests, and given also that the scales for the tests reflect these variations, one would expect that the highest possible scaled scores on the various tests would vary substantially and systematically from test to test.   As a result, it would be possible for an able student of physics to earn a higher scaled score than an equally able student of a subject that is generally chosen by less able students, simply because the scale for the physics test permitted it.   In order, then, to equalize the opportunities for high scores among the different subgroups of candidates, the maximum score of 800 is imposed across the board for all tests, and the test specifications for each of the Achievement Tests are so written as to ensure, as nearly as possible, that a score of 800 may be achieved on every test and on every form.   Similarly, the minimum score of 200 is imposed across the board for all tests; any raw score that would ordinarily "scale out" below 200 is reported as 200.   However, because precise discriminations in the vicinity of 200 are not as often necessary on the Achievement Tests, there is no corresponding effort to ensure that a score of 200 is possible on every test.   The 200 and 800 limits simply mean that scores are not reported *beyond* those limits.   The principal reasons for having them are: (1) as already indicated, to minimize *gross inequities* across the test offerings, and (2) to make it clear that the tests can discriminate adequately only within a limited range of scaled scores.

NECESSITY FOR REFINED SCORES

For the most part, the processes of equating and scaling the scores for College Board tests produce effects of relatively small magnitude, and thus they could be regarded as only a pedantic refinement.   But that is not so.   They are a refinement of a basic sort, introduced to ensure that no student will be put either to an advantage or a disadvantage simply because he happened to have taken an easier or more difficult form of the test or because he happened to have taken the test with a less able or more able group of students.

Fundamentally, what is desired is a measurement of ability that will serve the best interests of the students, a measurement that is not only accurate and relevant for them, but also, and just as importantly, equitable.   It is to this goal of equity that the "refinement" of equating and scaling is directed.

PROBLEMS

1. Why does the ETS make use of different test forms at different administrations of a test? What are the problems this practice causes?

2. In "equating" the scores on two different forms of a given test, how do the test constructions overcome the problem that each of the two forms is used by a *different* group of students?

3. Sam Johnson has received a scaled score of 600 on the verbal section of the SAT. Is this a good score? Explain your answer.

4. In applying the procedure used in "equating" forms the author says: "Here special care must be taken to avoid differential contextual effects." What does he mean by "differential contextual effects"?

5. Susan Smith just took the SAT's and found them surprisingly easy. She is certain that her raw score will be relatively high. Can you make a prediction about her scaled score? Why or why not?

6. In order to allow what it considers fair comparisons of groups of students who take different achievement tests, ETS scales achievement test scores against the expected scores of the entire body of SAT-takers. Explain the scaling procedure employed. In what sense does the procedure seem fair? In what sense does it seem unfair?

7. Consider Figure 1.
    (a) Estimate the most frequently achieved hypothetical score on the physics test for all candidates in the SAT standardization group.
    (b) Estimate the most frequently achieved score of those actually taking the physics test.

8. The author claims the student can't "beat the game" by making strategic choices about which achievement tests to take. Do you agree? Explain your answer. (Hint: Refer to Figure 1.)

9. The physics achievement test has been characterized as one which is taken by a "more able" group of students. Let us say that the subject A achievement test is usually taken by a "less able" group of students. Is it possible to get an 800 on both tests?

10. Notice the scaling procedure of question 6 allows an able student of physics (taken by relatively high SAT students) a higher scaled score than an equally able student of subject A (taken by relatively low SAT students). Explain how ETS "corrects" for this inequity.

STATISTICS, SPORTS, AND SOME OTHER THINGS

Robert Hooke   *Westinghouse Research Laboratories*

SPECTATOR SPORTS provide more than just observation of athletes who perform with admirable skill.   There is, for example, the drama of a young quarterback trying to lead a professional football team for the first time in front of 70,000 onlookers or that of a veteran pitcher calling on his experience to augment his dwindling physical resources in a crucial game of a close pennant race. Because these dramas are truly "live" and unpredictable, they are much more fascinating to some people than the well-rehearsed performances of the stage.

Not every moment in sports is dramatic, of course, but throughout any contest between professionals, the spectator is privileged to watch a group of people carrying out their jobs almost in full public view, to see how they meet their problems, and to see how they react to their own successes and failures.   Baseball and football provide especially good opportunities for such observation because each of these games consists of a sequence of plays, as opposed to the fairly continuous action of basketball, hockey, soccer, and racing sports.   The spectator sees more than strikeouts and home runs, completed passes and interceptions.   He sees a manager gambling on a hit-and-run play, or a quarterback deciding whether to run or pass for the first down that he desperately needs.   The fan has opinions on what his team should do in various situations, and he comes to decide that its manager or coach is a good strategist or a poor one.

Management in professional sports has many similarities to management in business and industry.   Some managers and coaches are smarter than others, and some make use of more advanced methods than others do.   This is true in sports, as it is in business, in spite of the folk wisdom of the sports pages that often maintains that all managers and coaches are pros and about equally good.   Some can get a great deal out of inferior personnel, but none can overcome more than a certain amount of incompetence among the people who work for them.   Some are natural gamblers, some are always conservative, and only a few are intelligent enough to be one or the other depending on what the circumstances call for.   Sports managers are different from business managers mainly in that their actions are so much more visible. Because of this visibility we should all be able to learn by watching them as they make their various moves in the goldfish bowl of professional sports.

STATISTICS AND MANAGEMENT

What does all this have to do with statistics?   The real concern of statistics is to obtain usable quantitative information, especially about complex situations that involve many variables and uncertainties.   "Usable" means that its purpose is to help us to improve our behavior in the future, that is, to help us learn how to extract from these situations more of whatever it is that we're trying to get.   Some managers make good use of statistics and some don't; this is true whether they manage factories or baseball teams.

Suppose, for example, that we are manufacturing rubber tires.   An expert will no doubt be able to detect from the example that I know nothing about the tire business; in fact, I chose this example because I've never been in the tire business and hence it will not implicate real people.   At some point in the process, let's suppose that we have a mass of liquid rubber that will ultimately be turned into tires.   Being aware that this batch may possibly have been improperly prepared, we would like to test it in some way so that, if it is defective, we can throw it away without wasting money processing it into defective tires.   Unfortunately, the true test of a tire is a road test, and we can't road test a batch of liquid, so we must perform some test that we think is relevant, such as a viscosity measurement.   This measurement will take time and cost money, and sooner or later someone will raise the question "Is it worth the money we're spending on it?"   This is always a good question, and it usually leads to much heated debate.   The debate will include arguments based on intuition, experience, laboratory tests, and scientific theory; each has its place in the process of seeking after the truth, but they are basically predictors, and the only way to be sure of what will happen in the field is to see what actually happens in the field.   This means that we should collect statistics.   After measuring a batch, we should follow it through the manufacturing process and see what is the quality of a sample of tires made from this batch.   We repeat this on another batch, and so on.   After a while we can establish the relationship between tire quality and viscosity, and we can use this to determine whether to continue with the test, taking into account the testing costs, the cost of the manufacturing process that follows the test, the value of a good finished tire, and so on.   Now you may say, of course, that this method isn't infallible because it involves sampling and, hence, sampling error (see the essay by Neter) and because the process may change unexpectedly, and so on.   But this method gets us as close to the truth as we can come, and this final objection merely says that you never have it made, even if you use statistics.

THE STRATEGY OF BUNTING

The student of management behavior can find many instances of this type of problem in sports, and if he is smart, he can profit from the mistakes that are made visibly on the diamond or the gridiron.   Take, for example, the sacrifice bunt in baseball.   There are those who swear by it and there are those who seldom use it.   They engage in passionate arguments as to whether it is a good strategy.   As we shall see, statistics can't settle the issue once and for all, but it can shed a great deal of light on the problem, and most of the argument could be eliminated if people would look at some of the facts.

The sacrifice bunt is a play that is used to advance a base runner from first base to second, or from second to third, normally sacrificing the batter, who is thrown out at first.   Many managers use the sacrifice bunt routinely, and they refer to their behavior as "percentage baseball," as if they knew the percentages, which, apparently, most of them do not.   The routine is that you bunt if there is a man on first or second, nobody out, and your team is only slightly ahead, tied, or not "too far" behind.   One or more runs behind is considered too far for the visiting team, and two or more runs behind too far for the home team, the difference coming from the fact that the home team bats last and can afford to "play for a tie."

Why does the manager decide to bunt?   Ultimately, of course, he does it to win more games.   At the moment of doing it, he is trying to increase his chance of getting one additional run while giving up some of his opportunity to get several.   The theory is simple.   It takes at least two singles or a double to score a man from first, while a single alone will usually score him from second; if he is already on second, we note that if we could get him to third with only one out, he could score on any hit, error, wild pitch, passed ball, long fly, balk, or slow grounder.   Proponents of bunting are fond of quoting this list, but it contains some fairly rare events, and this raises the real questions: when we use the bunt, by *how much* is our chance of scoring one run increased, and *how much* do we sacrifice in terms of possible additional runs?   Again, the only way to get an answer to this question that is relevant to real major league players playing under pressure of real games is to take statistics from actual games.   There is no way to provide realistic conditions for an experiment, and theory (see Cook 1966, Hooke 1967) is of dubious value.

Although records of games played exist in the archives of organized baseball, turning these into usable data is a major task which, if it has been done, has not been made public to my knowledge, except in Lindsey (1963). Lindsey discussed records of several hundred major league games played in 1959 and 1960, and he produced some very interesting statistics, some of which are shown in Table 1.

To see how we, as armchair managers, would use this table to decide about bunting, let's look at the first two lines.   (For the moment, we'll think only of average cases, but no good statistician dwells on averages alone, so we'll discuss special situations later.)   We start, say, with a man on first and no outs.   The table says that this situation was observed 1728 times (occasionally, perhaps, more than once in the same inning).   In a proportion of these cases equal to 0.604, no runs were scored during the remainder of the inning; that is, in 1044 cases no runs scored, and $1044/1728 = 0.604$. This means also that the proportion of times at least one run scored from this situation is $1 - 0.604$, or 0.396.   We use these proportions as estimated probabilities of the various events; thus near the end of a tight game, the

TABLE 1.   Relation of Runs Scored to Base(s) Occupied and Number of Outs

BASE(S) OCCUPIED	NUMBER OF OUTS	PROPORTION OF CASES NO RUNS SCORED IN INNING	PROPORTION OF CASES OF AT LEAST ONE RUN SCORED IN INNING	AVERAGE NUMBER OF RUNS IN INNING	NUMBER OF CASES
1st	0	0.604	0.396	0.813	1728
2nd	1	0.610	0.390	0.671	657
2nd	0	0.381	0.619	1.194	294
3rd	1	0.307	0.693	0.980	202
1st, 2nd	0	0.395	0.605	1.471	367
2nd, 3rd	1	0.27	0.73	1.56	176

Source: Lindsey (1963).

number 0.396 measures the average "value" of having the situation of a man on first and no outs. For earlier parts of the game, the value is more closely related to the number of runs that are scored in an inning, on the average, starting from this situation; this is given in the fourth column as 0.813 for the situation in question.

Now if we make a sacrifice bunt that succeeds in the normal way, the runner on first will move to second and there will be one out. Is this a better situation than we had? In the sense of average number of runs scored, it is decidedly worse; the first and second line of Table 1 show that the average number of runs scored from the man-on-first-no-out situation is 813 per thousand, but from the man-on-second-one-out situation, it is only 671 per thousand. On the average, then, a normally successful bunt loses 142 runs per 1000 times it is tried. But what about the last inning of a tight game when we only care what has happened to the probability that at least one run will be scored? This figure has dropped from 0.396 to 0.390; these numbers are so close that their difference is readily explained by chance fluctuation from the sample. So we conclude that although the probability of scoring at least one run is increased by moving the runner to second with no additional outs incurred the increase is almost exactly canceled when an additional out occurs, as usually happens in a bunt play.

Conclusion. On the average, bunting with a man on first loses a lot of runs. On the average, it doesn't increase the probability of scoring at least one run in the inning. Here we've assumed that the batter is always out at first, but, of course, he is sometimes safe, thereby increasing the efficacy of bunting. It is probably more often true, however, that the front runner is thrown out at second, a disaster to the team that chose to bunt. It would appear that bunting with a man on first early in the game should be done only when it so takes the defense by surprise that the chance of the batter's being safe is substantial. Even late in a tight game there is no visible advantage to such bunting unless special circumstances prevail.

Now let's think of the problem of the man on second with nobody out. The table tells us that he will score (or at least somebody will score) in all but 381 cases out of 1000, that is, in 619 cases out of 1000. *If* we can move him to third by sacrificing the batter, we can raise the 619 to 693. (Note that we lose 214 runs per 1000 tries doing this, but let's again consider the case where it is late in the game and we need only one run.) Here there is indeed something to be gained by a successful bunt play, but it's time to face reality: the bunt play doesn't always work. How often it works depends on a lot of things, and we don't have statistics for an average result, but let's see how we would use them if we did.

If the batter bunts the ball a little too hard, the defending team happily fires the ball to third base and the lead runner is put out, leaving the offensive

302 PART THREE: OUR SOCIAL WORLD

team with a man on first and one out, their probability of scoring at least one run having gone from 0.619 to 0.266, the latter figure coming from the complete table in Lindsey's paper. The typical manager does not admit the possibility of such an event. After it happens he dismisses it with the remark "These young fellows don't know how to bunt like we used to." I know this remark was being made before any of today's managers were making their first appearance as professional players, and it was probably originated in the nineteenth century by the first nonplaying manager. The remark is merely an excuse for not studying the problem, but let's not be too hard on the baseball managers; we have pointed out already that the moves they make in plain sight are duplicated by other kinds of supervisors in less visible circumstances.

As we said above, we don't have statistics for the results of a bunt try with a man on second, so we'll make up some, trying to be as realistic as possible from unrecorded personal observations over the years. Here they are:

(1) 65% of the time the runner moves to third and the batter is out (normal case).

(2) 12% of the time the runner is put out at third, and the batter is safe at first.

(3) 10% of the time the runner must stay at second and the batter is out, for example when the batter bunts a pop fly, or strikes out.

(4) 8% of the time the batter gets on first safely, that is he gets a hit, and the runner also advances.

(5) 5% of the time the bunter hits into a double play, that is, he and the runner are both thrown out.

Now to compute the overall probability of scoring at least one run, we simply multiply and add according to the rules of probability. The reader who doesn't know these rules can do it this way: start with 1000 cases. In 650 (i.e., 65%) of these we have result (1), namely a man on third and one out. The table says that he will score 69.3% of the time, so we take 69.3% of 650, and get 450. That is, in 450 cases the bunt succeeds as in (1), and a run ultimately scores. Now in 120 cases the outcome is as in (2), and Lindsey's complete table says that a score then occurs 26.6% of the time. So we take 26.6% of 120 and get 32. Add this to the 450 and keep going. What we get for all five cases, using Lindsey's complete table where necessary, is

$$0.693(650) + 0.266(120) + 0.390(100) + 0.87(80) + 0.067(50)$$
$$= 450 + 32 + 39 + 70 + 3$$
$$= 594.$$

In other words, we will get at least one run in only 594 cases out of 1000. Before the bunt our chances were 619 out of 1000, so we have shot ourselves down. Of course, if our hypothetical data in (1) to (5) above are too pessimistic, the correct result will be a little more favorable to bunting, but it would appear that any realistic estimates will lead to the conclusion that bunting is not profitable on the average.

The intelligent use of statistics requires more than just a look at the averages. The above data and accompanying arguments show that bunting, used indiscriminately as many managers do, is not a winning strategy. This doesn't mean that one should never bunt, however. The man at bat may be a weak hitter who is an excellent bunter, and the man following him may be a good hitter; the batter may be a pitcher whose hitting ability is nil, but who can occasionally put down a good bunt; or the other team may clearly not be expecting a bunt, so that the element of surprise is on our side to help the bunt become a base hit. In any of these cases the bunt can be a profitable action. The role of statistics is to show us what our average behavior should be. In general, if the average result of a strategy is very good, we should use it pretty often. If the average result is poor, we should use it sparingly, that is, the special circumstances that lead to it should be very, very special. There are those who say that statistics are irrelevant and that they treat every case as a special case. This is probably impossible, and if such people would examine their behavior over a long period of time, they would probably find it quite statistically predictable. Incidentally, if one takes the point of view that surprise is the whole thing, that is, that the objective is to be unpredictable, then a randomized strategy is indicated; this is elaborated in any book on mathematical game theory.

THE STRATEGY OF THE INTENTIONAL PASS

Another strategic move in baseball is the intentional base on balls. The opposing team has a man on second, say, with one out, and we decide to put a man on first intentionally, either to try to get a double play or to have a force play available at all three bases. Is it worth it? Lindsey's table shows that before the intentional walk the probability of scoring at least one run is 0.390, but afterward it is 0.429. Clearly, on the average, the intentional pass is a losing move; followed by a double play it's great, but followed by an unintentional walk it can lead to a calamity, and the latter possibility is part of the reason for the numerical results just quoted. Widespread use of the intentional pass seems to be based on sheer optimism, as the statistics appear to show that the bad effects, from the point of view of the team in the field, definitely outweigh the good ones, on the average. What about special cases? If the batter is a good one, to be followed by a poor one, then the data don't necessarily apply, and the intentional pass may be a good thing. It probably should seldom be used early in the game, though, unless the following batter is a weak-hitting pitcher because it causes the average number of runs to go up from 671 per 1000 to 939 per 1000 owing to the additional base runner, and it is doubtful that there are many special cases that are so special as to outweigh this fact.

COLLECTION AND USE OF DATA

Figures such as those in Table 1 are obviously of little value unless they are based on a rather large number of cases. It isn't at all obvious, though, how large the number of cases should be. Mathematical statistics answers questions about how large sample sizes should be, but the questions must be specific. We can't, as we are sometimes asked to do, say that 100 (or 1000 or 2000) is a good all-around sample size. If, however, we are asked to find the probability of at least one run resulting from a man on first with no outs, we can, with certain reasonable simplifying assumptions, deter- mine how large the sample must be so that we can be 90% sure, for example, of being within 0.005 of the correct answer. Table 1 shows in the last column the sample size that was used to produce the data of the earlier columns. For an individual keeping records as a pastime, this represents a major effort. We would think that baseball people, engaged in a competition in which a few extra victories can make a difference of a great deal of money, would go to the trouble to collect even larger samples. They wouldn't want to go too far in this direction, however, because information tends to become obso- lete. Changing rules, playing fields, and personnel cause the game to change slightly from year to year. Sometimes scoring is relatively low for a few years, and then it increases for a few years. Data gathered in one of these periods of time may not be altogether valid as a basis for decisions in another.

Data of the sort we have been talking about here are sometimes called *historical* as opposed to *experimental,* or *controlled.* The distinction is impor- tant in many areas. For example, if statistics are produced showing that smokers have lung cancer with much higher frequency than nonsmokers, this *historical* fact *in itself* does not demonstrate that smoking increases the lung cancer rate. (After all, children drink more milk than adults, but this is not why they are children.) The problem is that there may be other variables that, for example, help cause lung cancer and also influence people to become smokers (see the essay by Brown). Nevertheless, the historical statistics on cancer were very suggestive and led to various experiments in laboratories which have strengthened most people's belief in a causal relationship. We can make good use of historical data, in other words, but we must be careful about inferring cause-and-effect relationships from them.

No doubt because of frustrations in trying to draw conclusions from histori- cal data, statisticians developed the science and art called the *design of experi-*

ments. If we can do a properly designed experiment, we are in a much better position to draw valid conclusions about what causes what, but the possibility of a designed experiment is not always open to us. When we can't experiment, we must do what we can with available data, but this doesn't mean that we shouldn't keep our eyes open to the faults that such data have.

CONCLUSIONS

So what have we learned from our look at sports statistics? We have learned these do's and don'ts:

(1) Don't waste time arguing about the merits or demerits of something if you can gather some statistics that will answer the question realistically.

(2) If you're trying to establish cause-and-effect relationships, do try to do so with a properly designed experiment.

(3) If you can't have an experiment, do the best you can with whatever data you can gather, but do be very skeptical of historical data and subject them to all the logical tests you can think of.

(4) Do remember that your experience is merely a hodgepodge of statistics, consisting of those cases that you happen to remember. Because these are necessarily small in number and because your memory may be biased toward one result or another, your experience may be far less dependable than a good set of statistics. (The bias mentioned here can come, for instance, from the fact that people who believe in the bunt tend to remember the cases when it works, and vice versa.)

(5) Do keep in mind, though, that the statistics of the kind discussed here are averages, and special cases may demand special action. This is not an excuse for following your hunches at all times, but it does mean that 100% application of what is best on the average may not be a productive strategy. The good manager has a policy, perhaps based on statistics, that takes care of most decisions. The excellent manager has learned to recognize occasional situations in which the policy needs to be varied for maximum effectiveness.

PROBLEMS

1. Refer to Table 1. In how many cases with a man on third and one out, did no runs score?

2. Suppose second base is occupied and there are either no outs or one out. In how many of such cases in Table 1 are no runs scored in the inning?

3. Suppose there are runners on first and second, no outs, and it is early in

the game. Assuming the batter is out and the runners advance one base, do the figures in Table 1 suggest a bunt? Explain your answer.

4.   Suppose there are runners on first and second, no outs, and it is the last inning of a tight game. Assuming the batter is out and the runners advance one base, do the figures in Table 1 suggest a bunt? Explain your answer.

5.   Suppose the statistics for the results of a bunt try with a man on second are 70%, 13%, 9%, 5%, 3% respectively instead of 65%, 12%, 10%, 8% and 5% assumed by the author. Would bunting then be profitable on the average in this situation? Explain your answer.

6.   When might a sacrifice bunt be a wise move in a situation where, on the average, it is not?

7.   (a) Distinguish between *historical* and *experimental* data.
      (b) Why didn't Lindsey conduct a controlled experiment?

8.   Use the following additional statistics from Lindsey and the outcome percentages given in the text. Assume there is a man on second and one out. The batter attempts a bunt.

Base occupied	No. of outs	Probability that no runs score in the inning
1	2	.886
2	2	.788
3	2	.738
1,3	1	.367

   (a) How many times (out of 1000 cases) will at least one run score?
   (b) How does possibility (5) (bunter hits into double play) enter your calculation?

REFERENCES

E. Cook. 1966. *Percentage Baseball.* Cambridge, Mass.: MIT Press.

R. Hooke. 1967. Review of Cook (1966). *Journal of the American Statistical Association* 62:688–690.

G. R. Lindsey. 1963. "An Investigation of Strategies in Baseball." *Operations Research* 11:477–501.

VARIETIES OF MILITARY LEADERSHIP

Hanan C. Selvin   *State University of New York, Stony Brook*

WORKERS ON an assembly line, students in a third-grade classroom, and soldiers in an army training camp do different kinds of work in radically different settings, but they have in common one important social relationship. They all spend a good part of their day in close contact with lower-level leaders, such as foremen, teachers, and company-level officers, both commissioned and noncommissioned. From both individual experience and empirical research we know that the behavior of workers, students, soldiers, and others in subordinate positions, at work and afterward, is significantly affected by the actions of their leaders.

The empirical study reported here shows how the actions of company leaders in twelve U.S. Army training companies affected the *nonduty* behavior of several hundred soldiers undergoing basic training. The unraveling of these effects of leadership was unusually complex. Unlike the student and the worker, who usually are subject to only one leader in the course of a working day, the trainee had two company-level commissioned officers (the Commanding Officer and the Executive Officer) and two company-level "non-coms" (the First Sergeant and the Field First Sergeant). There was constant turnover in these positions: during his training cycle the typical trainee had seven company-level leaders.

An additional source of complexity in this study is the way in which we constructed descriptions of the "leadership climates" of the companies. For this study, it seemed better to rely on the trainees' description of their leaders in a questionnaire that they filled out at the end of their basic training, rather than on judgments by superiors or outside experts, as is often done in evaluating how well an organization achieves its goals. Accordingly, each trainee rated each of his leaders on fifteen different questions, ranging from how well the leader inspired confidence to whether he punished the men at every opportunity.

The sheer bulk of these data is impressive: an average of 150 men in each of 12 companies rated an average of 7 leaders on 15 questions. Multiplying these figures together ($150 \times 12 \times 7 \times 15$) yields a total of about 189,000 separate ratings of company leaders. The major statistical problem was to boil down this mass of data into descriptions of the leadership climates of the companies.

Part of this statistical "boiling down," or *data reduction* as it is usually called, consisted of such simple procedures as computing averages. Another large part, much more complex and more illuminating, was a statistical procedure called *factor analysis,* which played a central part in measuring the leadership climates. These factor-analytic procedures not only exemplify a powerful technique, but also can be applied whenever several people can give independent judgments about someone with whose behavior they are familiar. Examples are teachers as described by their students, students as described by several teachers, and mental hospital patients as seen by the ward staff. Finally, although most studies of teaching effectiveness in colleges and universities rely on ratings of the teachers by their students, they do not typically go on to the kind of analytic clarification that this procedure would afford.

THE IDEA OF FACTOR ANALYSIS

The three primary colors (red, yellow, and blue) when suitably combined, yield thousands of different colors. Similarly every bit of matter can be analyzed into some combination of the hundred-odd chemical elements. These two facts, familiar to all adults from their school days, are parallels in the realm of physical science to what the statistical procedure of factor analysis can sometimes do with such social phenomena as opinions, votes, and symptoms of mental illness. Factor analysis is, in short, a way to discover or construct from a larger group of observed characteristics, or *items*, a small set of more general characteristics, or *factors*, various combinations of which will produce each of the observed patterns of items.

THE BACKGROUND FOR THE DATA

To explain this work, let us start with the gathering of the data at Fort Dix, New Jersey, in the spring of 1952 by two physicians, Arthur M. Arkin and Thomas M. Gellert, then on the staff of the Mental Health Consultation Service (a central psychiatric facility to which soldiers were referred from dispensaries located near their companies). Over a period of several months they began to notice patterns in their records. Some companies had higher rates of accidents, other companies suffered more psychosomatic illnesses, and still other companies had greater proportions of men going AWOL for short periods. Because all companies followed essentially the same program of training, lived in identical barracks, and ate the same food, the staff members speculated about the kinds of factors that might be responsible for the differences they had observed. They reasoned that differences in the nature of the leadership among companies might account for the differences in rates of accidents, of psychosomatic illnesses, and of going AWOL.

Further reflections and some pilot studies soon led to the development of the two questionnaires that are the basis of this study. One, the "behavior questionnaire," asked each trainee to report the frequency of 24 kinds of nonduty behavior, such as going to the PX for food between meals, having sexual intercourse, going to the movies, and seeing the Chaplain. This questionnaire also asked for the trainee's age, education, and marital status.

The second questionnaire dealt with the company-level leaders that the trainee had had during the sixteen weeks of basic training: Commanding Officer (C.O.), Executive Officer (Exec.), First Sergeant (1/Sgt.) and Field First Sergeant (F-1/Sgt.). In general, the C.O. and the F-1/Sgt. worked directly with the trainees, and the other two leaders usually remained in the company office, or orderly room, and had less contact with the men.

APPROACHES TO THE ANALYSIS

Three elements combined to shape the analysis of the leadership data: the nature of the data as described above, the properties of the available statistical methods, and my training as a sociologist. At the outset of the analysis, there was a choice between two essentially different problems: the *psychological* problem of trying to explain a particular event (say, why Pvt. John Doe got drunk on his first weekend pass) and the *sociological* problem of variations in *rates* of behavior in different social units (why, for example, did Doe's company have a higher proportion of men getting drunk than did any other company?).

I chose to work on the second problem, both because of my training as a sociologist and because the data lacked the detailed psychological information on each individual soldier needed to learn why he behaved as he did. Once made, this decision helped to shape the answer to the second basic question of the study: how to describe the leadership of each company. It gradually became clear to my assistant (E. David Nasatir) and me that the data had to be put together in two different ways. First, we wanted to describe the leadership behavior of all of the leaders in a company, not simply that of the C.O.; we expected the nonduty behavior of the trainees to be affected by the overall leadership climate of the company. (We were able to show that each leader contributed something of his own to that climate and that his actions were not simply copies of the actions of the C.O.) Second, all of the leadership data we had were embodied in the responses of the individual trainees, so it was necessary to combine the responses in some way for two reasons. We wanted to find the common elements in the evaluations of leadership in a company, not the idiosyncratic perceptions of one or a few trainees, and, consistent with our sociological orientation, we wanted to focus on how the trainees in each company, as a group, saw their leaders.

The central statistical task is thus to describe the entire set of company-level leaders as seen by the entire set of trainees in each company, in other words, to reduce the 189,000 ratings of the 82 leaders by the 1800 trainees on the 15 questions to a small set of descriptions of each company's leadership climate.

THE IDEAL STUDY AND THE REAL STUDY

It will clarify the statistical reasoning to put the questionnaires aside for a moment and ask how one would go about describing leadership climates if one had unlimited resources of money, trained personnel, and time.

Ideally, perhaps, one would assemble a group of trained observers—or even one "omniscient observer"—and ask them to live with each company for a significant part of its training cycle. These observers would watch, record, and evaluate the behavior of the leaders and somehow produce a concise description of each company's leadership climate.

Even if everything were ideal, this would be extraordinarily difficult. For one thing, the observers would have to be everywhere, watching everything, and yet not interfering with the training activities or affecting the nature of the leadership. No, a corps of observers would not do, but if we could depart altogether from a realistic observational situation, at least to the extent of thinking of what an ideal arrangement might be, we would like to have an omniscient observer, a kind of observational superman who could see every interaction, describe it, and combine it appropriately with all of the thousands

TABLE 1. How Pvt. Doe Answered Question 15 of Leadership Questionnaire

	C.O.	Exec. Off.	1st Sgt.	Field 1/Sgt.
15. If you were ordered into combat and you could choose the men who would be your leaders use the No. 1 for those men in your unit you would like MOST to lead you; No. 2 for those men whom you would like LESS to lead you; and the No. 3 for those men you would like LEAST to lead you if at all.	1 3	2	2 1	3 3

of others that he observes. Such an omniscient observer does not exist, but we were able to create an approximation to his observations statistically by basing the descriptions of the leaders' behavior on the experience of the trainees. To see how this was done, consider question 15 (see Table 1) of the leadership questionnaire filled out by Pvt. John Doe of company X.

During his 16 weeks of training, Doe had seven company-level leaders: two C.O.'s, one Executive Officer, two First Sergeants, and two Field First Sergeants. The numbers at the right in Table 1 are his ratings of each officer and noncom as a combat leader. Doe apparently thought that the first C.O. would have made a good combat leader, for he gave him the highest rating, 1. His unwillingness to follow the second C.O. into combat is indicated by the low rating of 3.

Every trainee in Doe's company rated the same company-level officers on this question. For the sake of illustration, assume that there were 100 trainees in this company and that their ratings of the first C.O. as a combat leader were those shown in Table 2. The average of these ratings is 1.70, so this is the rating that the first C.O. was assigned on combat leadership.

TABLE 2. Ratings Given to First C.O. by 100 Trainees

RATING (1)	NUMBER OF TRAINEES (2)	(1) × (2)
1	50	50
2	30	60
3	20	60
	100	170

$$\text{Average} = \frac{170}{100} = 1.70$$

We can now turn away from the trainees and take each average rating as a characteristic of the leader being rated. Thus the first C.O. in the illustrative example would be said to have a rating of 1.70 as a combat leader. In other words, the average ratings received by a leader may be considered as his (*perceived*) attributes.

In the leadership questionnaire each leader was rated, as in the foregoing illustration, on the extent that he:

(1) Influenced the lives of the trainees
(2) Commanded the respect of the trainees
(3) Was a "sucker for sob stories"
(4) Was a "good Joe" one minute and "mean as Hell" the next
(5) Could create a real fighting spirit against the enemy
(6) Acted in such a way that the trainees were afraid of him
(7) Could not be depended on to keep his promises
(8) Created a feeling of confidence in the trainees
(9) Told the trainees when he thought that an order from higher headquarters was unfair or silly
(10) Displayed a real interest in the trainees without babying them
(11) Treated the trainees "like dirt"
(12) Gave more breaks to his favorite trainees than to others
(13) Seized every opportunity to punish his men
(14) Tried to have his men excused from "dirty details" ordered by higher authorities
(15) Would be preferred as a leader in combat

This use of average ratings, instead of the original ratings by each of the trainees, yields an impressive reduction in the amount of data. Instead of some 189,000 individual ratings, there are now only 1230 averages (82 leaders rated on 15 questions). Even more important than the quantity, however, is the quality of these statistically derived data. The original ratings of each leader show a great deal of variation, with misperception, failure to follow instructions, facetiousness, and errors of processing all distorting the true ratings.

COVARIATION OF RATINGS

The quality of these average ratings appears most clearly when we see how much the characteristics of the leaders that *should* vary together *do* vary together. To show this, we must introduce a numerical measure of this joint variation. We chose the *coefficient of correlation*, invented by Sir Francis Galton in the 1880s to measure how much various physical characteristics, such as height, are inherited. If the height of a son can be predicted *exactly* by a mathematical equation for a straight line using the height of his father and if tall fathers give rise to tall sons, then the value of the correlation coefficient is 1.0, the largest value that this coefficient can have. If the height of a son can be predicted exactly from the height of his father, but tall fathers produce short sons, then the value of the correlation coefficient is —1.0, its largest negative value. And, if there were no relation between the heights of fathers and the heights of sons, the correlation coefficient would be 0. (See the essay by Whitney for a further description of the correlation coefficient.)

In real data on individuals, values close to 1.0 or —1.0 are rare. Thus the correlations between pairs of ratings given to any one leader by the men in his company seldom were higher than 0.30. These are, of course, the correlations between the responses of the individual trainees to the leadership questions, before the computation of averages. For example, a trainee who rated a particular leader high as a combat leader might be almost as likely to rate him low in displaying an interest in the trainees as he would be to rate him high on this second trait.

The situation is altogether different for the average ratings. For example, a leader who has a high average rating on instilling a fighting spirit in his men (question 5) almost always has a high average rating on commanding their respect (question 2); the correlation between the averages on these two characteristics is 0.82. Similarly, a leader who punishes at every opportunity (question 13) usually produces fear (question 6); the correlation in this case is 0.84. The size of these correlations between averages is striking: of the 105 correlations in the leadership data, 49 are numerically greater than 0.50, 28 are numerically greater than 0.70, and 13 are numerically greater than 0.80.

On the other hand, we might expect competence and coercion to be negatively related—that, by and large, leaders who were judged to be competent would be less likely to be judged coercive. The data only partially bear out this expectation. The correlations between the average scores on inspiring respect and the averages on the two questions that measure coerciveness (6 and 13) are moderately negative (—0.28 for the question on fear and —0.45 for the one on punishment), but the corresponding correlations between the averages on the question on instilling a fighting spirit and the averages on the measures of coerciveness are so close to zero (—0.01 and —0.16) that they indicate that there is no appreciable relation.

Even though there are only four items in the analysis in the previous paragraph, the discussion was a bit complicated. Part of this complexity might be removed by a better choice of words, but there is a limit to the complexity that words can clarify. Imagine the complexity in trying to relate all 105 correlation coefficients between pairs of averages, instead of only four!

USING FACTOR ANALYSIS TO DESCRIBE THE STRUCTURE OF LEADERSHIP

Factor analysis offers a way out of this complexity. This statistical procedure often makes it possible to untangle large sets of correlation coefficients; in brief it determines which items go together and which do not. Moreover, it expresses this structure of relations numerically, so that we can tell *how much* of what kinds of order there is in the set of correlations and how these simpler orders fit together.

Before turning to the leadership data it is important to say a few words about the goals of factor analysis. In psychology, the field where factor analysis was invented and has been most used, it is customary to speak of the statistically derived factors as "underlying," "basic," or "fundamental" variables and to use the verbs "discover" or "uncover" to describe the process used to calculate the factors. This language suggests that psychologists and statisticians have invented a statistical procedure for discovering scientific truths, much as chemists discovered the 100-odd fundamental chemical elements. I prefer to use a different set of terms. I shall speak of "constructing new variables," or factors, from combinations of the original items. The statistical procedures are the same; only the shades of meaning attached to them differ.

It turns out that the 15 original questions in the leadership data can be combined into 3 new variables or factors, which we labeled "positive leadership," "tyrannical leadership," and "vacillating leadership." For example, a leader who received averages close to 1 on the question of willingness to follow into combat and on similar questions that make up the factor of positive leadership would get a high score on that factor. The factor-analytic computations thus lead to a set of scores for each leader on the 3 factors, scores that, to a considerable extent, can replace his scores on the 15 original variables. That is, if we know a leader's scores on these factors, then we can estimate his average ratings for the 15 original questions with a high degree of accuracy. The statistical relations between items and factors suggest the names for the factors. Thus a leader who scores high on the factor of positive leadership is one who creates confidence in the trainees, is able to instill a fighting spirit in his men, is interested in them, and is one whom the men would like to have leading them in combat. Similarly, leaders who receive high scores on the second factor, tyrannical leadership, are likely to be seen as producing fear in the trainees, as punishing them at every opportunity, and as treating them "like dirt." High scores on vacillating leadership go to leaders who play favorites, punish at every opportunity, and are "Good Joes one minute and mean as Hell the next."

The construction of the 3 factors from the original 15 items was entirely a statistical operation, based only on the numerical correlations among the 15 items. Neither the wording of the questions nor the analyst's expectations entered into these computations. These nonstatistical considerations enter only in choosing the names for the factors, and even these choices are relatively unimportant when one has access to all the significant numerical results.

VERIFYING THE MEANING OF THE FACTORS

Instead of dreary columns of numbers, let us look at other, perhaps more meaningful evidence that these factors really do express the trainees' perceptions of their leaders. The evidence comes from the unsolicited comments that many trainees wrote on the leadership questionnaires. For example, one trainee wrote of an officer who turned out to have a particularly high positive-leadership score:

> I think that our commanding officer, Capt. ————, was a great leader, he held the respect of all the men and was just about everyone's choice to lead them in combat if we ever saw action.

And a First Sergeant who happened to receive a conspicuously low score on this dimension elicited the remark:

> . . . he is the most unsympathetic character that I have ever encountered in my life also sneaky . . . . I don't see how he ever earned his stripes for he has the mental capabilities of a mongoloid.

Similarly, an officer with a very high score on tyrannical leadership drew this comment:

> The C.O. beat men until they ran to the I.G. (Inspector General). Very few of us got passes during basic. We never got breaks in our marches because the C.O. was either trying to set a record or win some money.

Finally, of the leader who had the highest score on the factor of vacillating leadership, one trainee wrote:

> If [he] wouldn't lie to the men so much and stop trying to make major . . . this soldier hates his guts for the way he treated me and the rest in basic training.

The last quotation may seem almost as indicative of tyrannical as of vacillating leadership. Indeed, we shall shortly see that there was a high correlation between these two factors.

LEADERSHIP CLIMATE OF COMPANIES

The computation of factor scores simplifies the data a good deal; instead of there being, for each of 82 leaders, scores on 15 items, there are only the scores on the 3 factors. One more important step remains: to combine the scores for each leader in a company into measures of the *leadership climate* of that company. At first glance it might appear that one could take a simple average of the leadership factor scores for the leaders in each company. There are two reasons, however, for not doing this. First, the leaders did not all serve the same length of time; some were with their companies for the entire 16 weeks, but others served as little as 2 weeks of the training cycle. Second, the leaders also varied in the extent of their influence on the trainees. In general, C.O.'s and Field First Sergeants had more influence than did leaders in the other two positions. And, of course, the personal qualities of the leaders also made some of them more significant than others.

Fortunately, one of the questions on the leadership questionnaire made it possible to measure the relative influence of the leaders:

(1) The four men listed on the right side of this paper are all important in the life of a trainee. Place the No. 1 in the column under the name or names of the men who had the MOST influence in your life as a trainee; the No. 2 in the column under the name or names of the men who had LESS influence and the No. 3 in the column under the name or names of the men who had the LEAST influence or NONE AT ALL.

The average score received by each leader on this question can serve as a measure of his perceived relative importance in determining the leadership climate of his company. Incidentally, the Field First Sergeant had the most influence just as often as the C.O., thus bearing out the point made earlier, that there is more to the effects of leadership than rank alone.

It seems obvious that the dimensions of leadership climate should be the same as the three factors of leader behavior, provided that the scores on these factors can be modified to take into account the variations in length of service and in importance to the trainees. A procedure for doing this uses a modified, or *weighted,* average; each leader's factor scores are given more or less weight according to his length of service and his relative importance. Thus, in the "indexes of leadership climate" for the company, a leader who served all 16 weeks would have his factor scores counted twice as heavily as a leader who served only 8 weeks. Similarly, leaders with high "influence scores" would have their three factor scores weighted more heavily in the indexes of leadership climate than would leaders with low "influence scores."

Computing these weighted averages of the factor scores yields three indexes of leadership climate for each company, one for each factor of leadership. For this study, it suffices to condense these indexes into only two values, "high" and "low" (actually, relatively high and relatively low). A further simplification comes from the high correlation between the indexes of tyranny and vacillation. With only one exception, companies high on tyranny were also high on vacillation. With only 12 companies, it was impossible to separate tyranny from vacillation.

TABLE 3. Types of Leadership Climate

INDEXES OF LEADERSHIP		LEADERSHIP CLIMATES	NUMBER OF COMPANIES
Positive	Tyrannical and Vacillating		
High	High	Paternal	1
High	Low	Persuasive	6
Low	High	Arbitrary	3
Low	Low	Weak	2
		Total	12

When this is done, there are only four different types of leadership climate, corresponding to high and low values on the first two indexes of leadership climate, as shown in Table 3.

The statistical techniques of averaging, correlation, and factor analysis have made it possible to distill these four types of leadership climate from the 189,000 separate ratings of leaders. Simply in the sense of reducing a mass of virtually indigestible data to a set of straightforward types, this is impressive. Data reduction alone was not the point of this research; rather, it was to study the effect of leadership on nonduty behavior. The value of this statistical analysis thus lies in finding how much difference these types of leadership climate make in the patterns of nonduty behavior. The gross differences in rates of different kinds of behavior between leadership climates are seldom larger than 10 percentage points, but they are remarkably consistent. There is space here only to sketch these effects; for further details the reader may consult Selvin (1960, especially Chapters 5 to 7).

EFFECTS OF LEADERSHIP ON BEHAVIOR

Because of a change in the behavior questionnaire during the gathering of the data, it was not possible to compare the frequencies of different kinds of nonduty behavior in the "paternal" climate with the rates in the other three climates. The remaining three climates—"persuasive," "weak," and "arbitrary"—can be thought of as spanning the continuum from competent, democratic, and considerate leadership to incompetent, coercive, and unsympathetic leadership. By and large, these differences in type of leadership correspond to the differences in frequencies and patterns of nonduty activities. The "persuasive" climate has the lowest rates on many of the nonduty activities, the "weak" has intermediate levels, and the "arbitrary" has the highest; or, to put it quantitatively, comparing the rates in the three climates with the rates for all trainees taken together, the rates in the "arbitrary" climate

TABLE 4. Proportion Reporting Getting Drunk at Least Once During Basic Training, by Leadership Climate

"Persuasive" climate	25%
"Weak" climate	36%
"Arbitrary" climate	34%

are higher than the rates for all trainees in 13 activities, the rates in the "weak" climate are higher in 9 and the rates in the "persuasive" climate are higher in 5 activities.

As an illustration of the type of relation found in this study, consider the effect of leadership climate on the incidence of drunkenness. In answer to the question "How many times during basic training did you get really drunk?" the figures shown in Table 4 were obtained. The maximum difference in this table, 11 percentage points between "persuasive" and "weak" climates, is typical of most of the differences between leadership climates in this study. They were usually no larger than 10 percentage points. This may not seem like much of a difference. Does the smallness of this relation mean that leadership has little effect on nonduty activities? Or does leadership have a larger effect, one that somehow does not appear in these figures?

The latter conjecture seems to be correct. The effects of leadership differences bear unequally on different kinds of men, some showing great differences in their rates of specific nonduty activities from one leadership climate to another and others seeming almost immune to differences in leadership. Thus consider the same relation between leadership and drunkenness, but this time examined separately for single and married men (see Table 5).

Compare the two columns with each other and with the figures in the preceding table. Among the single men, leadership climate has only a small effect on rates of drunkenness; the difference between the highest and lowest is only

TABLE 5. Proportion Reporting Getting Drunk at Least Once During Basic Training, by Leadership Climate and Marital Status

	SINGLE MEN	MARRIED MEN
"Persuasive" climate	30%	14%
"Weak" climate	38%	32%
"Arbitrary" climate	33%	36%

8 percentage points. Among the married men the picture is altogether different. The difference between the highest and lowest rates is 22 percentage points, almost three times as much.

Similar findings hold for most of the activities in this study and for the other two individual characteristics on which data were gathered, age and education. The effects of leadership are felt most among the older, married trainees who had not graduated from high school, and they are felt least among the younger, single high-school graduates. Statisticians express relations like these by saying that the leadership climate and individual characteristics *interact* in their effects on behavior; the effects of leadership climate on behavior depend on the background of the trainee, and, correspondingly, the effects of background on behavior vary from one kind of leadership climate to another. In short, the types of leadership climate constructed by the elaborate statistical procedure described in this chapter not only make sense; they also make a difference.

OTHER APPLICATIONS OF STATISTICS IN EVALUATING INDIVIDUALS

The method of describing leadership climate by a combination of statistical procedures appears to be applicable to a wide range of situations in which an individual (a leader, a doctor, a patient, or even an inanimate object like a book, picture, musical performance, or other aesthetic object) is rated independently on a number of variables by a group of judges, each of whom is well acquainted with the individuals he is rating. Perhaps the most significant extension of this work would be its application to other kinds of military units, both in training and in combat. Such studies should also examine what this study chose to ignore, the effects of leadership on the performance of assigned duties as well as nonduty behavior.

PROBLEMS

Note: Table 1 is incomplete.

1. The object of this study was to make some statement about off-duty behavior of enlisted men. Why did the author feel it was important to determine leadership climate?

2. Explain the idea of factor analysis.

3. The author draws a distinction between a *psychological* and *sociological* problem. Explain this distinction.

4. At what stage in the data reduction were the coefficients of correlation obtained?

5. Consider the author's statement: "Imagine the complexity in trying to relate all 105 correlation coefficients between pairs of averages, instead of only four!" Why are there 105 such correlation coefficients? (Hint: There are 15 questions.)

6. What would be the meaning of a correlation of −.77 between two questions? A correlation of .15?

7. Suppose the author had relied on his prior expectations of correlation to determine the three factors. How would this affect the nature of the study? Would it necessarily affect the outcome of the study?

8. In Table 3, why are "tyrannical" and "vacillating" combined into one category?

9. How were the terms "paternal", "persuasive", "arbitrary", and "weak" chosen to describe the various leadership climates?

10. Which leadership climate tended to have associated the most desirable off-duty behavior?

11. Consider Table 5. True or false: For each leadership climate, a greater percentage of single than married men reported getting drunk at least once. Explain your answer.

12. What does a statistician mean by the term *interact*?

REFERENCES

Raymond Bernard Cattell. 1952. *Factor Analysis: An Introduction and Manual for the Psychologist and Social Scientist.* New York: Harper. Probably the clearest account of the principles of factor analysis.

Harry H. Harman. 1960. *Modern Factor Analysis.* Chicago: University of Chicago. This is a very technical, but encyclopedic, book.

Hanan C. Selvin. 1960. *The Effects of Leadership.* New York: Free Press.

THE CONSUMER PRICE INDEX

Philip J. McCarthy   *New York State School of Industrial and Labor Relations*
Cornell University

ALMOST EVERYONE worries about the prices he pays for all kinds of things. Perhaps the best way for an individual to find out how the price level changes is to follow in newspapers the Consumer Price Index (CPI) of the Bureau of Labor Statistics, a part of the U.S. Department of Labor. The individual consumer sees the CPI as a good measure of price changes in goods and services that he purchases. He reacts to newspaper statements such as "In September of 1969, the average urban family must spend $12.93 for the same amount of goods and services that could be obtained for $10.00 in 1957–59." He wonders whether or not his income has increased sufficiently to compensate for this increase in prices.

Unions and management pay particular attention to the CPI because they know its value will play a critical role in wage agreements and that a 7% annual increase, for example, in the CPI will lead to a demand for at least

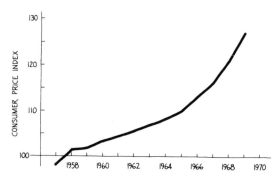

FIGURE 1
*Consumer Price Index, annual
averages (1957–59 = 100)*

a 7% increase in wages. Furthermore, in January of 1969 there were 2.55 million workers whose wages were covered by contracts containing *escalator* provisions, that is, provisions calling for automatic changes in wage rates in accordance with specified changes in the CPI.

Finally, economists concerned with the fiscal and monetary policies of the U.S. Government use the CPI as one of the principal indicators of the existence of an inflationary spiral in which higher prices lead to higher wages, which, in turn, lead to a greater demand for goods, thus raising prices again, and so on. During recent years, most writings on the state of the U.S. economy have emphasized the fear of inflation as evidenced by the increasing values of the CPI. The reason for this fear is illustrated clearly in Figure 1, which shows annual average values of the CPI for 1957–69. These values have been rising at an increasing rate since about 1964.

Although statistical studies of prices and living conditions in the U.S. were conducted in the late nineteenth and early twentieth centuries, the first complete "cost-of-living" indexes were published by the Bureau of Labor Statistics in 1919. They referred to 32 large shipbuilding and industrial centers, and arose through an agreement between the Shipbuilding Labor Adjustment Board and labor chiefs that one of the factors to be considered in settling labor disputes was that of "adjusting wages to the higher cost of living resulting from the war."

Since that time, the CPI has broadened in scope and increased in importance. This has led to many professional appraisals of the CPI, among them one by an advisory committee of the American Statistical Association in 1933–34 and one by the National Bureau of Economic Research in 1959–60. These appraisals have influenced major revisions in the CPI, made at approximately ten-year intervals since 1940, as well as many minor revisions.[*]

[*] See Footnote at end of article.

PROBLEMS IN CONSTRUCTING THE CPI

The typical consumer is quite aware of changes that occur in the prices of goods and services that he purchases regularly. He knows when the price of gasoline increases by one cent per gallon, or when the price of milk is increased by one cent per quart. Furthermore, he is able to predict with reasonable accuracy the impact of these price changes on his monthly or yearly budget, and he is therefore in a position to estimate the extra income (including an allowance for taxes) he must obtain in order to compensate for these increases. Price changes relating to less frequent and more sporadic purchases—clothing, doctors' services, appliances, automobiles, and homes, for example—are not as visible to the individual consumer, and it is much more difficult for him to assess their effects on his budget. Thus he may well know that it costs him more to live, or less to live, during the current year than in the preceding year, but he would find it almost impossible to provide an exact value for this change in his cost of living. His problem would be further complicated by changes that might occur in such matters as family status (births and deaths), family desires (color TV instead of black and white), and family purchasing patterns (chicken instead of steak). And yet it is exactly this evaluation—for the nation as a whole, for separate regions and cities of the nation, for different classes of expenditures (food, clothing, housing, transportation, and the like), and on a monthly basis—that the CPI provides. This essay is concerned primarily with the contribution of statistics to the construction of the CPI.

If the individual consumer actually wished to keep sufficiently detailed records to measure the change in prices of the items he purchases from, say, January of 1968 to January of 1969, he might proceed as follows. During January of 1968, he would keep a record of every purchase and would then summarize these purchases in terms of quantities (numbers of quarts of milk, number of pairs of shoes, number of haircuts, and the like) and the unit price of each item. In effect, he defines a "market basket" of goods and services and obtains the cost of this market basket in January 1968. Suppose this cost is $750.00. In January 1969, he would price exactly the same market basket of goods and services. Let us assume that the cost then was $800.00. Therefore, this consumer's personal CPI for January 1969 with January 1968 weights (quantities) and with January 1968 as base ($= 100$) is $100 \times 800/750 = 107$. The CPI is effectively computed in the same way, but the operation is much more complicated, since it deals with a large population rather than a single consumer.

Even if we assume that the family status and desires of our consumer did not change from January 1968 to January 1969—an assumption that can seldom be exactly true—there are many practical problems and tantalizing questions that would plague those responsible for the construction and interpretation of this index.   In particular,

(1) Even for a single consuming family, the number of different items purchased during a month or year is extremely large, and the continual pricing of such a list would be costly and time-consuming.   Preparation of a CPI for groups of families obviously magnifies this problem to a wholly impractical size.

(2) The average family has available to it a host of vendors of goods and services, and prices vary from vendor to vendor.   In January of 1969 should one attempt to return to the same places where purchases were made in 1968?   And what about stores that have gone out of business or stores that have come into existence in the interim?

(3) Prices vary from day to day.   Must our consuming family return to the same store at the same time to compute the cost of its market basket in January of 1969?   And how does one take into account the facts that some families take advantage of sales and others do not and that some families shop on a day-to-day basis while others shop less frequently?

(4) Suppose that a major purchase, such as an automobile, was made by our family in January 1968.   The fixed market basket still contained this item in January 1969 even though such a purchase was not made at that time.   How should this be handled?

(5) The goods and services available to the consumer change from time to time.   Items contained in the original market basket may not exist in the stores when we return to price them at a later date, or they may exist only at an improved or lowered quality level.   What do price changes mean under these circumstances?

(6) We also observe that our consuming family—even though its composition, status, and desires are assumed not to have changed—may be able to "beat" a rise in prices by appropriately altering its purchasing patterns. Suppose, for example, that they notice that the price of steak has risen sharply and decide that the desires of the family are as well satisfied by chicken as by steak.   If chicken is substituted for steak, the family's satisfaction level will not change, and yet the amount spent for food may actually decrease in spite of what may be a general increase in the price level.   In effect, we can think of replacing the "fixed-contents" market basket with a "fixed-satisfaction" market basket.

The above problems relate to both measurement and concept.   Thus we may find that a particular item of merchandise has changed in quality, and yet it may be a most difficult task to measure this change in quality and to translate it into a dollar value.   The Bureau of Labor Statistics does attempt to account for these changes in quality in constructing the CPI.

Item 6 above is at a deeper conceptual level. Most economists would prefer a "constant-utility" or "constant-satisfaction" or "welfare" type of price index to the "fixed-market-basket" type of index that is currently produced by the Bureau of Labor Statistics. In other words, what change in expenditures must a consuming family make from one time to another in order to maintain a constant level of satisfaction, with the recognition that the contents of the market basket can be changed in order to accommodate changes in prices, changes in products and the like? Although steps are being taken to move the CPI in the direction of such a "constant-utility" index, the practical and conceptual problems are difficult to overcome, and progress has been slow. As a matter of fact, in September 1945, the official title of the index was changed from "The Cost of Living Index" to "The Consumer Price Index" in order to emphasize the distinction between these two approaches. In this essay we shall not treat these conceptual problems, but rather we shall focus on the role of statistics as it helps to solve CPI measurement problems.

SAMPLING AND THE CPI

Many of the above problems become more manageable if attention is shifted from the individual consuming family to a group, or *population,* of consuming families. Thus, even though some members of the population did purchase automobiles in January 1968, other members of the population did not make such a purchase, and similarly for January 1969. An automobile can then be introduced into the market basket with a weight that will reflect the effects of changes in its average price, averaged over the population of families. This shift in emphasis from the individual consumer to a population of consumers, where differences among the members of this population are known to exist, means that the construction of the CPI requires statistical sampling (and analysis) from the population of consuming families.

Moving in this direction also forces one to think in terms of the *population of goods and services* available to all members of the population of consumers and in terms of the *population of outlets* at which all of these goods and services may be purchased. Furthermore, it is manifestly impossible to study every consuming family, and to price each item of consumption in all the outlets where it can be purchased. Hence it becomes necessary to select samples from each of these populations and to draw inferences from the samples to the entire populations. The methods of statistics can assist and guide these steps.

The Consumer Price Index has never attempted to measure the changes in the prices of goods and services for all families and individuals living in the United States. Rather, because of its traditional use in collective bargaining between labor and management, its scope has been restricted to urban families. More specifically, the *population of consuming families* covered by the index consists of all urban families (including single workers) for which 50% or more of the family's income comes from wages or from salaries earned in clerical occupations and for which at least one member of the family unit works for at least 37 weeks during a year.

In determining the contents of an average, or "representative," market basket of goods and services for this defined CPI population of consuming families, it is impossible to study every family. Hence a sample, or a portion of this population, must be chosen. Because a list of the members of the population is not available, complex methods of statistical sampling must be employed.

In brief, a sample of urban communities is selected first, and then a sample of families is taken from each of the selected communities. In the 1959–64 revision, this was accomplished through an original selection of 50 cities. The 12 largest cities in the U.S., plus one city from Alaska and one city from Hawaii, were automatically included in this sample of cities. The remaining urban areas of 2500 and over were placed in homogeneous groups according to size and geographic location and 36 cities were chosen from these groups in accordance with probability methods of selection. An additional six large cities were added to the sample later, primarily so that individual city indexes could be published. This sample of 56 cities serves not only as a basis for studying the expenditure patterns of consuming families, but also as a basis from which to obtain the prices that must be used to determine the current cost of the CPI's market basket of goods and services. The CPI will continue to be based upon this sample of 56 cities until the next major revision takes place.

Within each of the 56 chosen cities, a sample of consuming units was selected and interviewed during the period 1960–61. These units were drawn in accordance with the tenets of statistical sampling theory; the goal was to choose samples in such a way that objective measures could be devised for assessing the likely size of deviations between averages for the sample and corresponding (unknown) averages for the whole population. This particular survey was called the *Consumer Expenditure Survey* (CES) because its primary purpose was to collect data relating to family expenditures for goods and services used in day-to-day living. Information was obtained through lengthy personal interviews with family members. Among the items recorded during this interview were the following:

(1) A complete account of receipts and disbursements for the preceding calendar year.

(2) The estimated value of goods and services received free.

(3) The characteristics of housing occupied by both home owners and renters.

(4) An inventory of major household furnishings owned.

(5) A detailed listing for a seven-day period of expenditures for food and beverages, household supplies, and tobacco.

Altogether, intensive interviews were conducted with 9476 consuming units. Of these, 4860 interviews were with members of the defined population of consumers for the CPI. The remaining 4616 units did not satisfy the CPI restriction to the population of consuming families, although the data obtained from them are of value for other purposes.

Data collected in the CES interviews with members of the CPI population were used for a variety of purposes in determining the structure of the current CPI. First of all, these data determined the complete contents and total cost of each family's market basket in the survey year. The number of items of expenditure in all market baskets totaled about 1800. The items were classified into five major groups: food, housing, apparel and upkeep, transportation, and health and recreation. Each of these groups was further subdivided, so that the final classification scheme consisted of 52 expenditure classes. Some examples of expenditure classes are: meats, eggs, fuel and utilities, housekeeping services, footwear, auto repairs and maintenance, hospital services and health insurance, and reading and education. In effect, the consumer's market basket was divided into 52 compartments. The total cost of each compartment was then determined for all sample consumers selected from within a particular city, and these total costs were expressed in relative terms; for example, at December 1963 prices, meats were estimated to account for 4.45% of the total cost of the nationwide market basket. Different market baskets are used in different cities to allow for differences in such characteristics as climate and the availability of different foods. Thus we use many average market baskets rather than a market basket that applies to a single consuming unit.

As observed earlier, the CES interviews provided a market basket filled with some 1800 items that consumers had purchased. It would be impossible to price all of these items in their almost infinite variety each month, even recognizing that this pricing need be carried out only in the 56 index cities. Hence another sampling problem arises. Again the theory of statistical sampling was used to select a sample of items from each of the 52 expenditure classes. The final sample for the 1959–64 revision contained 309 items. The sampling approach allows the contents of the market basket to change. For example, if one item disappears from the market, replacements may be made by further sampling. The compartments and their original weights, however, remain fixed through time.

Once a sample of items has been selected and its members specified in detail, another, more difficult sampling problem must be faced. Within each of the 56 index cities, prices of the sample items must be obtained on a monthly basis from the outlets patronized by members of the CPI population of consuming units. It is impractical to obtain price quotations from all possible outlets, and so a sample of outlets must be selected to serve as sources of price information.

Among the problems that have to be considered in developing a sample of outlets are the following:

(1) Ideally, we would define a separate population of outlets for *each* item to be priced and would select a sample from *each* of these outlet populations. This approach would be too costly because even a sample of only four or five outlets for each item would require that 1500 outlets in a city would have to be priced. Furthermore, it is difficult to compile lists of outlets on an item-by-item basis because of merchandising patterns. For example, department stores sell a tremendous variety of items, including clothing, appliances, and furniture; tire and automobile accessory stores also well appliances and toys. Hence compromises must be made in developing a sample of outlets.

(2) The sampling problems are quite different for different items. We can obtain the price rate for electricity by merely visiting the local utility company or the price of newspapers by calling the local publishers. On the other hand, the price of items such as meats and fresh produce vary widely from one grocery store to another, and a fairly large sample of price quotations are required in order to determine the average price of a grocery item with any degree of precision.

(3) There is no clearcut way of identifying the particular outlets that are patronized by the population of consuming families to which the CPI is supposed to apply. The Bureau of Labor Statistics recognizes that the shopping patterns of consuming units now range over a wide geographic area, even though the residences may be confined to a city. Thus the final sample attempts to give proper representation to the downtown and neighborhood areas of the central cities, as well as to suburban areas where so many shopping centers have been developed in recent years.

The field operations of monthly pricing are intricate. Not only must a large field staff be trained and supervised, but one must also obtain the cooperation of store managers, and make provision for businesses that cease operations or come into existence during the ten year period that ordinarily elapses between major revisions. The magnitude of these endeavors is indicated by the fact that food prices are collected in 1775 stores each month and that the Bureau is in touch with about 16,000 outlets for the pricing of nonfood commodities and services.

There is a final sampling problem, not always recognized as such, associated with the CPI. Although it is published monthly, it does not refer to any definite date during the month. The pricing operation has to be almost continual, and it is therefore necessary to choose a sample of points in time at which the prices are obtained. This sampling is not carried out in as formal a manner as are the other sampling operations. Nevertheless, every attempt is made to ensure, for example, that sale and nonsale days for food are represented in their proper proportions, and that a similar balance is maintained for other items, such as newspapers and theater admissions, whose prices may change periodically.

One important goal in statistical design and analysis is to have an objective measure of the precision of sample analyses. Although this goal has not been fully realized in the complex setting of the CPI, substantial progress was made in this direction during the 1959–64 revision, and crude estimates of sampling error have been obtained. Since 1967, these estimates are given in *The Consumer Price Index,* a monthly bulletin published by the Bureau of Labor Statistics. A recent issue of this bulletin states that ". . . any particular (month-to-month) change (in the CPI) of 0.1 percent may or may not be significant. On the other hand, a published change of 0.2 percent is almost always significant, regardless of the time period to which it relates." All indications are that the sampling operations are reasonably well under control, and that uncertainty in the value of the index due to sampling is relatively small compared to the uncertainties arising from other aspects of the process, for example, the effects of quality changes on the index.

SUMMARY

The production of monthly values of a Consumer Price Index by the U.S. Bureau of Labor Satistics is a highly complex undertaking that involves problems of *basic economic theory* (e.g., choice between a price index or a constant-satisfaction index), *measurement and quantification* (e.g., of changes in the quality of items purchased by consumers), *sampling statistics* (definition of, and selection of samples from, a wide variety of populations), and *operations* (e.g., training and supervision of price reporters).

The Index is concerned with a *population of consuming families* and with the *population of cities* in which these families live. A sample of cities serves two purposes. First, from within the selected cities a sample of consumers can be chosen from which it is possible to determine average expenditure patterns. Second, prices are collected in the selected cities to determine the value of the current CPI. The Consumer Expenditure Surveys also define a *population of goods and services* for which consuming families spend their income. From this population a sample must be taken for current pricing purposes. Within each of the index cities there exist *populations of outlets* at which items can be purchased and samples must be chosen to represent these populations. Finally, there is a *population of times* within a month at which price quotations can be obtained, and this population also must be sampled.

It is certainly true that no one is completely satisfied with the CPI in its present form. Improvements can and probably will be made in many parts of the Index, for example, in the basic data on consumer expenditure patterns, in the sampling of outlets and in the collection of price data from these outlets, in the preparation of indexes for a wider variety of subpopulations, and in techniques for handling quality change problems. There will be continuing pressure also to move the CPI more in the direction of a constant-utility type of index. It may even happen that some completely new approach to the construction of the index may be developed, possibly through the use of newer mathematical techniques. No matter what its form, however, the CPI undoubtedly will remain one of the main indicators of the state of the U.S. economy.

PROBLEMS

1. Refer to Figure 1. In 1965 how much would the average urban family have paid for the same amount of goods and services that could have been obtained for $10.00 in 1957–1959?

2. Explain the difference between a "constant-utility" and a "fixed-market-basket" type of index. Which type would you prefer the Bureau of Labor Statistics to use?

3. What is the purpose of dividing up cities into homogeneous groups before choosing the cities to be included in the price survey?

4. Why are all 12 of the largest cities automatically included in the price survey?

5. Suppose the market basket used in computing the CPI in January 1968 cost $750. How much more would it have cost in February 1968 before you concluded that a statistically significant price increase had occurred?

6. Suppose a group of rural congressmen pushed through a bill requiring the Department of Agriculture to find out the cost of living for farm families.
   (a) Could the Department refer the Congressmen to the CPI?
   (b) If not, what steps would it have to take to implement this law?

7. Suppose an anthropologist decided to construct a CPI for a remote community of Northern Albertan gold prospectors which he was studying. In January 1975 he took a survey and found that the prospectors spent on the average:
   (a) 100 ounces of gold for flour, which cost 4 ounces a sack;
   (b) 20 ounces for burros, which cost 100 ounces on the average;
   (c) 40 ounces of gold for whisky, which costs 5 ounces per pint.

In January 1976, the anthropologist finds that flour has gone up to 6 ounces of gold per sack; burros sell for only 80 ounces on the average; and whisky costs 10 ounces a pint.

What is the CPI for January 1976 (January 1975 = 100)?

REFERENCES

Ethel D. Hoover. 1968. "Index Numbers: II. Practical Applications." David Sills, ed., *International Encyclopedia of the Social Sciences,* vol. 7, pp. 159–165. New York: Macmillan and Free Press.

Philip J. McCarthy. 1968. "Index Numbers: III. Sampling." David Sills, ed., *International Encyclopedia of the Social Sciences,* vol. 7, pp. 165–169. New York: Macmillan and Free Press.

Erik Ruist. 1968. "Index Numbers: I. Theoretical Aspects." David Sills, ed., *International Encyclopedia of the Social Sciences,* vol. 7, pp. 154–159. New York: Macmillan and Free Press.

U.S. Department of Labor. *The Consumer Price Index: History and Techniques.* Bulletin No. 1517. Washington: U.S. Government Printing Office.

*In 1970 the U. S. Bureau of Labor Statistics started preparations for making a major revision of the Consumer Price Index. Among other activities, Consumer Expenditure Surveys were carried out during 1972 and 1973; a new sample of cities was selected; and improved methods of sampling were devised for choosing samples of items and samples of outlets at which current prices would be collected. The new CPI will first be published in February, 1978. A general description of the revision is given in

Julius Shiskin. 1974. "Updating the Consumer Price Index—an Overview." *Monthly Labor Review.* 97, No. 7 (July 1974), pp. 3-20.

and a description of the Consumer Expenditure Survey is given in

Michael D. Carlson. 1974. "The 1972-73 Consumer Expenditure Survey." *Monthly Labor Review.* 97, No. 12 (December 1974), pp. 16-23.

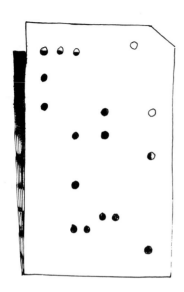

HOW TO COUNT BETTER: Using Statistics to Improve the Census

Morris H. Hansen   *Westat Research, Inc.*

THE IMPORTANCE OF CENSUS RESULTS

In 1790, Thomas Jefferson gave to George Washington, our first President, the results of the first census of the United States of America. Every ten years since then, as provided in the Constitution, the decennial census has determined for the nation essential information about its people.

The basic constitutional purpose of the census is, of course, to apportion the membership of the House of Representatives among the states. From the beginning, the census has had many important purposes beyond the constitutional one. The development of legislative programs to improve health and education, alleviate poverty, augment transportation, and so on, is guided by census results. They are used also for program planning, execution, and evaluation. Now the distribution of billions of dollars a year from the Federal Government to the states, and from the states to local units, is based squarely

332

upon census counts. (For example, if New York City is underenumerated by 2%, the result will be a loss of about 150,000 people and thus a loss of about one and a half million dollars a year in funds, or about $15,000,000 until the next census is taken.)   Private business uses the census for many purposes, for example, plant location and marketing.   Much social and economic research would be essentially impossible without census information.

The importance of the census to current problems such as poverty, health, education, civil rights, and others has brought about requests for shifting to a five-year, rather than a ten-year, census in order to keep information more nearly up to date.   These requests, which are now under congressional review, have come from governors of states, mayors of large cities, scientists, businessmen, and many others.

THE JOB OF PLANNING AND TAKING THE CENSUS

The need for a fast and reasonably accurate census is fairly obvious.   What may not be so apparent are the major problems involved in taking a census and making the results available promptly, and in adapting the census questions to serve current needs.   In each census, some questions have been changed in response to new needs, but certain basic information has been consistently required.   Questions in the 1970 population census included age, sex, race, marital status, family relationship, education, school enrollment, employment and unemployment, occupation and industry, migration, country of origin, income, and other subjects.   In conjunction, a census of housing, with additional questions, was taken at the same time.

The job of organizing and taking the census is a major administrative and technological undertaking.   Even though most of the questionnaires were filled out by the respondents themselves in a "census by mail" and the application of modern computers and other advanced technology has eliminated a lot of paperwork, the taking of a census requires the recruiting and training of about 150,000 people, most of them for only a few weeks of work.   Once specific goals are set in terms of questionnaire content and desired statistical results, the massive job of organization and administration begins.

The system for canvassing and for collecting, receiving, processing, and summarizing the vast numbers of completed questionnaires must be planned. Specialized electronic and paper-handling machines designed and built at the Census Bureau automatically read the information recorded by respondents or enumerators.   These complex machines first photograph on microfilm and then scan and read microfilm copies of the census forms that have been filled out mostly by the respondents themselves.   The magnitude of the job is difficult to comprehend.   For the 1970 census, approximately a quarter of a billion pages (counting each side of a relatively large sheet as a page) were handled. The results were recorded on magnetic tape, and then computers examined these results.   In the process, the forms were edited for certain types of incompleteness or inconsistency, and adjustments were made automatically or special problem areas were identified for further manual investigation.   The later steps of tabulation and printing for publication also were accomplished on electronic computers.

The approach to the job of taking a census differs from that of designing a totally new system, in that the census has been taken many times before, and the background and experience of the past serves to guide current efforts.

The availability of extensive past experience, however, has a disadvantage. There may be long traditions that have come to be regarded as essential, but that, in fact, only represent ways in which the job has been done in the past.   For example, the tradition of taking the census by an enumerator canvassing an area and personally asking the questions of any responsible member of the household had been long established.   This approach was regarded as proven by long use to be the only reasonable way to elicit information.   Because the concepts in some census questions are difficult, it was thought that only a trained enumerator could ask the questions and elicit the proper information.   But this view did not recognize the difficulties in training and controlling an army of temporary interviewers.   Nor did it recognize that the responses obtained in the census interview situations were sometimes based on a misunderstanding of the questions or were spontaneous without the opportunity for a considered reply.   Furthermore, the interviewer himself and his conceptions or misconceptions could importantly influence the response.

In the nineteenth century, many potential advantages of prior experience were lost, for the Census Bureau was not created as a permanent and continuing agency until 1902.   It then became far more feasible, with a continuing staff, to benefit from lessons and experience of prior censuses.   The situation for the 1900 and earlier censuses is illustrated in the following quotation from *The History of Statistics* (Cummings, 1918) :

> Mr. Porter gives the following account of his experience, which must have been essentially that of every Superintendent of the Census.

> The Superintendent in both the last two censuses [1880 and 1890] was appointed in April of the year preceding the enumeration, but when I was appointed I had nothing but one clerk and a messenger, and a desk with some white paper on it.   I sent over to the Patent Office building to find out all I could get of the remnants of ten years ago, and we got some old books and schedules and such things as we could dig out . . . I was not able to get more than three of the old men from this city . . . .   I knew most of the old census people. Some of them were dead and some in private business . . . .   But little over sixty days were allowed for the printing of 20,000,000 schedules and their distribution, accompanied by printed instructions to the 50,000 enumerators all over the country, many of them remote from railroads or telegraph lines . . . .   Now

to guide us in getting up these blanks, we had only a few scrapbooks that someone had had the forethought to use in saving some of the forms of blanks in the last census. He had taken them home, a few copies at a time, and put them into scrapbooks. The government had taken no care of these things in 1885, when the office was closed up. Some of them had been sold for waste paper, others had been burned, and others lost.

In addition to showing the potential gains from continuity and learning from the past this quotation suggests the great complexity of the censuses in the latter half of the nineteenth century. At that time, there was little in the way of a continuing statistical program in the federal government and as a consequence the decennial census was loaded with a range of questions that proved difficult if not impossible to collect through decennial census inquiries—hence the great number and variety of questions and forms. Many of these types of information are now collected through sample surveys or compilations from administrative records.

THE USE OF STATISTICS IN PLANNING AND TAKING THE CENSUS

Statistical concepts and methods have provided fundamental improvements to the census over the past 30 years. We might say that these improvements form a technological explosion. Part of this explosive change has come from introduction of the large-scale electronic computer, but even more of the change has come from statistical advances and the application of statistical studies. (In fact, development of the high-speed computer and modern statistical methods were both substantially motivated by census problems, and were in part carried out by census scientists or with census support.)

The Introduction of Sampling as a Tool for Census-Taking. Sampling was first used in collecting census information in the 1940 census. A series of questions was added for a 5% sample of the population. [Roughly speaking, this meant asking every twentieth person the additional questions (Waksberg and Pearl 1965).] This was a major advance, as the tradition for a century and a half had been universal coverage for every question. The questions asked of the 1940 sample included one on wage and salary income—income had not previously been a census question—one on usual occupation (as distinguished from current occupation; the 1940 census was planned during the Depression, when unemployment was very high and there was frequently a difference between a person's usual occupation and the occupation at which he was currently working), and several other questions.

The art and science of modern statistical sampling were evolving at the time of the 1940 census, and at the same time there was increasing public acceptance of sampling. It was possible to proceed with greater knowledge and confidence about what a sample would produce than would earlier have been possible. Thus, for a particular size and design of sample, statisticians could establish a reasonable range for the difference between a sample result and that obtained from a complete census. Suppose, for example, that a city had a total population of 100,000 persons of whom 30,000 were employed and earned wages or salaries, and suppose that 10,000 of these received wages and salaries of less than $2000 in 1939 (the year preceding the census). There would be approximately 5000 people in a 5% sample from that city. The estimate of the number receiving less than the $2000 wage and salary income would, with a very high probability, when estimated from the 5% sample, lie within the range 9300 to 10,700. This kind of accuracy was sufficient to serve many important purposes and, in fact, was as much accuracy as could be justified in the light of the less-than-perfect accuracy of the responses to the question on income. Not only could estimates be prepared from the sample of what would have been shown by a complete census, but in addition, the range of probable difference between the sample estimate and the result of a complete census could be estimated from the sample! Sampling theory also guides in designing samples to achieve a maximum precision of results per unit of cost.

In considering the advantages of the use of sampling, it may appear to some that the main work involved in taking a census is the time it takes in going from door to door and that, once some questions have been asked at a household, the cost of adding questions would be small whether they were added for a 5% sample or for the total population. Such a presumption is far from true. Suppose, for example, that an additional question about whether a person has a chronic illness adds an average of 20 seconds of work for each person counted in the census. With some 200,000,000 people in the population, this would add more than 1,000,000 hours of work and perhaps $4,000,000 to the cost of the census. Thus, obtaining the added information for, say, five questions from only a 5% sample instead of from all persons can produce needed and highly useful results at a fraction of the cost for complete coverage. Finally, the use of a sample permits tabulation and analysis much sooner than complete coverage.

Starting with the forties, sample surveys on a wide range of subjects were introduced so that continuing and up-to-date information would be available between censuses. For example, the Current Population Survey is a sample of the population conducted monthly by the Bureau of the Census. The Survey collects information each month on employment, unemployment, and other labor force characteristics and activities of the population. It also serves to collect information on other subjects, with different supplemental questions in various months. In one month each year almost the full range of population census questions is asked. In other months information may be requested on recreational activity, housing, disability, or other subjects. Sample surveys include a continuing health survey, and surveys covering retail trade, business and personal services, the activities of governmental units, and so on. These surveys have large enough samples to provide national information, some information for regions of the nation, and even information for large states and metropolitan areas. They cannot, however, provide information for the many individual cities and counties, and for relatively small communities within the cities and counties. To obtain that kind of fine detail, very large samples are needed, samples such as those taken as part of the decennial census.

In the 1950 census, the use of sampling was extended to some questions that in earlier censuses had been collected from all persons. For these questions, a 20% sample was used. This sampling was extended in the 1960 census to most of the items of information. In the 1960 and 1970 population censuses, only the basic listing of the population, with questions on age, sex, race, marital status, and family relationship, was done on a 100% basis.

The following question is often raised: If sampling is so effective a tool, why not take the whole census with a sample? Isn't it a waste of effort to do a complete census? In response, I always point out that the primary purpose of the census is not to obtain national information, but to provide information for individual states, cities and counties, and for small areas within these. The results obtained by converting the whole census to a relatively large sample (perhaps including 20% of the population) *would* be adequate for most purposes. Such a sample, however, would not apportion representatives in the state legislatures in the same way as a complete census. Similarly, there could be important differences in the distribution of vast amounts of funds to thousands of individual small areas. Even very small differences in the total counts by states may decrease by one the number of congressmen from one state and increase by one the number from another— and they can alter the allocation of funds to the states by billions of dollars. Also, in some states, the legal status of many communities depends on the exact size of the official population count; for example, a city with 10,000 or more people can issue bonds. Hence a complete census is needed for total population counts and for some other basic population data, but the great bulk of the information may be collected from a relatively large sample. Much smaller samples can be, and are, used to collect data on items needed only for larger areas such as individual large cities, metropolitan areas, and states.

The Use of Sampling and Experimental Studies to Evaluate and Improve Census Methods and Results. Substantial steps to evaluate and improve census methods were taken beginning in the forties and have been continued and greatly extended since then. Statistical studies have been made of various aspects of the census. One such study was made by repeating the census enumeration, in a well-designed sample of areas, shortly after the original census enumeration, using the same procedures as in the initial census. Thus we were able to see something of how much alike two censuses taken under the same conditions and procedures would be.

Studies of these types show high consistency and accuracy of response for questions such as sex, age, race, and place of birth, but they show higher degrees of inconsistency and inaccuracy in responses to the more difficult questions relating to occupation, unemployment, income, education, and others. The information from such studies guides both in improving the questions in the next census (by, for example, showing which questions cause trouble and need rewording), and in interpreting the accuracy of census results when the questions are put to specific uses.

Studies that compare alternative methods and procedures within the framework of well-designed and randomized experiments have been exceedingly important for learning about the effectiveness of various procedures and in comparing their costs and their accuracies. Such comparisons have been made, for example, of various types of questionnaire designs or of other variations in procedures. For example, one study reexamined the census coverage and questions in a sample of areas, but used more highly trained enumerators, more detailed sets of questions, and other such expensive improvements.

These rather wide-ranging studies led generally to the conclusion that some of the methods earlier regarded as the way to achieve major improvements would not be effective in relation to their cost (although some worthwhile improvements in questionnaire design and procedures were accomplished). We find, for example, that simply spending much more time and money on training an army of temporary enumerators and paying them hourly rates instead of piece rates, making questions more detailed, and insisting on personal interviews for each adult respondent would all add greatly to the cost of a census, but would not make corresponding improvements in its accuracy. There appear to be intrinsic limits to what can be done with a vast temporary organization.

The Surprising Effect of Enumerators. One study, however, led to surprising conclusions and then to a basic improvement in census procedures. It had long been known that enumerators can and do influence the answers they obtain—presumably, unconsciously most of the time. But the *magnitude* of this enumerator effect was not known. Hence a large statistical study was carried out as part of the 1950 census to measure the magnitude of enumerator effect.

The study plan (in a simplified description) was based on areas divided into 16 work assignments (areas small enough so that 2 work assignments could be canvassed by a single enumerator) and 8 enumerators. Two of the 16 work assignments, chosen at *random,* were given to the first enumerator; two of the remaining ones, also chosen at random, were given to the second enumerator, and so on. Of course the whole 16-fold experiment was repeated many times throughout the country.

The random choice of work assignments was essential here in order to interpret the results in a useful way. (Random choice means choice by a method equivalent to writing the numbers 1, 2, . . . , 16 on identical cards, shuffling or mixing them thoroughly, and then picking first one, then another, and so on.)

Another essential feature of the plan was that each enumerator had two work assignments and that there were a number of enumerators. That way, good estimates could be obtained for the variability introduced by a single enumerator (roughly, the differences in performance by the same enumerator that were not attributable to differences in areas) as compared with the variability stemming from differences *among* the enumerators.

Further details of this path-breaking analysis cannot be given here, but we can summarize the results. Far greater differences between enumerators were found than had been anticipated, not so much on items such as age and place of birth, but on the more difficult items such as occupation, employment status, income, and education. For those items, in fact, a complete census would have as much variability in its results (because of enumerator effect) as would a 25% sample if there were no enumerator effect!

What to do about this? One approach might be to expend far more resources on the selection, training, and supervision of enumerators. But the other studies mentioned above had shown that this was not feasible under the conditions of a national census, in which temporary enumerators are hired, trained, and do their job in only two or four weeks.

Another possible answer was to eliminate the need for the enumerator by leaving the carefully designed questionnaires with the respondents and asking them to fill them out and mail them. (Enumerators would be needed only when respondents' returns were incomplete or where the respondents asked for help.) This method was tested in further studies and found to work quite well. It was used in the 1960 census for the larger census forms and was a great success, in terms of both cost and added accuracy. Hence, in 1970, this method was used still further: most of the population received and returned the census forms by mail.

Thus, in the process of using statistics to improve the census, the completeness of coverage has been improved, the accuracy of the items of information collected has been increased, and the time taken to publish the results has been decreased by about half. Now most of the data are available for public use within six months to one and a half years following the completion of the collection of the data, while in the 1950 and earlier censuses the corresponding time for making available the same basic information was roughly one to three years.

With all of this, costs (in terms of equivalent dollars, adjusted for changes in salary rates and other costs) have been reduced. The results have been achieved by the application of sampling and other statistical methods to census data collection in the field, to the processing of the information, and to the study of methods for evaluation and improvement.

PROBLEMS

1. Give several reasons why a census is taken.

2. Why is it desirable in some instances to take a five percent sample as opposed to a complete census? What does the five percent sample lose?

3. Comment on the advantages and disadvantages of using enumerators in a census.

4. Answers to questions regarding sex, age, race and place of birth seem to be more reliable than those regarding occupation, unemployment, income and education. Give two possible reasons for this. Do you think answers to mailed questionnaires would be more accurate than interviews by enumerators on these questions? Why?

5. (For group work). Take a small random sample on some issue, for example, opinions of the local newspaper. Report on the problems and situations that arise.

6. Suppose the Census Bureau decided to add a battery of 6 questions on energy consumption to the 1980 Census. About how much money would the Census save if, instead of asking everyone the questions, they asked them of only a 1% sample?

7. Suppose the 1980 Census uses mailed questionnaires instead of enumerators, and 40% of the returned questionnaires have no answer marked to a question on mental health. Would the Census be justified in counting only the marked responses and publishing the result as a 60% sample? Justify your answer.

8. Someone in the Census Bureau suggests using telephone interviews instead of mailed questionnaires for families with telephones. He thinks that this technique might lead to a more accurate census.

   (a) What arguments can you think of which support or refute his idea?

   (b) Design an experiment to determine whether his idea is correct.

   (c) What other considerations (besides accuracy) would determine whether or not the telephone technique would be adopted?

REFERENCES

John Cummings. 1918. "Statistical Work of the Federal Government of the United States." John Koren, ed., *The History of Statistics*. New York: Macmillan. Pp. 678–679.

Morris H. Hansen, Leon Pritzker, and Joseph Steinberg. 1959. "The Evaluation and Research Program of the 1960 Censuses." *1959 Proceedings of the Social Statistics Section, American Statistical Association.* Washington: American Statistical Association.

Morris H. Hansen and Benjamin J. Tepping. 1969. "Progress and Problems in Survey Methods and Theory Illustrated by the Work of the United States Bureau of the Census." Opening address at Symposium on Foundations of Survey Sampling, April 22–26, 1968, University of North Carolina, Chapel Hill. Norman L. Johnson and Harry Smith, Jr., eds., *New Developments in Survey Sampling.* New York: Wiley (Interscience). Pp. 3E(1)–3E(11).

Ann Herbert Scott. 1968. *Census, U. S. A.: Fact Finding for the American People 1790–1970.* New York: Seabury.

Joseph Waksberg and Robert B. Pearl. 1965. "New Methodological Research on Labor Force Measurements." *1965 Proceedings of the Social Statistics Section, American Statistical Association.* Philadelphia: American Statistical Association. Pp. 227–237.

INFORMATION FOR THE NATION FROM A SAMPLE SURVEY

Conrad Taeuber   *Associate Director, Bureau of the Census*

THE DOORBELL at the Robert Brown house rings as the Brown family is finishing lunch. Nick Brown, the oldest son of the family, goes to the door. "Mom," he calls back, "it's Mrs. Smith, the Census Bureau lady." This is not the Census lady's first visit to the Brown's, so no further introduction is needed. "Well, tell her to come on in," says Mrs. Brown, and Mrs. Smith is invited to join the Browns for coffee.

Courteously, but without wasting time, Mrs. Smith verifies that all members of the Brown family are still living at home and that no one is there to visit or stay. Assured that there are no changes, she begins her questions:

Mr. Brown, what were you doing most of last week, working or something else? How many hours did you work last week at all jobs? For whom did you work? What kind of business or industry is this? What kind of work were you doing? Mrs. Brown, what were you doing most of last week, keeping house or something

342

else? Did you do any work at all, not counting work around the house? Did you have a job or business from which you were temporarily absent or on layoff? Have you been looking for work?

Mrs. Smith turns to Nick, who says he's been looking for work:

What have you been doing to find work? Why did you start looking for work? How many weeks have you been looking for work? Have you been looking for full-time or part-time work?

And similar questions are repeated for each member of the Brown household who is 14 years old or older.

Every month, about 50,000 families in the nation are interviewed in this way. Yours may have been one of them, although that is unlikely. Why is this program carried out? To understand this, we turn from the specific interview in the Brown household to the broad national scene.

A newspaper headline in the fall of 1969 announced "U.S. Jobless Rate Advances to 4 Percent, Highest Since '67." The same newspaper went on to observe, "Administration aides greet rise as sign of restraint on boom spurring inflation. Former Vice President [Humphrey] is reported as being critical, but a Treasury official said no change in tax policy is yet under consideration." The reporter had learned that the development was greeted by Administration officials as a welcome sign that their policies, aimed at ending inflation by slowing business expansion, were effective.

The rise in unemployment that month was spread throughout the work force. Unemployment rates increased for almost every category of workers— adult men, adult women, teenagers, Negroes, and Caucasians. The ones most seriously affected, however, were men in the 20 to 24 age group, blue collar workers, construction, trade and manufacturing employees, and agricultural workers.

These facts and many more gave the nation a major indication of the state of our economy in September 1969. Similar figures are available every month and they provide the Congress and the Administration with essential information to help in charting the course of the economy. The figures are widely published. Both the Administration and its critics find in these figures information about the levels of employment and unemployment and the persons who are most directly affected by changes in employment. At a time when there is serious concern with the possibility of inflation and with the consequences of efforts to slow down the rate of economic growth, the statistics on unemployment are carefully watched. Responsible officials want to be sure that the policies are working as desired and that the Government has an early warning if the effects are so great that action is needed to try to avoid dangerous levels of unemployment or recession.

Such statistics, and many others, are essential tools in the nation's efforts

to guide its economy.   Information is needed on how many people are working, how many are unemployed, whether the workers are on overtime or
on short time, whether the unemployed are primarily married men, single
young men or women, and whether unemployment is hitting harder at young
or older ages and at black or white.   There is a need also for information
on how many workers are on temporary layoffs, how many are young persons
looking for their first jobs, and how many are women looking for jobs when
their husbands have been thrown out of work.

Unemployment may mean many different things, depending on who is
unemployed and for how long.   The unemployed worker may be simply between jobs, with a short break before he takes on the new one.   He may
have been laid off because a plant has been shut down or relocated, or simply
because a plant has temporarily reduced its work force.   He may be a young
person testing the labor market, looking for a job for a short time, or trying
to decide whether to get a job or to stay in school.   She may be a housewife
who would like some additional work to help out with the family expenses,
to buy some special item, or to finance the schooling of the children.   She
may be a housewife needing a job in order to supplement a husband's inade-
quate income or to make up for the fact that he has no income.   The unemployed worker may be a man who finds it difficult or impossible to leave
his home area where there is no job in order to go to some other location
where one may be available.   He may be an older worker whose skills have
become obsolete and who faces continuing unemployment unless he can be
retrained.   Among three million unemployed, there are people in many different situations.   Because of its concern with the welfare of all Americans
and its commitment to a policy of full employment, the U.S. nation has
developed a wide variety of measures to help meet the problems of unemployed workers.   It needs a continuing source of reliable information on the
actual situation at a given time and on the changes from one month and one
year to the next.

We now have much more up-to-date information than we did in the
early thirties when the nation was deeply concerned with the problems of unemployment.   Then everyone knew that there was an unbearably large number of unemployed, but hard information was lacking on how many there
were and what types of persons they were.   Different authorities made widely
differing estimates.   No one had a real count; people had to rely on guesses.
Though the Administrator of the Federal Emergency Relief Administration
could state emphatically that hunger is not debatable, even he did not have
a good measure of the size or nature of the problems that he had set out
to correct.   It was the early efforts of the Works Progress Administration
to get some reliable measure of changes in the levels of unemployment that
led to the establishment of the survey which now provides the official monthly
figures on employment and unemployment.

These figures, of course, do not settle the arguments about how much unemployment is normal, tolerable, or dangerous. Nor do they entirely satisfy the people who feel that a person on part-time work, but who would like full-time employment, should be counted as unemployed. Some think that 14- and 15-year-olds who want work should be counted as unemployed. There can be differences of opinion about the proper classification of persons on temporary layoffs, those who are not working because of some labor dispute, seasonal workers in the off-season, the "discouraged workers" who have given up looking for a job because they are sure none is available, and the persons who aren't sure but say they would like a job if an attractive one comes along. It is sometimes argued that the primary concern should be about married men with dependents and that social economic policy need not be seriously concerned about others who may be unemployed.

Statistics cannot settle such policy questions, but they can supply information that helps in identifying essential issues and in narrowing the debate to the policy issues. Persons advocating different policies can carry on useful discussions or debates only if they are agreed on certain basic propositions. If they can agree on the statistics, they can then match their different interpretations against each other. If, however, they do not agree on what the basic facts are, much of the argument may be fruitless. Even though they may disagree on precisely who should be included and who should not, they can narrow the policy argument if they can agree on how many and what kinds of people are included in the groups about which they disagree.

How do we know about the number of unemployed and whether they are young or old, men or women, black or white, blue collar or white collar workers? How do we know how many of them have just become unemployed and how many have been looking for work for three months or more? How do we know that unemployment in one month is more or less than it was in the preceding month or in the same month a year ago? There are important seasonal changes in some kinds of work—the harvesting of crops, canning, retail selling with its large need for temporary help before Christmas and Easter, and many others. We need to take all of these seasonal changes into account in assessing whether a change from one month to the next is really significant or whether it is simply a reflection of the normal seasonal development.

THE CURRENT POPULATION SURVEY

The basic source of such information, both in this country and in some others, is the Current Population Survey (CPS). In the U.S., every month some 1100 persons leave their housework or other duties and call at some 60,000 addresses to ask the occupants a number of carefully worded questions. The answers (usually there are some 50,000 of them) are promptly assembled, and from them, the nation has its monthly report on employment and unemployment.

Every month, in the week that includes the 19th, these interviewers begin their rounds. Each one has about 60 addresses on his or her list. The questionnaires are so designed that many of the answers can be recorded simply by filling in little circles. Some of the answers, however, must be written out, for example, the kind of work that each worker was doing. The interviews begin on Monday and always relate to activity during the previous week, that is, the week that includes the 12th of the month. Completed questionnaires go to a regional office of the Census Bureau, where they are given a quick review, and then promptly shipped to the Census Bureau's processing office in Jeffersonville, Indiana. There trained persons translate the answers from ordinary language into a language that the computer can read. If the interviewing begins on the 15th of the month, all of it is completed by the 22nd, and the computer receives all of the forms by the 29th. Overnight it does the work of combining the data into usable tables. Approximately two weeks after that month's enumeration began, the first statistics are available. They need to be reviewed to be sure all of the processing, including that done by the computer, has been done in accordance with the instructions. Within two days more, the results are ready for distribution to newspapers, radio, and television, which pass on the information to the public. Analysts in the Government and outside then have a new set of figures to analyze in order to arrive at proper policy decisions in the light of the changes revealed that month. This process is repeated month after month, as it has been in essentially similar form for more than 25 years.

Anyone using such statistics is likely to ask how reasonable it is to draw conclusions about all of the 60 million households in the country from a sample of a little more than 50 thousand (about 0.1%). To answer this natural question, we must look at the method by which the sample of households is drawn. If that method is highly restrictive (for example, if only families in major industrial cities were selected or only farm families), then it would obviously be inappropriate to reach conclusions about all the country from the sample. Similarly, if families were chosen by asking each interviewer to interview 50 families who are his friends or neighbors, we might properly doubt inferences from the sample. We want to be sure that the sample has appropriate numbers of rural and urban families, of high- and low-income families, and so on. All parts of the country should be represented in proportion to their share of the population.

The major assurance that the sample will reflect conditions for the entire country comes from the fact that the households are probabilistically selected from within homogeneous strata. Not only does this help to keep the process free from personal bias on the part of the persons choosing the sample, but it also permits accurate knowledge of the precision of the estimates that come from the sampling. This also means that it is possible to determine the sample size that will assure the degree of precision in the data needed to meet the requirements of public policy. The details of how the sample is selected are described below.

There are more than 60 million households in the U.S., and the problem is how to select an appropriate sample each month of about one out of every 1240 for interview. If there were an up-to-date list of all households in the U.S., one out of every 1240 of them might be selected in an appropriate manner, but there is no such list.

Because the information is to be collected by interviewers visiting the households, it is desirable to select the sample in clusters of neighboring households. This helps the interviewer get to all of them within the few days each month in which the interviewing must be done. In this survey, clusters of about six households are used.

Another requirement is that the sample should be selected in such a way that it is possible to estimate from the sample itself how much the results differ from those which would have been secured if every household in the country had been included. It is well known that the results from a sample are rarely precisely the same as those from a complete enumeration, but from a properly designed sample, we can measure the chance that these deviations are small enough so that the major findings from the sample can be used with confidence. The same type of consideration applies to the comparison of changes from month to month. For them, too, it is necessary to know whether a difference is real or is simply the kind of chance difference that could be the result of "sampling error."

The problem for the statisticians was to develop a way of drawing the sample from the entire country in such a way that an economical and reliable survey could be conducted each month. A measure of the degree of confidence to put in the figures also should be obtained from the sample.

It was decided to give each interviewer a fixed set of addresses at which to call. If no one is living at one of the addresses or if the persons found there do not actually live there, that address will not contribute data to the survey for that month. If a household moves away from an address between two of the monthly visits, it is dropped from the survey and the household which has moved to the sample address is included. The interviewer is responsible for completing an interview at each of the assigned addresses, or explaining why no interview was required or possible. If no one is found at home after several calls or if the residents at the assigned address refused to provide information, it is so noted. No substitutions for sample households are allowed.

The Current Population Survey, which is the source of this information, is carried out in 449 sample areas, which include 863 counties and independent cities.   They are located in every state and the District of Columbia. In all, about 60,000 residential addresses are designated for the sample each month; about 52,000 of them, containing about 105,000 persons 14 years of age and over, are actually interviewed.   The survey is limited to the civilian population and excludes members of the Armed Forces, as well as persons living in institutions, such as prisons, long-stay hospitals, and so on.   Most of the 8000 addresses for which interviews are not secured are for housing units that are vacant, units that have been converted to nonresidential use, units whose usual occupants are temporarily away, or those that are temporarily occupied by persons who actually live somewhere else (for example, persons occupying a home on their vacation).   Answering the questions is voluntary, but fewer than 2% of the persons interviewed refuse to answer.

The first step in selecting the households is to select a sample of the counties or equivalent governmental units.   These are then subdivided into subareas, and a selection is made among these subareas.   Within each of the selected subareas, a sample of addresses is selected for interviewing.[1]   In making the final selection of the specific addresses to be used, two different procedures are utilized.   In urban areas it is generally possible to work with addresses that give specific house numbers and streets and even apartment numbers. When such lists are available, a cluster of about 18 consecutive addresses is selected from the census enumeration districts (ED).   Every third address within the cluster is taken for the current sample; the remaining 12 addresses are saved for use in future samples.   Arrangements are made to take into account new construction since the last census, chiefly by checking building permits.

In rural areas and other areas where such addresses are not available, *area sampling* is used.   The sample EDs are subdivided into small land areas with well-defined boundaries.   Insofar as can be determined from available information, each such area segment has about six housing units.   If it is not possible to define area segments of that size, larger segments may be defined.   These six addresses are drawn by a systematic sampling of all housing units.

It is desirable to have a household in the sample for consecutive months, and for the same months in successive years in order to secure measures of month-to-month and year-to-year change.   To avoid overburdening the households who cooperate in the interview, it has been arranged that interviews are conducted at a sample address for four successive months, then that address is omitted for eight months, and after that, it is interviewed again for four consecutive months.   This rotation occurs in such a way that each month one-eighth of the sample addresses are entirely new, one-eighth consists of addresses that are starting on the second round of four interviews, and three-

---

[1] See the appendix to this essay for a more detailed description of the sampling procedure.

fourths were interviewed in the preceding month. Thus one-half also were interviewed in the same month a year before.

PRECISION AND ERROR CONTROL

Modern sampling theory makes it possible to measure the size of the fluctuations arising from the sampling process, for the probability of including any unit in the sample is known. Of course, as in any survey, there are also other, nonsampling sources of error that must be investigated and controlled. Interviewers must be carefully trained both in the techniques of interviewing and in the content of the questionnaire they are using. Every effort must be made to assure that respondents understand the questions and provide correct answers. Controlling such a series of interviews requires attention to all possible sources of error, and taking appropriate steps to control them. In the case of the Current Population Survey, there is continuing training of the interviewers, careful review of their work each month, periodic observation, and a program of reinterviews by supervisory personnel.

On the average of twice a year, a subsample of the addresses assigned to each interviewer is visited a second time by a supervisor and the same questions are asked again to make sure that the correct information has been obtained. The interviewers do not know when their work will be checked or which addresses will be selected for the reinterview. The supervisors do not know at which addresses they are to reinterview until after the initial interviews have been completed. If the information secured at the two interviews differs, an effort is made to find out which of the two answers is correct and why the differences occurred. The reinterview program serves as a basis for further training of the interviewers and gives a measure of the quality of the survey in general.

All steps in the office processing of the interviews are similarly kept under constant control. The preparation of the estimates in which the results of the sample are projected to the entire population also requires application of modern statistical principles. The results are published with a measure of the sampling variability of each of the major figures.

QUESTIONS ASKED

Selecting the sample properly and training and supervising the interviewers carefully to make sure that they interview at the designated addresses and report the replies accurately would not be adequate if the questionnaire itself were not well designed or if the questions to be asked were left to the interpretation of each interviewer or each respondent. A question such as "Were you unemployed?" would give results of little value, for there would be wide differences in the interpretations placed on such a question. Instead, over the years a battery of questions has been developed to secure information on what a person actually did during the survey week and to classify him or her on that basis. If he reported that he was working, he is asked how many hours he worked during the week, and because people are likely to report some standard number such as 40, he is also asked whether he worked any overtime or lost any time or took off any time. He is asked also to state what his job was and for whom he worked, as well as the kind of business or industry in which he worked.

If he did not work during the survey week (i.e., the week preceding the interview) he is asked whether he had a job or business from which he was temporarily absent or on layoff. If he was absent, he is asked why and whether he was paid for the time off. If he was not working, but looking for work, he is asked to indicate what he has done during the preceding four weeks to find work, such as checking with an employment agency, with employers, or with friends and relatives, placing or answering advertisements, and so on. He is also asked why he started looking for work, whether because he lost his job, quit his job, left school, or wanted temporary work. There are questions to ascertain how long he has been looking for work, whether he was looking for full-time or part-time work, and whether there was any reason, such as illness or school attendance, why he could not have taken a job last week. There are questions on when he last worked for pay, and the kind of work he did on his last job.

If he is not looking for work, he is asked when he last worked for pay, why he left his last job, and whether he intends to look for work of any kind in the next 12 months. He is also asked why he is not looking for work, whether because he believes that no work is available in his line of work or in the area in which he lives, that he lacks the necessary training or skills, that employers think he is too old or too young, or that there is some personal handicap that stands in his way. Other possible reasons include family responsibilities, the inability to arrange for child care, ill health, a physical disability, or the fact that he is going to school. The answers to these questions provide a basis for meaningful classification of persons as employed or unemployed. Information on the respondent's sex, age, color or race, marital status, relationship to the head of the household, and years of schooling is also available to assist in providing some meaningful classification of persons as employed, unemployed, or not in the labor force.

The published results provide a monthly measure of national employment and unemployment. From time to time a few questions are added to the questionnaire. These special questions, in addition to those that are asked every month, provide the basis for much of the information about important social trends in the years between the major censuses.

Because the sample continually reflects the growth or decline of population in the sample areas, it provides a basis for estimating the major geographic shifts of the population. Between 1960 and 1968 the total population of the central cities of our metropolitan areas remained almost the same, but there was a major change in the makeup of the population of these cities, including a net loss of about 2.1 million Caucasians and a gain of about 2.6 million Negroes. The continuing movement of people to the suburbs is clearly reflected in the statistics from this survey, with about four-fifths of the national growth occurring in the suburban areas. The survey has also reflected a slowing down of interregional migration during the sixties, compared to the higher rates of migration in the fifties.

The annual report on the number of families and persons in poverty comes from this same survey, for once a year each person in the survey is asked to report the amount of income he received during the preceding year. The number of persons living in poverty declined from nearly 40 million in 1959 to about 25 million in 1968. The percentage of the population living in poverty declined from 22 in 1959 to 13 in 1968. There were differences in regard to income and poverty between Caucasians and Negroes, as shown by these figures. In 1959, more than half the Negroes were classified as living in poverty; in 1968, the comparable figure was 35%. Average family income has been increasing. In 1968, it amounted to $8600, which represented a real gain of about 3.5% over the previous year, even after allowing for the increased level of prices.

An important measure of educational attainment is the proportion of young adults who have completed high school. Between 1960 and 1969 the proportion of white men 25 to 29 years old who had completed at least four years of high school rose from 63 to 78%. For black men in the same age bracket, the percentage went up from 36 to 60% in the same nine years.

Americans are a mobile people, and the survey provides a measure of that mobility. One person in five changes his residence in the course of a year; most of these moves are within the same county. There has been very little change in this rate of mobility over the last 20 years for which information is available.

The survey also indicates that in homes with children that include only one parent (usually the mother) the family income is likely to be relatively low. In 1969, 89% of white children and 59% of black children were living with both parents. In recent years, there has been little change in this percentage among whites, but some decrease among blacks.

Such facts about life in the U.S. are available annually because the families included in the survey are willing to answer the questions put to them by the interviewers. A census covering the entire population is taken only once in 10 years. Between censuses, we now have statistics that reveal changes for the nation as a whole and for the major regions. A survey of this size, however, cannot provide figures of the same reliability for states, individual cities, or for metropolitan or smaller areas.

From time to time, the Bureau of the Census is asked to add some other questions to the questionnaire. As a result, it has been able to supply statistics on the number of persons who smoke, and the proportions of young men and women and older men and women who do so, as well as the numbers who quit smoking. The most recent survey showed that about 2.7 million persons had quit smoking between 1966 and 1968. Some persons began smoking during that time, but the total number reporting that they smoke dropped by about 1 million. Information has also been supplied on the number and proportion of persons who have had immunization for polio, smallpox, diptheria, and other communicable diseases.

In view of the public interest in the percentage of persons who vote, questions have also been asked about whether the individual had voted and, if not, whether he had been registered to vote. The survey found shortly after the 1968 election, that 68% of all persons of voting age reported that they had voted. Men had higher voting participation rates than women. Northern blacks had a higher voting rate than Southern whites. Persons over 65 and those under 35 had lower voting participation rates than persons between 35 and 64; persons with higher educational levels and those with higher incomes tended to have higher voting participation rates. Unskilled workers had lower voting participation rates than persons in the occupations that required higher levels of training.

THE ROLE OF THE SURVEY

This survey, which involves the willing cooperation of more than 50,000 households each month, has proven to be an important source of information about the U.S. This information is needed by the Government in planning its economic and other policies. It is also needed by many other agencies and by private organizations concerned with employment and unemployment, educational levels, poverty, incomes, health, and living conditions generally. Through the application of modern sampling theories and through a system of carefully developed training and supervision, it has become possible to provide this and much other information on a timely basis and at a fraction of the cost of conducting a complete census. Congress and other policy makers and administrators look to this source for up-to-date information, knowing that the results are reliable.

Such a survey is not a substitute for a census, which provides information for each state, county, and city, and for smaller areas, but it does provide essential information between censuses and is an indispensable source of data needed for a continuing appraisal of important developments in the nation.

APPENDIX: PROCEDURE FOR SELECTING SAMPLES FOR THE CURRENT POPULATION SURVEY

The first step in selecting the addresses to be visited was to determine the counties in which the interviewing was to be done. This was accomplished by combining all of the counties in the U.S. into Primary Sampling Units (PSUs). Each of the Standard Metropolitan Statistical Areas (SMSA) was taken as a PSU. An SMSA is a city of 50,000 or over, plus the county in which it is located and adjoining counties that are closely tied to it economically and socially as determined by certain specified criteria. Outside the SMSAs, counties were grouped into PSUs (small groups of counties that are sufficiently compact so that a sample of households within a unit could be visited without undue travel cost). Whenever possible, a PSU was constructed to include both urban and rural residents of both high and low economic levels and, to the extent feasible, a variety of occupations and industries. When the current sample was selected in 1962, there were 1913 PSUs, including the 212 SMSAs defined in the 1960 census.

The next step was to combine these PSUs into 357 strata. Each of the SMSAs with 250,000 or more residents was treated as a single stratum. The other strata, in general, consisted of sets of PSUs as much alike as possible in various characteristics. These included geographic region, density of population, rate of growth between 1950 and 1960, the proportion of the population in 1960 that was not white, the principal industry, type of agriculture, and so on. Except for the strata in which an area represented itself (such as the larger SMSAs), the strata were so arranged that their populations in 1960 were approximately equal.

If a PSU was a stratum by itself, it automatically fell into the sample, and thus there were 112 strata that consisted of only one PSU. Within the other 245 strata, PSUs were selected for this sample at random in such a way that the probability of each being drawn was directly proportional to its 1960 population. Thus, within a stratum, the chance that a PSU with 100,000 persons would be drawn was twice as great as that of a PSU with only 50,000 persons.

The next step was to select the sample households within the designated PSU. The sampling rate within each PSU was determined in such a way that the overall chance of an address being included is equal to one in 1240. In this way, it is possible to reflect changes, such as new construction or demolition of housing units or changes in the characteristics of the population of the area that have occurred since the last census was taken. Within each designated PSU, the first step was to select a sample of the census enumeration districts (ED) that were used in 1960. These small administrative units contained approximately 250 households in the 1960 census. The EDs were arranged in geographical order to make sure that the sample EDs will be spread over the entire PSU. The probability of selection of any one ED is proportionate to its 1960 population.

PROBLEMS

1. Explain why a large battery of questions are asked on unemployment rather than just the question, "Were you unemployed?"

2. Briefly explain how households are selected for inclusion in the CPS.

3. List at least three advantages which the Bureau of the Census gains by sampling 60,000 households rather than interviewing every household in the Current Population Survey.

4. Are the figures reported monthly for unemployment among black males aged 20–24 likely to be just as accurate, more accurate, or less accurate than the figures reported monthly for unemployment among all workers? Why?

5. Why doesn't the Census just sample employers and find out how many workers are employed, instead of carrying out the CPS to determine an unemployment index?

6. The Census uses rotation sampling instead of taking a completely fresh sample every month. Why? Do you think that the Gallup Poll should use rotation sampling? What about the Nielsen television ratings survey?

HOW CROWDED WILL WE BECOME?

Nathan Keyfitz   *University of California, Berkeley*

ALL STATISTICAL facts concern the past. The Census of April 1970 counted 205 million of us, but we did not know this until November, despite the census emphasis on speed, pursued with ingenuity and with much new electronic equipment. Stock-market prices and volumes are hours old before they appear in the evening paper. Statistics of plans or intentions are only an apparent exception. No one can ever gather data directly on the future.

Yet the actions that statistics serve to guide can occur only in the future. The local telephone company wants to know how much this town will grow in population over the next few decades. Its interest is not abstract curiosity, but contemplated construction of new lines out toward a certain suburb. The investment might occur in the next two or three years, and the service given by the investment along with the income derived from it would be spread over 30 years. If the town does not grow as much as expected, the construction would be wasteful. If the growth is in the direction of a different suburb, then lines will be idle on one side of the town and too often busy on the other side. School authorities, the bus company, a textile manufacturer, all similarly need statistics on the future for the conduct of their business, and these are nowhere to be collected until the future has become past and it is too late.

With producers of population statistics all working on the near side of *now* and users all concerned with the far side, it is lucky that even in times of rapid change, some continuities are to be found between past and future. Population projection rests on these continuities.

The continuities are not be found in simple totals. We know that the number of people in the U.S. does not increase evenly from year to year, and still less does the population of one town or one age group increase evenly. The age classes especially have fluctuated erratically in recent decades. Today the U.S. includes an exceptionally large proportion of young people 10 to 25 years of age, the result of the baby boom of the forties and fifties. They have crowded the high schools and colleges, and they are seeking jobs and entry into graduate schools across the country. But during the sixties, births fell sharply, and the number of pupils entering elementary schools leveled off.

Yet we can say something about the future. At the end of the seventies, schools and the labor market will be reached by the wave of what may be called the nonbirths of the sixties. But, though kindergartens and public schools will slow their expansion in the seventies, they may have to accelerate it again in the eighties to accommodate a new generation—children of the children born in the postwar baby boom. How such things can be projected with some confidence is our subject.

The approach, or model, that we shall build for projection serves other purposes than prediction. It is especially valuable for judging the effects on population growth of a possible change or a proposed policy.

PROJECTION WITH CONSTANT BIRTH AND DEATH RATES

The trick in projection is to seek elements that remain nearly constant through time. The increase in total population from year to year plainly does not qualify, but certain *rates* do remain more or less the same, and on these we rest our analysis of the future. For example, the proportion of people aged 30 who die each year is likely to remain much the same in 1960, 1970, and 1980. These death rates are constant enough that some fairly reliable predictions can be hung on them, and we proceed to the exploitation of this constancy.

Our projection of population into the future includes three parts:

(1) The statistical data of a baseline census from which work starts
(2) Effect of death
(3) Effect of birth

Demographers ordinarily recognize five-year age groups, to the end of life, for men and women separately, and they have a computer do the arithmetic. To show the procedure without being swamped in numbers, we consider here girls and women only, and these just up to age 45. Moreover, we need only consider three age groups, each of 15 years' width. For purposes of this illustration, three numbers describe the population at any one time.

We can make a fairly complete analysis for these three groups, and show the whole worksheet. The census of April 1, 1960, counted 27.4 million girls under 15 in the U.S. It showed only 17.7 million between 15 and 29 years. An intermediate number, 18.4 million, were between 30 and 44. (This article follows the census in always counting people at their age last birthday.) Those under 15, born between 1945 and 1960, constitute the baby boom; the next older group, born between 1930 and 1945, are survivors of the meager crop of depression babies; the oldest, aged 30 to 44, were born between 1915 and 1930, when birth rates in the U.S. were higher than in the thirties, but lower than in the fifties.

Now these three numbers can be written one below another in an array known as an age distribution; see Table 1.

So much for the counts made in 1960, our point of takeoff into the future. We now need to know how death and birth will act on this starting distribution. (Migration, which demographers usually take into account in making projections, is probably going to be relatively small and not likely to affect our conclusions seriously, so we shall ignore it.)

Let us start with death, but look at its positive side: the people who do not die, but survive into the next period. The question is, how many of the 27.4 million girls under 15 years of age counted in the 1960 census may be expected to survive to 1975? We have at hand a *life table*, as such collections of survival probabilities are called, that gives the proportion of girls under 15 who survive for 15 years as approximately 0.9924. This life table was calculated from deaths in the U.S. in 1965, and it would not be very different if calculated for any other recent year. Hence the expected number of survivors 15 years later of the 27.4 million counted in 1960 would be

TABLE 1. Age Distribution
of American Girls and
Women, 1960

AGE	MILLIONS OF GIRLS AND WOMEN
0–14	27.4
15–29	17.7
30–44	18.4

27.4 multiplied by 0.9924, or 27.2 million. These girls would be 15 to 29 years of age in 1975.

No such multiplication can give the *exact* numbers in 1975. Individuals survive or die at random, and even if 0.9924 were the probability for each separate girl 0–14 years of age, a few more or a few less than 27.4 × 0.9924 million could survive in the particular years 1960–75. If the U.S. were subject to serious epidemics, chance events each affecting large numbers of people, then the variation from year to year would be substantial. Because, in fact, death and survivorship act like events affecting each of us more or less independently, the multiplication is permissible, though even then the result could be made wrong by a war or epidemic on the one hand or a medical breakthrough on the other. We shall suppose that the chance of survival does not change very greatly over the period of the projection.

In the same way the proportion surviving 15 years among girls 15 to 29 in 1960 is estimated at 0.9826, and hence the projected number aged 30 to 44 in 1975 would be 17.7 × 0.9826 = 17.4 million. The projections to this point stand as shown in Table 2. Our next task is to fill the upper cell on the right, which requires an estimate of the number under 15 in 1975. (Remember that, to keep things manageable and simple, we are neglecting women 45 or more—of course, only for present simplicity, as the wives of some of us will remind us.)

All of the girls under 15 years of age in 1975 will have been born since 1960, and we need to estimate not how many girl births take place in the 15 years, but how many of these births survive to 1975. We know, also from the 1965 experience, that, on the average a woman 15 to 29 can expect 0.8498 surviving girl babies by the end of a 15-year period. We have counted girl babies only for this purpose because a female model is what we are constructing, and we have deducted deaths among the babies so as to come up with girls under 15 who will be alive in 1975. There were 17.7 million women aged 15 to 29 in 1960, and their contribution to the total girls under 15 in 1975 is expected to be 17.7 × 0.8498 = 15.0 million.

TABLE 2. Projected 1975 Population of American Girls and Women

AGE	MILLIONS OF GIRLS AND WOMEN	
	1960	1975
0–14	27.4	?
15–29	17.7	27.2
30–44	18.4	17.4

Children will be born also to the women 30 to 44 years of age; on the average, these women will have 0.1273 girl babies alive at the end of the 15-year period. The contribution that these make to the total girls under 15 in 1975 is expected to be $18.4 \times 0.1273 = 2.4$ million. (The actual calculation was made to more decimals than shown here.)

Finally, children will be born before 1975 to girls under 15 in 1960, a large proportion of whom will become of childbearing age during the 15 years. On the average (again at 1965 rates), they will have 0.4271 surviving girls. This average, like the others above, is taken over many different cases; it includes the girls too young to become mothers, those who will be old enough but not yet married, and those who will marry but not have children. The expected contribution here is $27.4 \times 0.4271 = 11.7$ million.

To find the total number of girl children under 15 surviving in 1975 we must add the numbers reached in the three preceding paragraphs: $11.7 + 15.0 + 2.4 = 29.1$ million in all. Figure 1 shows schematically what is happening. (Because so few children are born to women over 44, we can afford to ignore them. Our simple model will give almost the same rate of increase of the population as more elaborate models.)

By repeating exactly the same argument, except that we now start with the 1975 projected population, we obtain the age distribution in 1990; any

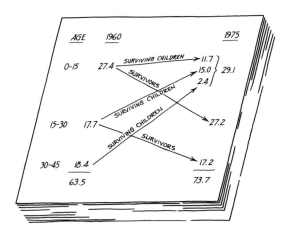

FIGURE 1

Calculation of 1975 population of girls and women under 45 years of age (figures are in millions)

TABLE 3.   Millions of Girls and Women Under 45 Years of Age in the U.S. if Birth and Death Rates Remain at the 1965 Level

AGE	1960	1975	1990	2005	2020	2035	2050	2065
0–14	27.4	29.1	37.7	44.1	54.3	65.0	79.0	95.3
15–29	17.7	27.2	28.9	37.5	43.7	53.8	64.5	78.4
30–44	18.4	17.4	26.7	28.4	36.8	43.0	52.9	63.4
Total	63.5	73.7	93.3	110.0	134.8	161.8	196.4	237.1

number of additional 15-year cycles may be calculated similarly.   Table 3 shows the resulting numbers up to 2065.

WAVES OF MOTHERHOOD

The first age group, girls under 15, increases less than two million between 1960 and 1975, while the women 15 to 29 increase by almost 10 million.   The 15 to 29 group in 1975 are the babies born between 1945 and 1960, the postwar baby boom, and as these succeed the depression babies in any group we expect its number to rise rapidly.   Women 30 to 44 actually become fewer during this first 15-year period, even though the 0 to 44 population as a whole is growing.

Because most children are born to mothers 15 to 29 years of age, we can expect a new baby boom, an echo of the first one, at the time when the babies of the fifties themselves pass through childbearing age, and indeed the under-15s grow by 8.6 million from 1975 to 1990 according to Table 3.

In fact, the depression and boom will keep echoing to much later times, supposing, as we do throughout, that childbearing practices remain fixed.   But the table also shows that as time goes on the irregularity of the 1960 age distribution steadily lessens.   At the end of 105 years all ages are increasing at very nearly the same rate.

That the several ages ultimately increase at the same rate can be seen by dividing each 2065 figure in Table 3 by the corresponding 2050 figure.   In Table 4, this ratio is shown to be about 1.2 for the three age groups and the total.   By carrying the projection further, we could have had these ratios as close to one another as we wanted; in fact, further calculation shows that they all would converge to 1.2093.

This ratio may be called intrinsic, or the true ratio of natural increase.   It can be shown to depend not at all on the 1960 age distribution with which the process started, but only on the rates of birth and death, and it is the most informative single summary measure of that set of rates.   It tells us that any population that is subject to our particular birth and death rates

TABLE 4. Increase of Age Groups of Girls and Women
in the U.S. from 2050 to 2065

AGE	2050 (MILLIONS)	2065 (MILLIONS)	RATIO, 2065 TO 2050
0–14	79.0	95.3	1.206
15–19	64.5	78.4	1.216
30–44	52.9	63.4	1.198
Total	196.4	237.1	1.207

over a period of time will sooner or later settle down to an increase in the ratio 1.2093, which is to say by about 21% per 15-year period. Under the operation of the projection, applying the assumptions we have made, *a stable age distribution* is sooner or later attained in which all the irregularities of 1960 due to boom and depression have been forgotten. Age distributions tend to forget their past when persistently pushed forward by the method developed above.

Let us find numerically the component of population growth that increases in the same ratio in every cycle, a mode of increase spoken of as *geometric*. If we divide each of the numbers shown under the year 2065 in Table 3 by 1.2093, we get back to an estimate for 2050; if we then divide again by 1.2093 we get back to 2035, and so on. To get back to 1960 we would divide by the seventh power of 1.2093, written $(1.2093)^7$ and equal to 3.78. Carrying out the division gives 95.3/3.78 or 25.2 million for age 0 to 14, and similar calculations for the other ages provide what we may call the stable equivalent for 1960; see Table 5.

Table 5 shows the set of numbers that, increasing in the constant ratio 1.2093, would sooner or later exactly join the track of our projection in each age group. If we multiply the stable equivalent by the fixed number 1.2093 to obtain the geometric track, and subtract this from the projection of Table

TABLE 5. Main Component
of Female Population in the
U.S., 1960

AGE	STABLE EQUIVALENT (MILLIONS)
0–14	25.2
15–29	20.7
30–44	16.8
Total	62.7

TABLE 6.   Departures of Projected
Population in Table 3 from Geometric
Progression in Millions

AGE	1960	1975	1990	2005
0–14	2.2	−1.4	0.9	−0.6
15–29	−3.0	2.2	−1.4	0.9
30–44	1.6	−2.9	2.1	−1.4

3, we obtain Table 6. For example, for girls 0 to 14 in 1975, we have 29.1 − 25.2 × 1.2093 = −1.4. Our analysis has separated the prospective population change into two parts, one a smooth geometric increase, the other a series of waves that are departures from the geometric.

These departures gradually diminish in amplitude. For 1960, we have 2.2 million as a measure of the temporary "excess" of the 1945–60 babies. The −3.0 million are the deficiency of the depression babies, and 1.6 million, again an excess, relate to the twenties. Note that by 1990, each of these has an echo, of the same sign but on the whole of smaller amount.

The tendency of the waves to diminish in amplitude is related to women having their children over a range of ages. If all children were born to mothers of the same age, the waves would steadily *increase* in amplitude. With such concentration any irregularity in the age distribution caused, for example, by a war or depression would not only continue echoing through all later generations, but become magnified. In the U.S. today, women prefer to have their children around age 25, whereas our grandmothers spread theirs from about 20 to 45. The new style, associated with the effective use of birth control, could mean diminished stability.

In this analysis of the U.S. population we have gone from the facts of the 1960 census, through various more or less realistic calculations concerning 1975 and even 1990, into a kind of fantasy as we proceed far into the future. The early part of the projection can within limits be useful for practical purposes; the later part is so dependent on various *if's* that one would be very foolish to count on it. The biggest doubt attaches to the birth rate. It may seem that birth is as individual a matter as death, and therefore births across the country ought to be independent of one another, yet in fact high and low birth rates spread like epidemics across the country.

Why then do we bother with the fantasy of such far-out projections referring to the distant future? The answer is that they help us understand the present. We ascertain the meaning of the present rates of birth and death by calculating what they *would* lead to if they continued for a hundred years or more. Let us see why this is even more important in study of the birth and death rates of developing countries than of a developed one like the U.S.

GROWTH OF DEVELOPING COUNTRIES

The task is in some ways easier for developing countries because they do not have a history of changing birth rates. It is true that their death rates have been falling, and where this occurs for young children only, it is the equivalent of a rise in the birth rate: as far as the population mathematics is concerned, a fall in infant deaths has the same effect as a rise in births. In fact, however, deaths have been falling at nearly all ages, and births are relatively unchanged. The fact is that age distributions are already more or less in the condition we called stable, and which could be attained by the U.S. only in the course of several generations of fixed rates. Because of past uniformly high birth rates, developing countries tend to grow much faster than the U.S. Moreover, they show a simple geometric increase, with all ages rising uniformly. The sort of waves that we have been studying do not occur for them.

Let us concentrate then on the geometric component, and take Malaysia as an example. In the mid-sixties, Malaysia was growing in the intrinsic ratio defined above of 1.59 per 15 years. This corresponds to an annual rate of increase of the 15th root of 1.59 or 1.031, that is, about 3.1% per year, against about 1% for the U.S. To convince ourselves of this we could multiply 1.031 by itself 15 times, that is calculate $(1.031)^{15}$, and we would find the result to be just under 1.59.

We can see the long-term prospect more clearly by translating into doubling times. How long does it take a country that is increasing at the rate r % per year to double in population? The equation to be solved for the unknown time t is $[1 + (r/100)]^t = 2$. The solution is obtained by taking logarithms of both sides and comes out very near to $t = 70/r$, where r is expressed as percent increase. This rule applies to money lent out at interest, and financiers use it because they are very interested in doubling times. The same sort of rule works for the half-life of a piece of radium or other substance under radioactive decay.

As an example of a geometric projection of a population, suppose that Malaysia's rate of 3.1% per year were to go on for about $70/3.1 = 23$ years. This would carry it from the present 10 up to 20 million people. Suppose it went on for another 23 years; this would mean another doubling. At the end of 115 years at this rate, the population would have doubled five times, which means multiplying by 2^5 or 32; Malaysia would contain 320 million persons. By the end of 230 years, it would have doubled ten times and would contain 2^{10} times as many as now, or 1024 times ten million.

No one could mistake such calculations for predictions of what will happen. In a sense they are the opposite; we might call them counterpredictions, for they show that in much less than 100 years, the birth rate will go down or the death rate will go up or both. Most demographers are optimistic enough to believe that the adjustment will be through the birth rate.

Other countries are today growing faster than Malaysia. Mexico's present 50 million population is increasing at about 3.5% per year, which, by our rule, would give it a doubling time of $70/3.5 = 20$ years. At this rate, it would be 100 million by the year 1990, 200 million by the year 2010, and 400 million by the year 2030. This again is a counterprediction; shortage of food, excess of pollution, and many other reasons would prevent it from coming true. The usefulness of the calculation is in showing that births must be reduced; anyone who makes a *principle* of permanent opposition to birth control in effect favors an increase of the death rate sooner or later. The most that opponents of birth control can argue is that it should be delayed a few years.

In diluted form, the same is true of the U.S. Our calculation showed that the geometric component, neglecting waves, was an increase of 21% for 15 years, or about 1.2% per year, according to births and deaths of 1965. That means a doubling in 60 years, quadrupling in 120 years, and so on. Contract the U.S. time scale by about three, and the future growth of the U.S. is the same as that of Mexico. And even our having three years to Mexico's one is partly offset by the greater damage to the environment caused by our more advanced industry.

RELIABILITY OF PREDICTION

The techniques presented in this article and obvious extensions of them are much used for predicting the future. They are used not because they are perfect, but because nothing better is available. Whatever continuities exist in birth and death rates are exploited by the makers of projections. From about 1870 to 1935 in Western Europe and the U.S., the birth rate and the death rate were both falling; projections could be made by the method outlined here, except that instead of using fixed rates, the past downward trend in birth and death was projected into the future. Such projections were acceptably accurate as long as the downward trend continued.

But these same countries reached a turning point in the forties. People married younger, and births rose rapidly. Moreover, couples varied the timing of their children as well as varying the total number. The fact that in a modern society couples plan their children, both in number and in timing, can be used to strengthen the predictions. Samples of young couples are surveyed to find what their childbearing intentions are, just as we ask intentions on buying houses and automobiles. The official estimates of the U.S. Bureau of the Census take account of these intentions.

The Census Bureau's projections, which use a vastly elaborated form of the method of this article, can be compared with ours. Theirs are more detailed than ours above have been, and they are also cautious enough to make a variety of projections rather than betting on just one. They end up with four numbers for each age, sex, and future year. For example, for 1990, their four numbers for girls 0 to 14 years of age range from a low of 29.5 million to a high of 41.8 million. Our Table 3 shows 37.7 million.

How well would past application of our model have foretold the 1970 population of the U.S.? If we had worked forward from the 1920 census total of 106 million, using exactly the procedure of this article, but applying it in five-year age intervals to all ages and to both sexes, we would have found about 185 million for 1970. If we had allowed for immigration less emigration of 200,000 per year, this would have brought us to 195 million against the 205 million actually counted. Something a little lower would have been found starting from 1950; starting from 1960, we would have slightly overestimated the census figure. An error of about 5% in estimates made up to 50 years ago is not bad, considering that the Bureau of the Census estimates its own actual count to be subject to nearly 2% error.

We would have done much worse starting in 1940, however; the method of this essay, plus about 6 million immigrants, would have produced a total of only about 160 million. Put another way that sounds even worse: the increase from 1940 to 1970 was about 62 million, and of it, we would have estimated less than 30 million. This is not a good score. The baby boom of the fifties was a historic event about as hard to predict in advance as the war that sparked it.

MODELS PERMIT EXPERIMENTS

We have discussed the population projection as a way of making predictions, and also as a way of making counterpredictions—calculating what would happen if present rates continued as a way of showing that they cannot continue. This last suggests what may be the most important use of the model of this essay, originally developed and still often applied to making predictions. This use is experimentation. Not only does our model answer the question "What would happen if the birth and death rates of the present time continue into the future?" but it answers a great variety of other important questions. What would be the effect on total population numbers if intensive research on heart disease was undertaken, and it reduced deaths from that cause by 90%? This could conceivably result from a research effort comparable to present investigations of outer space. But the effort could equally be put into reduction of infant mortality. Our model could compare the effects of these alternatives, taking account of the fact that the person dying of heart disease is of such an age that he will soon die of something else; the child saved from some lethal ailment, on the average, will have a long life ahead of him. A given fall in infant mortality increases the population by much more than the same fall in heart disease.

An example of a question that has been frequently asked, and to which our model provides a clear answer, is: what would happen if, starting now, each member of the population averaged one descendant? This means that each fertile couple would need to have somewhat more than two children, to allow for those who do not marry, for those who marry but are infertile, and for deaths in childhood. An average of about 2.3 children per married couple would constitute bare replacement, that is to say, would keep the population stationary at modern death rates.

But the stationary *level* at which it would keep it would be above that of the starting point. Any population that has been subject to birth rates higher than bare replacement in the past has a large proportion of girls and women of childbearing age. These will produce increasing numbers of children for about 50 years after the date at which the birth rate falls. The projection model developed in this article tells how high the population would rise if we drop to bare replacement numbers of children.

Application of the model shows that the U.S. would rise to about 270 million persons if bare replacement were adopted today, and Mexico would rise from its present 50 million to over 80 million. Most underdeveloped countries, such as Mexico, would increase by about 65% from the point at which they drop to replacement, and they would do this over 50 or so years. No country ought to fear that immediate adoption of contraception would freeze total population where it stands; a kind of momentum operates simply because of the favorable age distribution that results from past high fertility.

Former President Sukarno of Indonesia was against birth control because he thought Indonesia should have 250 million people. The present government applied the model described in this essay and found it will probably exceed 250 million even if the brakes are put on immediately; consequently it has formally sponsored a program of birth control.

CONCLUSION

We started this essay by developing a model to forecast the future. The model works for forecasts over short periods and over longer periods in which either the trend is steady or in which ups and downs offset one another. For forecasting major turning points it is of little use, but so is any other model so far developed.

While the model is moderately, but only moderately, successful for the purpose for which it is designed, it has the power to analyse hypothetical futures whose consideration is urgent. If the marriage age in India is raised to 20, what effect will this have on the birth rate? If 20% of couples aged

30 accept sterilization, how far will this take a given country towards zero population growth? How much long-run effect does the emigration from Jamaica have on its population increase? What kind of population waves would follow a sudden drop in the birth rate of an underdeveloped country? It is in answering questions such as these, of which examples have been given in the course of this essay quite as much as in making predictions, that the projection model presented here finds its use.

PROBLEMS

1. What is a "life-table"?

2. What assumptions must one make to be able to predict, with some accuracy, the future population size using a life table?

3. In Table 2, why cannot one use the same method to calculate the three entries in the last column?

4. What is meant by "diminished stability" as a result of the new style of childbirth associated with the effective use of birth control?

5. Give some reasons why the method of prediction used in the article might give biased results.

6. What is the usefulness of the proposed model of population growth? Describe a few questions for which the model can provide answers.

7. Using the stable equivalent for 1960 females in the age group 0–14 given in Table 5, we can get the stable equivalent for the year 2065 of the same age group by multiplying 25.2 by $(1.2093)^7$.

    a. What power of 1.2093 should 25.2 be multiplied by to get the stable equivalent for the year 1990? Calculate this number.

    b. Is the stable equivalent obtained for 1990 in part a. equal to 37.7 (the number given in Table 3)? If not, how big is the difference? Compare with the difference 0.9 shown in Table 6.

8. Verify the numbers in the column for the year 2065 of Table 3 using the method given in the text in connection with Table 2.

9. Verify the solutions for the doubling time $t = 70/r$ of the equation $(1 + (r/100))^t = 2$. If the growth rate of a country is 2%, what is the doubling time?

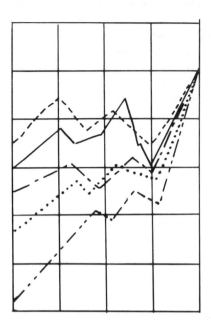

EARLY WARNING SIGNALS
FOR THE ECONOMY

Geoffrey H. Moore   *Bureau of Labor Statistics*

Julius Shiskin   *Office of Management and Budget*

DURING THE sharp business setback of 1937–38, the National Bureau of Economic Research (NBER) was asked by the Secretary of the Treasury, Henry Morgenthau, to devise a system that would signal when the setback was nearing an end.

At that time, quantitative analysis of economic performance in the U.S. was still in a rudimentary stage. The Government's National Income and Product Accounts, today the foundation of so much economic analysis, were just being established. A number of statistical series, including unemployment rates, were being developed or refined by public agencies trying to provide information that would be useful in fighting the Depression. Among business, labor, and academic groups, there was a surge of interest in obtaining better economic intelligence as a guide to policy making.

Under the leadership of Wesley C. Mitchell and Arthur F. Burns, the private, nonprofit NBER had since the twenties assembled and analyzed a vast amount of monthly, quarterly, and annual economic data on prices, employment, production, and so on as a basis for the NBER's major research effort to gain a better understanding of business cycles. From these statistical series, it selected 21 that, on the basis of past performance (dating as far back as 1854), promised to be fairly reliable indicators of business revival. This list was made available to the U.S. Treasury late in 1937 in response to its request and published in May 1938.

This, in brief, is the origin of the statistical indicators so widely used today in analyzing the economic situation, determining what factors are favorable or unfavorable, and forecasting short-run developments.

Since then, the availability, study, and use of economic indicators has been vastly expanded under the leadership of the National Bureau and other public and private agencies.

Especially during a period of change in business activity the indicators, and analyses based on them, are front page news. A drop or rise in the number of persons unemployed, a change in the rate of the nation's total output, a movement up or down in the indexes of consumer or wholesale prices, a rise or fall in new orders for goods—all of these command nationwide attention. Among professional economists interested in the future performance of the economy, indicators have become a major tool of economic forecasting. In this essay, we shall discuss what these statistical measures are and how they are used.

THE INDICATORS IN BRIEF

The indicators have been classified into three groups: those that provide advance warning of probable changes in economic activity, the *leading* indicators; those that reflect the current performance of the economy, the *coincident* indicators; and those that confirm changes previously signaled, the *lagging* indicators.

Coincident Indicators. The most familiar type of indicator, the coincident, measures current economic performance. Gross National Product, industrial production, personal income, employment, unemployment, wholesale prices, and retail sales are examples. Comprehensive in coverage, these indicators show how well the economy is faring in certain important respects. Their movements coincide roughly with, and indeed provide a measure of, aggregate economic activity. They tell us whether the economy is currently experiencing a recession or a slowdown, a boom or an inflation.

A number of government agencies cooperate in the gathering and compilation of these statistics. To produce the Wholesale Price Index, for example, the Bureau of Labor Statistics gathers by mail questionnaire each month about 7000 price quotations on more than 2000 commodities covering the output of manufacturing, agriculture, forestry, fishing, mining, gas and electricity, and public utility industries.

The Census Bureau conducts a monthly survey of 50,000 households to obtain information on employment and unemployment. This survey provides such indicators as the unemployment rate and the total number of persons employed. Indicators of the number of employees on nonagricultural payrolls and total number of man-hours worked are obtained from reports by 160,000 business establishments to the Bureau of Labor Statistics.

The Office of Business Economics of the Department of Commerce is the keeper of a number of statistical series of current economic performance. The most significant is the Gross National Product, which is the total of all goods and services produced by the economy, reported quarterly both in current dollars and adjusted for changes in the price level.

The Federal Reserve Board compiles the Index of Industrial Production, another important coincident indicator.

Leading Indicators. In view of the great impact that economic developments have upon our daily lives and upon long-term economic progress, there is intense interest in indicators that signal in advance a change in the basic pattern of economic performance. Examples are new orders for durable goods, construction contracts, formation of new business enterprises, hiring rates, and the average length of the workweek. These indicators move ahead of turns in the business cycle, primarily because decisions to expand or curtail output take time to work out their effects, while the factors that govern these decisions also take time to produce their influences. The early warning signals provided by leading indicators help to forecast short-term trends in the coincident series and help policy makers to take timely steps to avert, or at least to moderate, unfavorable economic trends.

Many of the leading indicators are produced by the same agencies or surveys that yield the coincident indicators. Thus, the average workweek of production workers, the hiring rate, and layoff rate are figures from the Bureau of Labor Statistics employment survey. The value of manufacturers' new orders for durable goods is compiled by the Census Bureau from the same survey that produces monthly statistics on inventories and shipments.

A number of leading indicators are compiled by private firms. For example, the index of the total value of construction contracts is prepared by the McGraw-Hill Information Systems Co. The number of new business incorporations is provided by Dun and Bradstreet, Inc., and the index of stock prices by Standard and Poor's Corp.

Leading indicators are used more and more widely by business, government, and academic economists in analyzing and forecasting business conditions. In a recent survey by the American Statistical Association, 261 forecasters were asked what principal methods they used, and 103 checked "lead indicators." One of the information sources on which business analysts depend is *Business Conditions Digest* (BCD), a monthly publication of the U.S. Department of Commerce that charts leading and other indicators.

The use of leading indicators has spread to individual states and to foreign countries. New Jersey is one of several states issuing monthly reports on the current position of leading indicators for the state. The Canadian and Japanese governments each have developed reports similar to BCD, and studies of leading indicators have been made for Great Britain, Germany, France, Italy, and Australia. As a result of this widening interest, we are more and more likely to read in newspapers, magazines, and business advisory publications about developments in the leading indicators.

Lagging Indicators and Others. Still another type of indicator is described as lagging. The fluctuations of these indicators usually follow, rather than lead, those of the coincident indicators. Examples are labor cost per unit of output, long-term unemployment, and the yield on mortgage loans.

Finally, there are important economic activities that have not behaved in a sufficiently consistent manner to be appropriately classified as leading, coincident, or lagging, but are nevertheless relevant to an overall appraisal of the current situation of the economy and prospective trends. Examples are government expenditures and the balance of payments.

CHARACTERISTICS OF INDICATORS

The selection of indicators has been guided by two considerations. First, does the measured process play a significant role in a widely accepted explanation of short-term economic fluctuations? While the recurrence of successive waves of rapid growth and slower growth or decline in business activity is generally acknowledged, many different explanations of the underlying causes have been advanced. Some economists lay primary stress on the relations between investments in inventory and fixed capital, on the one hand, and final demand, on the other (John Maynard Keynes, Paul Samuelson). Others assign a central role to the supply of money and credit (Milton Friedman). Still others look for clues in the relationships among prices, costs, and profits (W. C. Mitchell, A. F. Burns). All these factors undoubtedly influence the course of business activity, and some may be more important at one time than another, but there is no general agreement on which is the most important. Hence it is prudent to consider a variety of indicators that reflect all the processes.

The second consideration in selecting indicators has been their empirical record. How closely correlated are the fluctuations in a given series with those in aggregate economic activity? How consistent has been the timing record of a given series compared to aggregate economic activity; that is, does it consistently lead, coincide with, or lag aggregate economic activity? Specified criteria have been applied to the hundreds of economic series from which a list of indicators could be selected. These pertain to the economic significance of the series, its historical record, and various properties affecting its reliability as a current statistic.

In recent years, most of the research and testing in this field has been carried on by the National Bureau of Economic Research. Since publishing its first list in 1938, the Bureau has revised the list in 1950, 1960, and 1966. The most recent list appears in a book by the authors of this essay, published by the NBER in March 1967, under the title *Indicators of Business Expansions and Contractions*. Many of these indicators have been used for years in appraising economic conditions, but 13 were incorporated for the first time in the latest list. Among these new indicators are job openings at U.S. Employment Service offices, delinquency rates on installment loans, export orders for durable goods, and man-hours of nonagricultural employment. Up-to-date figures on these indicators and other related series are published each month in *Business Conditions Digest*.

An understanding of the role of the selected indicators in initiating or reacting to short-term economic fluctuations is aided by the two principles of classification used in presenting the current list of 88 indicators. First is a grouping by *cyclical timing*, as explained above, with 36 leading, 25 coincident, and 11 lagging indicators (16 are not classifiable by timing). The second type of classification is by *economic process*, in which series that pertain to different stages or aspects of the same process are grouped together.

All series are cross-classified according to these two principles. Thus, the cross-reference system shows, under the employment and unemployment category, five leading series representing marginal employment adjustments such as the hiring rate and the workweek (which varies with the amount of overtime worked), eight coincident indicators representing existing job vacancies and comprehensive measures of employment and unemployment, and one lagging indicator representing long-term unemployment. The capital investment category includes ten leading indicators representing the formation of business enterprises and new investment commitments, two coincident indicators representing the backlog of capital investment, and two lagging indicators representing current investment expenditures. Other economic process categories include production, income, consumption and trade; inventories and inventory investment; prices, costs and profits; money and credit; foreign trade and payments; and federal government activity. Together they provide a rather comprehensive view of the economic system.

Another innovation of the National Bureau publication is a method of assigning numerical scores, or *weights*, to each indicator, ranging from 0 to 100. The scoring plan covers six major elements: economic significance, statistical adequacy, historical conformity to business cycles, cyclical timing record, smoothness, and promptness of publication. The ratings throw into clearer perspective the characteristic behavior and limitations of each indicator, aid in their classification and incidentally suggest ways to improve them for purposes of short-term forecasting (see Table 1). For example, the low score of series 31 (change in inventories) for smoothness (column 6) is a warning that sharp erratic movements from month to month are to be expected and that it may take several months to discern a new trend.

The short, substantially unduplicated list of 26 principal indicators shown in Table 1 provides a convenient way of summarizing the current situation and outlook. This list includes 12 leading, 8 roughly coincident, and 6 lagging indicators. The series selected for each of these groups have high scores and cover a broad range of economic processes.

SUMMARY INDEXES

In studying the current economic situation, we can proceed by examining numerous and varied aspects of the economy so as to be sure that all the relevant points are covered or by examining just a few selected indicators that make it easy to grasp the overall trends. Most business analysts move back and forth from a detailed examination of many sectors to a broad view of the overall situation, and there is a feedback of information and insight from one view to the other. The indicator scheme just described makes it appropriate to ask whether a broader type of summary, combining indicators with similar short-term timing behavior, but pertaining to different aspects of activity, can be constructed.

The National Bureau studies make it possible to do this; that is, series that usually lead in the business cycle can be combined into one index, coincident series into another, and lagging series into a third. Similarly, within the leading group alone, series that represent orders or commitments for capital investment projects can be combined into one index, those representing inventory investment or materials purchasing into a second, and those representing sensitive flows of money or credit into a third. A suitable set of weights to be used in combining the series is provided by the scores referred to earlier.

In the sense that they are not expressed in a common unit such as dollars or tons, the series selected for inclusion in each of the indexes are heterogeneous. In the special sense that they measure related aspects of business change, are sensitive to business cycles, and experience similar timing behavior during cyclical fluctuations, they are homogeneous, however. In this respect, some of the best known aggregates are heterogeneous. For example, Gross National Product includes the change in inventories, a leading indicator; consumption expenditures, a coincident indicator; and investment expenditures, a lagging indicator.

TABLE 1. Scores for 26 Economic Indicators on 1966 NBER Short List

CLASSIFICATION AND SERIES TITLE	Average Score*	SCORES, SIX CRITERIA*					
		Economic Signifi-cance	Statistical Adequacy	Con-formity	Timing	Smooth-ness	Currency
Leading indicators (12 series)							
1. Avg. workweek, prod. workers, mfg.	66	50	65	81	66	60	80
4. Nonagrl. placements	68	75	63	63	58	80	80
12. Index of net business formation	68	75	58	81	67	80	40
6. New orders, dur. goods indus.	78	75	72	88	84	60	80
10. Contracts and orders, plant and equip.	64	75	63	92	50	40	40
29. New building permits, private housing units	67	50	60	76	80	60	80
31. Change in book value, mfg. and trade inventories	65	75	67	77	78	20	40
23. Industrial materials prices	67	50	72	79	44	80	100
19. Stock prices, 500 common stocks	81	75	74	79	87	80	100
16. Corporate profits after taxes, Q	68	75	70	77	76	60	25
17. Ratio, price to unit labor cost, mfg.	69	50	67	84	72	60	80
113. Change in consumer instal. debt	63	50	79	77	60	60	40
Roughly Coincident Indicators (8 series)							
41. Employees in nonagrl. establishments	81	75	61	90	87	100	80
43. Unemployment rate, total (inverted)	75	75	63	96	60	80	80
200. GNP in current dollars, Q	80	75	75	92	82	100	50
205. GNP in constant dollars, Q	73	75	75	91	58	80	50
47. Industrial production	72	75	63	94	38	100	80
52. Personal income	74	75	73	89	43	100	80
56. Mfg. and trade sales	71	75	68	70	80	80	40
54. Sales of retail stores	69	75	77	89	12	80	100
Lagging indicators (6 series)							
44. Unempl. rate, persons unempl. 15 + weeks (inverted)	69	50	63	98	52	80	80
61. Bus. expend., new plant and equip., Q	86	75	77	96	94	100	80
71. Book value, mfg. and trade inventories	71	75	67	75	66	100	40
62. Labor cost per unit of output, mfg.	68	50	70	83	56	80	80
72. Comm. and indus. loans outstanding	57	50	47	67	20	100	100
67. Bank rates on short-term bus. loans, Q	57	50	55	82	47	80	25

Source: "Indicators of Business Expansions and Contractions," 1967, New York: National Bureau of Economic Research, Inc.

* The scores run a scale from 0 to 100. For example, a series with a random relation to business cycles would be expected to score 0 for conformity and timing (columns 4 and 5). The average score is an average of the six criteria scores with smoothness and currency weighted one-half each.

374

The procedure used in constructing the indexes allows for the fact that some indicators, such as new orders, typically move in wide swings while others, such as the average workweek, experience narrow (but, nevertheless, significant) fluctuations. Each indicator is adjusted in such a way that, apart from its weight, it has the same opportunity to influence the index as any other indicator. The indexes themselves are adjusted in a similar manner, with the result that their swings are of the same order of magnitude on the average (namely, 1.0% per month) and can be compared readily. For example, if the most recent monthly increase in an index is 2.0, it is rising twice as fast as its average rate of change in the past; if the increase is 0.5, it is rising only half as fast as the historical average.

The index for the leading group is also subject to a further adjustment, designed to make its long-run trend the same as that of the index of coincident series. The major difference that remains is in cyclical timing, with the leading index typically moving first, the coincident index next, and the lagging index last, as can be seen in Figure 1, a type of chart currently published in *Business Conditions Digest*.

It is noteworthy that when the scoring system used to provide weights for the individual indicators is applied to the indexes themselves, the scores earned by the indexes are higher than those for any of the component indicators. In this sense, the indexes are a superior form of indicator. But to understand and interpret their movements, close study of the components is essential.

HOW THE INDICATORS HAVE PERFORMED

One of the uses to which the leading index can be put is to make explicit forecasts for short periods ahead of GNP, total employment, or other variables. Some tests of a simple method for doing this have been made, with promising results. For instance, when data for the third quarter become available, say in October, the percentage change in GNP (in current dollars) between the current calendar year and the next can be forecast by observing the percentage change in the leading index between the third quarter and the *preceding* fiscal year.

Annual forecasts obtained in this manner compare favorably in accuracy with what most forecasters have been able to achieve. During 1962-68, the average error in the GNP forecasts contained in the *Economic Report of the President* was 1.3 percentage points, while during this period the leading index method would have produced an average error of about 1.0 percentage points. In 1968, this superiority was not maintained because the forecast of increase in GNP obtained from the leading index was about 6%, the *Economic Report* (February 1968) put it at nearly 8%, but the actual change turned out to be 9%. For 1969, the leading index (using data through September

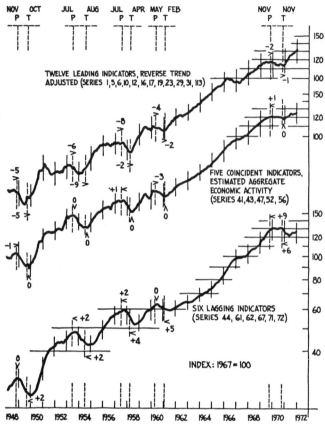

FIGURE 1

Composite indexes. Dates at top of chart are reference peak and trough dates when expansion ended and recession began (P) and when recession ended and expansion began (T). Numbers on chart are length of lead (−) or lag (+), in months, from index peak or trough to reference peak or trough. Source: U.S. Bureau of the Census

1968) forecast an increase in GNP in the neighborhood of 7%, significantly lower than the actual increase in 1968. The *Economic Report* (January 1969) also forecast 7%. The actual increase was about 7.5%. The leading index forecasts can be revised monthly as additional data become available.

The method clearly is no substitute for a carefully reasoned approach to the economic outlook.   It merely helps to summarize the information contained in the leading indicators regarding the near-term future course of GNP or other variables that are systematically related to the business cycle.   Hence it provides the forecaster with information useful in developing his actual forecast.   It can be used also as a standard against which to judge past efforts and might be of assistance in improving upon them, but it is not a recipe that will tell the forecaster everything he should know.

SUMMING UP

A few concluding observations may be helpful to those who would use these indicators as an aid in interpreting current trends.

(1) The change in pace of the leading indicators foreshadows increasing strength or weakness in aggregate economic activity.   Sometimes weakness in the leading indicators is followed by a recession, that is, an extended, substantial, and broadly diffused decline in aggregate economic activity, as in 1957–58.   At other times, however, a decline in the leaders is followed only by a slowdown in the rate of expansion of aggregate economic activity, as in 1962 and 1967 (see Figure 1).   Indeed, if the response of government policy to a decline in the leading indicators is prompt and vigorous enough, there may be no unfavorable developments in aggregate activity at all.

(2) Whether in any particular instance a decline in the leading indicators signals a retardation or a recession will eventually be determined by the movements of the coincident indicators, that is, by such measures of aggregate economic activity as total production, employment, income, consumption, trade and the flow of funds.   In view of the fact that economic upswings and downswings have several relevant dimensions (length of swing, how deep it swings, and how pervasive the effect), and that the data themselves are fallible and subject to revision, several months may elapse before a current decline can be reliably appraised.

(3) The lagging indicators should not be neglected.   They may provide evidence confirming a change in trend that has appeared earlier in the leading or coincident indicators.   Their value in this role is all the greater because the factors that make them lag also make them relatively impervious to erratic movements.   In addition, some of the lagging indicators have an important bearing on the *subsequent* behavior of the leading indicators.   For example, when unit labor costs, a lagging indicator, rise rapidly, they may exert a downward pressure on profit margins, one of the leading indicators.   In this connection, it is often the relation between the movements in lagging and coincident indicators (in this case costs and prices) rather than those of lagging indicators alone that is crucial.

(4) The ultimate objective of the work with indicators is the *prevention* of unfavorable developments, especially recession and inflation. The intent is not to compile good forecasting records; it is rather to develop warning signals which come sufficiently early to assure that effective preventive measures can be taken in time. Hence a successful record of forecasting recession and inflation would attest the failure of economic policy, while successful economic policy might well relegate many accurate forecasts to the limbo of apparent failure.

What can be said of the usefulness of business cycle indicators on balance? It seems clear from the record that the business indicators are helpful in judging the tone of current business and short-term prospects. But because of their limitations, the indicators must be used together with other data and with full awareness of the background of business and consumer confidence and expectations, governmental policies, and international events. We also must anticipate that the indicators will often be difficult to interpret, that interpretations will sometimes vary among analysts, and that the signals they give will not always be correctly interpreted.

Indicators provide a sensitive and revealing picture of the ebb and flow of economic tides that a skillful analyst of the economic, political, and international scene can use to improve his chances of making a valid forecast of short-run economic trends. If the analyst is aware of their limitations and alert to the world around him, he will find the indicators useful guideposts for taking stock of the economy and its needs.

PROBLEMS

1. From Table 1 find the leading indicator with the highest average score of all leading indicators. Do a similar thing for coincident and lagging indicators.

2. Explain the difference between leading, lagging and coincident indicators.

3. If you graph indicators from month to month, which type would tend to have the smoothest graphs—leading, coincident, or lagging indicators?

4. How do economists use the economic indicators described?

5. If you had a new index of economic activity how would you decide whether it was a leading, lagging or coincident indicator, or whether it could not be classified as any of these?

6. Have the indicators been changed since this article was written?

7. The Wholesale Price Index (WPI) comes out each month. Suppose we compute a new index—call it the Average Recent Wholesale Price Index

(ARWPI)—by averaging the values of WPI for the 6 previous months (so, for example, ARWPI for December = $^1/_6$[(WPI for Dec.) + (WPI for Nov.) + (WPI for Oct.) + (WPI for Sept.) + (WPI for Aug.) + (WPI for July)]).

    (a) Will ARWPI be a lagging, leading or coincident indicator? Why?

    (b) Will ARWPI be smoother or less smooth than WPI?

8. (a) If you wanted to know whether the business cycle was in a "down" phase or an "up" phase in the early 1950's, would you rather see unemployment figures or stock prices for this period?

    (b) If you wanted to know in which months of the 1950's the business cycle "changed directions," would you rather see unemployment figures or stock prices?

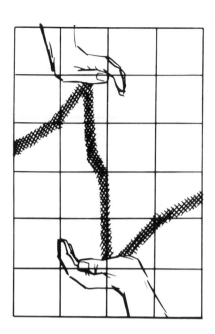

STATISTICS FOR PUBLIC FINANCIAL POLICY

Leonall C. Andersen   *Federal Reserve Bank, St. Louis*

ARTHUR BURNS,[1] Chairman of the Federal Reserve Board, sat back in his paneled office on Constitution Avenue, puffed his pipe pensively, and reflected on the forthcoming meeting of the Fed's Open Market Committee. What should they decide about purchase or sale of government securities in the open market, and about the many other instruments of policy available to the Committee?

Whatever the decisions might be, their consequences would affect wages paid to the steel worker in Gary, Ind., and the movie extra in Los Angeles; they would also affect the profits and the expansion plans of the largest steel company and the smallest corner grocer; and they would affect the prices paid for apples, automobiles, and xylophones by everyone.

---

[1] The names, for concreteness, are those of February 1971, when this was written. They may change, but the issues will continue.

At about the same time, Paul McCracken, Chairman of the Council of Economic Advisors, sat back in his office in the Executive Office Building, a less elegant, but more historic (and hard by the White House) office than that at the Federal Reserve. What should the Council recommend to the President about Federal spending and other fiscal actions? The Council cannot, of course, move in the independent and decisive ways open to the Federal Reserve, but its recommendations and arguments can be highly influential upon the views of the President, the Congress, and all Federal agencies dealing with economic matters.

Whatever the Council's recommendations might be, their implementation too would have effects on wages, profits, prices, and other economic quantities. My pocketbook and yours are linked inextricably to such decisions at the levels of great power and great wealth.[1]

The links, however, are complicated, they change over time, they have many interconnections, and some of their effects are very difficult to measure accurately and quickly. Economics is the science that tries to understand these complex matters.

FISCAL AND MONETARY ACTIONS

To discuss governmental economic policy at all, we must have some idea of goals and some idea of possible actions to try to achieve those goals. Three sorts of goals are widely accepted for the national economy:

(1) High employment
(2) Relatively stable prices
(3) Rising output of goods and services

It might be objected that these three goals contain mutually contradictory elements, and there have long been doubts (sharply increased recently) about working toward a rising output of goods and services, especially as it is usually measured by the Gross National Product (GNP).

Now what about actions? In this essay we shall consider two kinds of actions:

(1) *Monetary actions.* These are actions that change the amount of money and credit available to the economy. For example, a purchase of securities in the nation's money market (Wall Street) by the Federal Reserve will tend to increase the total available money and credit. (You should keep in mind that money includes the total of all checking accounts in banks.) Monetary actions are primarily under the control of the Federal Reserve System, and of the Treasury.

[1] A high official of the Office of Management and Budget recently quipped that his office is the only one in existence where 0.1 means one hundred million dollars.

(2) *Fiscal actions.* These are broad-scale acts of Government spending and taxation. They relate to the much-argued issue of government deficit financing.

In recent years a debate has been hotly raging over monetary versus fiscal actions as the more effective means of achieving important economic goals. The fundamental Keynesian viewpoint that has now become part of the mainstream (President Nixon has been quoted as saying "I am a Keynesian") concentrates almost exclusively on the direct influence of fiscal actions, primarily of government spending. Some economists, of whom Milton Friedman of the University of Chicago is perhaps the best known, argue for monetary actions as more effective and more predictable.

We shall here completely omit some aspects of this debate, in particular the difficult underlying economic theory and such institutional arguments as the ease of taking monetary actions relative to fiscal—the former can be effected by a single agency, while the latter require the whole political process of interaction between the Executive and Legislative segments of the government. (Fiscal proponents might rebut by emphasizing the democratic desirability of having major decisions go through the political process, even at the inherent cost of clumsiness and time lost.) Rather, we shall indicate how statistical methods have permitted insight into a direct empirical comparison of the two kinds of government action.

MEASUREMENTS AND PROCEDURE

An essay of this character, brief and for a wide audience, can hardly hope to provide more than a surface glimpse of a highly technical topic. Yet it is important, I feel, that such glimpses be given frequently and in many ways. Economics, like war, is too important to be left entirely to the specialists.

What we shall describe is a small part of a correlation study. (See the end of the essay for a discussion of the correlation notion).

As an indicator of economic well-being we shall use the Gross National Product. The GNP shows, in a word, the total goods and services sold in the market place during some specified period.

As a measure of monetary actions, we shall use the money stock, defined as currency and demand bank deposits held by nonbank individuals and firms.

As a measure of fiscal actions, we shall use government spending.

These three quantities are known each quarter, and the analysis is based, in fact, on quarter-to-quarter *differences.* Thus, suppose that in a certain quarter

(1) GNP increases by $400 million
(2) Money stock increases by $800 million
(3) Government spending decreases by $50 million.

Then in correlating the money stock and GNP, we would use the pair of numbers (800, 400) for that particular quarter-to-quarter change; in correlating spending and GNP, we would use the pair (−50, 400).

RESULTS

Scatter diagrams showing such relationships for the 68 quarters from 1953 to 1969 are shown in Figure 1. The major point to notice is that changes in GNP and those in the money stock are more closely related than changes in GNP and those in spending.

In fact the computed correlation coefficients are:

Between GNP and the money stock: 0.66
Between GNP and spending: 0.44.

The first of these is clearly larger than the second; this says that we can predict changes in GNP (by straight-line prediction) much better by using changes in the money stock than by using changes in government spending.

Another way of looking at this is in terms of the "scatter" or variability of GNP changes around the straight line that best fits the cluster of data in the left hand graph of Figure 1, as against the corresponding variability in the right-hand graph. The relevant quantities here are the squares of the correlation coefficients, 0.44 for changes in GNP and the money stock, and 0.19 for changes in GNP and government expenditures. If we measure

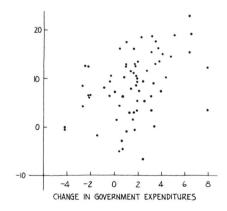

FIGURE 1

Comparison of correlation between changes in gross national product and those in money supply with correlation between changes in GNP and those in government expenditures (figures are in billions of dollars)

variability in terms of proportional decrease in a technical quantity called "variance," then changes in the money stock permit a 44% decrease in variability of changes in GNP, while changes in spending permit only a 19% decrease in that variability.

The full technical analysis was much more detailed and complex than that described above. For example, the full analysis carefully considered time lags: a change in spending now might not affect GNP for two quarters, or the effect of changes in money stock may require four quarters. The refinements, however, did not change the general result: *change in GNP is much more accurately predictable from change in the money stock than from change in spending.*

QUALIFICATIONS

For economic studies, unlike studies in most of the natural sciences, true experiments are generally impossible. We cannot play God, manipulating economic variables this way and that just to see how the economy responds. Economists are as a rule forced to do the best they can with data of the kind discussed earlier—data that arise in the natural course of our changing world.

Under these circumstances it is difficult to establish causation. To say that two quantities are positively correlated is not to say that artificially increasing one would increase the other. For example, imports of gold and the annual number of marriages are positively correlated over years (because both reflect economic health), but suddenly increasing gold imports by government action would scarcely be expected to change the number of marriages.

On the other hand, even when evidence is primarily correlational, people frequently do come to conclusions of causation. A current example is the relationship between smoking and health. When people ascribe causation based on a correlational study there is additional information—usually a reasonable theory of the mechanism behind the effect, an unusually high, consistent, and specific correlation, or some other piece of connective evidence.

A significant part of the debate between those favoring monetary actions and those favoring fiscal ones stems from the causation problem. Other aspects of the debate turn on the accuracy of the data and on possible artifacts of the data.

This debate, and similar ones, are bound to continue. One prediction can safely be made, however: the continuing debate will use statistical tools such as correlation coefficients. There is no choice in empirical quantitative argument. Statistical methods are essential, whether or not they are explicitly described.

A frequently used measure of how well one can predict one variable (e.g. GNP) from another variable (e.g. money stock or spending) by means of a linear equation is the *correlation coefficient**: the closer the correlation coefficient is to +1 or −1 (its largest and smallest possible values) the better the linear prediction will work. If the correlation coefficient is positive, an increase in the first variable is predicted to lead to an increase in the second variable; the two variables are then said to be *positively correlated*. Similarly, if a correlation is negative, an increase in the first variable leads to a predicted decrease in the second variable, and the two variables are then said to be *negatively correlated*.

PROBLEMS

1. Explain the difference between monetary actions and fiscal actions.

2. (a) Roughly how many of the 68 quarters from 1953 to 1969 had *smaller increases* in gross national product than the quarter which had the *greatest decrease* in money supply?

   (b) Roughly how many of these 68 quarters had *greater increases* in gross national product than the quarter which had the *greatest increase* in government expenditures? (Hint: use Figure 1).

3. Would you say that the correlational study in this essay is evidence in favor of fiscal actions over monetary actions? Why or why not?

4. (a) For a number of years in the United States, there was a high correlation between public school teachers salaries and liquor consumption. Do you think that the government could have discouraged liquor consumption by decreasing teachers' salaries?

   (b) Briefly explain the difference between correlation and causation.

5. Suppose the study described in this article had concluded that the correlation between GNP and the money stock was 0.1. How would this have affected the Federal Reserve's policies? What if the correlation was 0.9? −0.9?

6. Suppose you noticed that the number of marriages in a certain year increased by 5% from the previous year. Would you guess that U.S. gold imports increased, decreased, or remained about the same as the last year? Why?

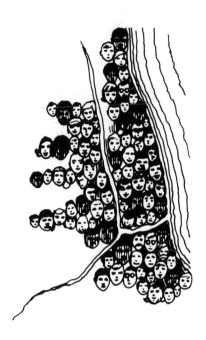

MEASURING RACIAL INTEGRATION POTENTIALS

Brian J. L. Berry   *University of Chicago*

DURING THE summer of 1966, Chicago's racial problems exploded in headlines across the country. The Rev. Dr. Martin Luther King, together with the Chicago Freedom Movement, initiated a series of marches and demonstrations in and around Chicago to highlight the problems of racial segregation in housing. At a rally held in Soldier Field, Dr. King said, "for our primary target we have chosen housing . . . we shall cease to become accomplices to a housing system of discrimination, segregation and degradation. We shall begin to act as if Chicago were an open city."

Reflecting the general conditions of racial unrest, rioting broke out on the west side in July, and National Guardsmen were called out to restore order. Violence and hostility continued throughout the summer, and in this unsettled context a series of meetings took place between the leaders of business, labor, government, civil rights, housing, industry, and religious groups. On December 6, 1966, they signed the Chicago "Summit Agreement" in which they accepted the responsibility for eliminating the dual black and white housing markets that they agreed existed in the city. The Leadership Council for Metropolitan Open Communities was created to carry out this mandate, and the U.S. Department of Housing and Urban Development (HUD) provided the funds to establish the fair housing service. The Center for Urban Studies of The University of Chicago was requested by HUD to monitor the Leadership Council's programs and to evaluate whether they were progressing towards the integration goal.

THE EVALUATION SCHEME

Several problems arose immediately, as the Center developed its methodology and work program. To perform the evaluation task properly it was important to have baseline information about the existing degree of racial segregation, to be able to forecast where existing trends were leading in the absence of the Leadership Council's programs, to have some measurement of the long-run ideal of truly integrated neighborhoods and to have estimates of how fast the ideal might be achieved if every change in residence occurring naturally within the metropolitan area were made so as to contribute to the ideal. Such "baseline" and "maximally-achievable ideal" forecasts were clearly essential if the evaluators were to be able to determine the size of the problem and then to assess whether the Leadership Council's programs were, first, making a difference in the sense that actual events were deviating from the baseline of existing trends and, second, were moving in the direction of the integration goal.

RESEARCH NEEDED

Each of the ingredients presented its own complexities. What, for example, was the degree of racial segregation of Chicago's neighborhoods in 1967, when the Leadership Council began its activities? The last federal census had been taken in 1960, and much change had taken place since that date. What changes might be expected in the spatial distribution of Chicago's black population if present trends and conditions continued? Census information is only available decennially, but we needed to know how changes unfolded on a weekly and monthly basis. Much time and effort had to be devoted to obtaining new kinds of information that permitted continuous monitoring of residential shifts in the Chicago housing market and updating of information on residential segregation. Equally complex was the problem of determining what the racial mixture of neighborhoods would be like in the ideal case, if all real estate transactions were completed in a colorblind housing market, one in which race did not affect the decisions of buyers, tenants, real estate dealers, sellers, or landlords.

MEASURING POTENTIALS

The immediate reaction was to think that this utopian state would exist when the black-white ratio in each neighborhood was the same as in the metropolitan area as a whole. A moment's reflection should indicate that such a uniform ratio is unreasonable, however. For a variety of historical reasons, educational levels and types of job training of blacks and whites now differ, and these differences will persist for many years to come, all the nation's efforts to the contrary notwithstanding. This inertia carries through into the job market, and gives rise to a higher proportion of blacks than whites in lower income categories. In consequence, more blacks than whites have to rent homes, and only a relatively small proportion of black families are able to afford to buy more expensive residences. Proportionately more black purchasers of middle-class homes have two wage-earners rather than one. Black families in lower income levels tend to be larger than white families, thus requiring larger homes or apartments. Further, a disproportionately large share of the poverty families in the central city is black, with female heads of large households looking to public welfare for support.

Somehow, any "ideal" that the evaluators developed had to be well grounded in such facts, recognizing the effects of inertia on existing differences between black and white homeseekers (as well as the fact that the neighborhoods of the city varied considerably in the kinds of housing they had to offer) on the potentials for integration in the next decades. The need to develop some insights into what might be achievable was also important because some observers, for example certain members of the real estate industry, had said that the residential segregation of black and white was the result of a system that takes exactly these differences in income and family type into account, and that there was little distinction, then, between actual conditions and what, ideally, might be achieved. Others, for example, the Leadership Council, argued that even if both income and family differences and the preferences of many families, both black and white, to live in racially homogeneous neighborhoods were taken into account, there was still much scope for increasing integration. The construction by the evaluators of an ideal map representing colorblind mixing subject to the effects of income, family, and other differences could, then, not only quantify the goal toward which the Leadership Council was striving; it could also help resolve the argument about the extent of change the Council might hope to achieve by determining the differences between actual and ideal.

The actual model developed was quite complex, taking many variables into account.   A simple example using only one variable, housing expenditures, will serve to illustrate how the evaluators approached this task of measuring racial integration potentials.   First, a table was prepared for each census tract in the metropolitan area, listing the number of homes available in each of a series of price and rental classes.   The proportion of the families demanding homes in each of these classes in the metropolitan area as a whole who were black was also calculated.   It was assumed that this proportion could be used as a measure of the probability that a homeseeker coming to the tract looking for a home in this category would be black if the housing market and individual homeseekers were completely indifferent to color.   Thus, if there were 50,000 families demanding homes of category A in the metropolitan area and 20,000 were black, the probability that a family selected at random from the 50,000 would be black is $20,000/50,000 = 0.4$.   If tract 1 has 1000 homes in category A, the expected number of black occupants is therefore $1000 \times 0.4 = 400$ on the average in the colorblind market, although there clearly could be random variations from this.   Similarly, if the metropolitan probability of a black family demanding a home in price range B, is 0.2 and there are 750 homes of this range in tract 1, the expected number of black occupants is $0.2 \times 750 = 150$.   The several expectations can be summed $(400 + 150 + \ldots)$, to yield the predicted number of black families in tract 1.   This, divided by the total number of homes in the tract, gives the expected black percentage, which usually differed from the actual black percentage currently observable in the tract.   The difference in percentages is called the percentage redistributive shift required if present segregation is to be changed to the ideal of integration.

The various percentages can be mapped to show spatial variations.   Figure 1, for example, maps the predicted black percentages computed for every census tract in metropolitan Chicago, using a more complex version of the model described above in which the expenditure categories were subdivided by household size, and each of the resulting categories was in turn subdivided into successively smaller groups on the basis of additional variables, including the sex of the household head.   In the map, different shades have been applied to the census tracts to indicate different percentage ranges.   The map reveals that few neighborhoods would have more than twenty percent of their residents black in this utopian case.   These are largely older central-city neighborhoods in which large apartments are available at low rents in deteriorating buildings, or areas in which there are new clusters of public housing.

Massive shifts in racial patterns would be required to achieve the levels of integration corresponding to the patterns of Figure 1.   These are shown in Figure 2, which maps the redistributive percentages—differences between the ideal and today's actual percentages—calculated in the model from which Figure 1 was derived.   To move from present racial patterns to the ideal would require, the map shows, 10 to 20% increases in the percentage of

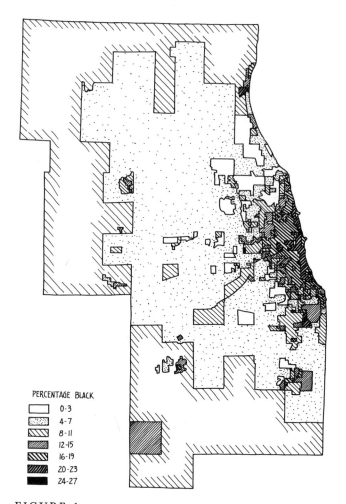

FIGURE 1

*Distribution of the black popu-
lation in a colorblind housing
market with expenditure standardization*

black households in most suburban areas and decreases of more than 60%
in the existing central-city ghetto neighborhoods.

How fast could the ideal of colorblind mixing be achieved under natural
moving rates? Analyses of residential mobility indicated that a minimum
of seven to eight years would be required if all relocations of families normally occurring within the metropolitan area, as they change residences to
adjust them to changing needs and income levels, were made so as to integrate

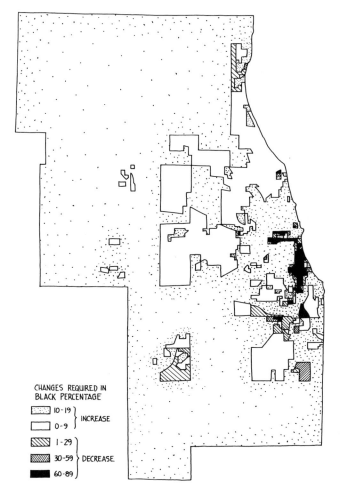

FIGURE 2

Redistribution of the black population to achieve expenditure-standardized equality

neighborhoods now segregated. Of course, this is the lower limit of the time required under the most favorable circumstances. In so far as families' changes of residence, made as they are to increase satisfactions with home and neighborhood, are inconsistent with increasing integration (as, for example, when a white resident of one of today's integrated communities receives a promotion and moves to an area in which the black percentage expected in terms of the model is lower), a much longer time will be needed. On the other hand, to the extent that both black and white prefer to live in racially

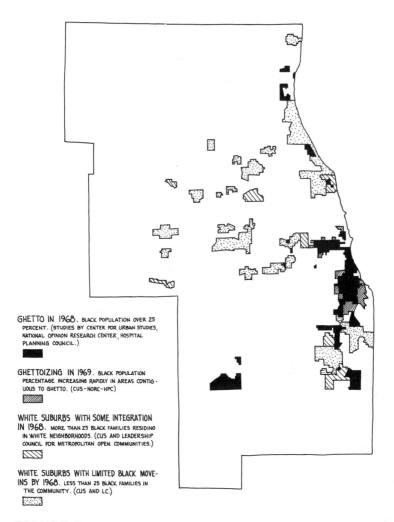

GHETTO IN 1968. BLACK POPULATION OVER 25 PERCENT. (STUDIES BY CENTER FOR URBAN STUDIES, NATIONAL OPINION RESEARCH CENTER, HOSPITAL PLANNING COUNCIL.)

GHETTOIZING IN 1969. BLACK POPULATION PERCENTAGE INCREASING RAPIDLY IN AREAS CONTIGUOUS TO GHETTO. (CUS-NORC-HPC)

WHITE SUBURBS WITH SOME INTEGRATION IN 1968. MORE THAN 25 BLACK FAMILIES RESIDING IN WHITE NEIGHBORHOODS. (CUS AND LEADERSHIP COUNCIL FOR METROPOLITAN OPEN COMMUNITIES.)

WHITE SUBURBS WITH LIMITED BLACK MOVE-INS BY 1968. LESS THAN 25 BLACK FAMILIES IN THE COMMUNITY. (CUS AND LC.)

FIGURE 3

Racial makeup and trends in metropolitan Chicago

homogeneous neighborhoods, Figure 2 may overstate the degree of integration desired.

THE EVALUATION

The primary goal in the Center's efforts was to provide a framework for evaluating the Leadership Council's programs by providing baseline data and measuring the goals that the Council aspired to achieve. This evaluation was still in progress as of mid-1970 although a first-year evaluation report entitled *Down from the Summit* was submitted to HUD by the Center in late 1969.

When this report was made, only a little progress in suburban areas could be reported, and in fact the central-city ghetto was characterized by increasing rather than decreasing polarization. Real estate agents' dual listings, the racial fears and prejudices of whites fleeing to suburbia, and the increasing feeling among blacks that integration is undesirable limited most additional housing available to blacks to neighborhoods contiguous to the existing ghetto. This in turn resulted in a wavelike pattern of ghetto expansion.

As for the Leadership Council, it could report that some suburbs had enacted fair housing ordinances and some now had a few black residents (Figure 3). Of 902 black families living in otherwise white suburban neighborhoods, 544 had moved in during the years 1967–68. Meanwhile, very few white families moved back into the central city where our maps showed that countervailing flows were necessary, except where public investment in urban renewal was involved. Clearly, massive efforts will be necessary in the future if society truly believes that integration is an important goal that must be achieved.

PROBLEMS

1. What were the types of information the Urban Studies Center felt it needed before it could begin to evaluate the success of the Leadership Council's integration program?

2. What is meant by "maximally-achievable ideal forecast"? How does it enter into the evaluation process?

Consider the table:

Home Category		Metropolitan Demand (families)		Number of Homes in
		Total	Black	Selected Census Tract
$10,000	$20,000	200,000	100,000	100
$20,000	$40,000	400,000	150,000	300
$40,000	$60,000	200,000	50,000	500
$60,000	$80,000	100,000	20,000	200
$80,000	$100,000	50,000	10,000	100

Using the technique outlined by the author, answer questions 3 through 5.

3. How many black-owned homes in the $40,000–$60,000 range would you expect in the selected tract with a colorblind market?

4. What *percentage* of the homes in the $20,000–$40,000 range would you expect to be *white-owned* in the selected tract with a colorblind market?

5. What would be the total expected number of black-owned homes in the selected tract with a colorblind market?

6. Suppose the total black population of Chicago is 30%. According to Figure 1, most of the northern quarter of Chicago would have 4–11% black residents in a colorblind market. Explain this concept of integration.

7. (a) If there were no racial or other biases at work in the housing market, what relationship would you expect between the racial percentage in a given neighborhood and in the city as a whole? Explain.

(b) What does the author call the situation in which only economic biases exist?

8. Refer to Figure 2. Estimate the percentage of the city's area which would require an increase in its black population to achieve integration.

9. Suppose a census tract had the following composition:

	No. of units available	% white Chicago families seeking housing in this category
Housing category A	500	50
Housing category B	300	90
Housing category C	200	60

What percentage of this census tract would be white, assuming a colorblind housing market? How would the study characterize this census tract? (Hint: Refer to Figure 3.)

10. Notice the author intimates that *current* residential mobility rates were used in estimating how long it would take to achieve the "ideal" integrated arrangements. If total or even partial success were achieved in the establishment of a colorblind market though, would you expect *current* overall mobility rates, for blacks at least, to go up or down? Why?

11. Describe the overall results of the Urban Studies Center evaluation of the Leadership Council's program. Might the success (or lack of success) found have to do with other factors besides the activities of the Leadership Council? If you say "no", why not? If you say "yes", name a few such factors.

CENSUS STATISTICS IN THE PUBLIC SERVICE

Philip M. Hauser   *University of Chicago*

ON AUGUST 28, 1963, the Chicago Board of Education passed a resolution creating an Advisory Panel of five members to study the problem of segregation in Chicago's public schools (see Reference). The Board's resolution followed an agreement to effect an out-of-court settlement in the suit *Webb versus the Board of Education of the City of Chicago* in which the Board was being sued to remedy the adverse consequences of segregation in the schools.

In its resolution, the Board stated that "without design on the part of the Board of Education or the school administration, there are schools under the jurisdiction of the Board which are attended entirely or predominantly by Negroes"; and that "there exists public controversy as to the racial composition of such schools, and the psychological, emotional, and social influences that may be brought to bear on the pupils in such schools and any harmful effects thereof on educational processes." The Board assigned to the Panel the following task: "to analyze and study the school system in particular regard to schools attended entirely or predominantly by Negroes, define any problems that result therefrom, and formulate and report to the Board as soon as may be conveniently possible, a plan by which any educational, psychological, and emotional problems or inequities in the school system that prevail may best be eliminated." The resolution contained a provision to the effect that on the receipt of the Panel's report "the Board shall promptly take such action as it may determine is appropriate or required to work toward a resolution of any problems involved and any inequities found to exist."

The Panel, fully appointed by October 30, 1963, was confronted with the exploration of a problem that, while restricted to the City of Chicago, was actually a nationwide problem of major importance: the problem of racial segregation in the schools. It remains a major source of cleavage and conflict on the domestic scene, constituting a major element in the nation's "urban crisis." The report of the Panel, submitted to the Board of Education on March 31, 1964, constitutes an interesting example of the use of census statistics in an effort to analyze this crucial national problem as manifest in Chicago and to make recommendations for its resolution.

Census statistics were utilized by the Panel in four distinct ways: first, to place the problem of segregated patterns of residence and schooling in its historical and national context; second, to show the relationship between school and residential segregation at the time of the investigation; third, to differentiate the socioeconomic status of neighborhoods by race in Chicago to see how such differences affected the educational process; and fourth, to project the trends discernible in the census data to anticipate the future. Each of these uses is described in what follows.

HISTORICAL AND NATIONAL CONTEXT

In a section of the Panel's report entitled "Demographic and Social Background," census statistics were analyzed for the nation and for Chicago to provide the historical and national context for consideration of the problem of racial segregation in the schools. The use of the census data in relation to other data and in the framework of sociological and demographic interpretation is best demonstrated by actually quoting from the report:

> Although Negroes have resided in the United States for over three centuries, they have not, in general, been able to enter the main stream of American life until the present generation. In 1860, 92 percent of all Negroes in the nation resided in the South. . . . By 1960, 73 percent of the Negroes, a higher proportion than of the white population, lived in cities; and 40 percent lived in the North and West. Half the Negro population, in 1960, resided in the central cities of metropolitan areas; that is, central cities having fifty thousand or more population.

The Negro population in Chicago, 44 thousand or 2 percent of the total population of the City in 1910, increased over eighteen-fold to reach a total of 812 thousand, or 23 percent of the population by 1960. . . . The growth of the Negro population has been of unprecedented magnitude and speed for any one ethnic or racial group in the history of the City.

As the Negro population has increased in Chicago, as in other metropolitan central cities, the white population has spread outward into surrounding suburbia. . . . In 1960, Negroes made up 23 percent of the population of the city but only 14 percent of the six counties that make up the Standard Metropolitan Statistical Area.

The entree of the Negro to Chicago paralleled that of the early white immigrant groups in at least three respects: . . . [they] entered the city in the inner slum areas in concentrated or segregated fashion; they did the dirty work with the lowest pay; they were greeted with suspicion, hostility, and prejudice, and subjected to discriminatory practices.

The concentration or segregation of Negroes, like that of white immigrants, was not only the result of external pressures, but also of internal forces. Negro in-migrants, like white immigrants before them, usually came to live with relatives, friends, and people from the same town or area of origin. . . . [I]t must be anticipated that enclaves of Negroes will continue to exist on a voluntary basis for a long time even after all economic and social barriers to integration have been removed.

As a result of residential concentration, the Negro population, like white immigrants before them, find their children attending *de facto* segregated schools. . . .

. . .

In 1960, non-whites (97 percent Negro) made up 23 percent of the total City population, but they constituted 34 percent of the population of elementary school age and 27 percent of the high school age population. In 1963, according to the October 3 Board of Education headcount, non-white pupils made up 54 percent of the elementary school pupils in the public schools and 36 percent of the high school pupils. (Parochial and private schools include a disproportionate number of Chicago's white students.) (*Report,* 1964, pp. 4–6)

The census data, analyzed over time for the U.S., for metropolitan areas, and for Chicago, enabled the Panel, as indicated above, to place the problem of segregation in the schools in the broad context of national historical developments.

THE RELATION OF RESIDENTIAL AND SCHOOL SEGREGATION

Statistics from the 1960 Census were used to construct maps showing the areas of Negro residential segregation. Into these maps were plotted the location of the various types of schools: elementary, "upper-grade centers" (grades 7 and 8), and high schools by designation as "segregated white schools," "segregated Negro schools," and "integrated schools," as determined by school

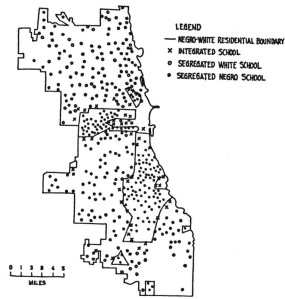

FIGURE 1

Location of integrated and segregated elementary schools and elementary school branches in Chicago

enrollments. An example of this procedure is shown in Figure 1 (*Report,* 1964, p. 57) for elementary schools.

It is clear from the map alone that the segregated Negro schools were located in the segregated Negro residential areas, that the segregated white schools were in the segregated white residential areas, and that the "integrated schools" were clustered at the boundaries between the white and Negro residential areas. It was clear also from this combined use of census and school enrollment data that school integration could be achieved, as long as segregated residential patterns persisted, only by massive movements of white and Negro pupils. Certainly integration could not be achieved with the continuation of "neighborhood schools," a fact that led to great controversy over the neighborhood school concept. Finally, it could be seen that more integrated schools could be achieved by more admixture of white and Negro pupils at the peripheries of the white and Negro residential areas.

The use of census data in conjunction with data from the schools permitted the Panel to reach the following significant conclusion:

> Negro children and teachers and other staff in Chicago Public School System are, by and large, concentrated in predominantly Negro schools located in pre-

dominantly Negro areas in the City. This *de facto* segregation in the public schools is not unique to Chicago. It is a pattern common to many central cities with relatively large Negro populations in the metropolitan United States, even when there is no *de jure* segregation; that is, legally enacted provision for segregated schools.

This conclusion, in turn, led to some of the more important recommendations of the Panel on "student open enrollment patterns" which could increase the number of integrated schools (*Report*, 1964, p. 27 ff.).

SOCIOECONOMIC STATUS OF NEIGHBORHOODS AS A FACTOR IN SCHOOLING

The Panel was faced with the need to ascertain not only the background and extent of school segregation but also the assessment of "any harmful effects" resulting from such segregation. It was decided, therefore, to obtain such measurements as were possible of "ability" and "achievement" in the white, Negro, and integrated schools, respectively, and to do this separately for neighborhoods of different socioeconomic status. By analyzing the data separately for different socioeconomic levels of neighborhoods, it would be possible to see whether the differences in ability and achievement were entirely a matter of race or whether the socioeconomic background of the population also played a part.

To classify neighborhoods by socioeconomic status the 1960 Census data were employed. Many other studies had previously demonstrated that "median years of schooling" (that is, the number of years of schooling above which and below which half of the population fall) was an excellent index of the socioeconomic status of a neighborhood. Neighborhoods were classified by using "years of schooling" of the population as reported by the census. The city's neighborhoods were divided into three groups: those of "high education status," "medium education status," and "low education status." For each category of neighborhood and for white, integrated, and Negro schools measurement of "ability" and of "achievement" were examined as shown in Table 1 (*Report*, 1964, p. 86).

The scores shown in Table 1 are "stanine scores," that is, scores based on nine classes of individual scores. A higher score means a higher ability or achievement. The average stanine score for the city was 5.

The Panel was aware, of course, that there is no "culture free" measurement of mental ability. That is, all tests of mental ability involve the use of language or other forms of expression acquired by the person in the process of socialization—the process that transforms the newborn infant into a member of society. All tests of mental ability are, therefore, in some sense also a test of achievement. The Panel believed, however, that the mental ability test could be used, with caution, as an indicator of what the pupil could

TABLE 1.   Median Stanine* of Ability and Achievement Test Scores by Race, Composition of School, and Socioeconomic Status of Neighborhood.

NEIGHBORHOOD	WHITE		INTEGRATED		NEGRO	
	Achieve†	Ability#	Achieve†	Ability#	Achieve†	Ability#
High education status	6	7	5	6½	5	5
Medium education status	5½	6	4½	5	4	4
Low education status	5	5	4	4	3	3½

* Medians are based on seven to nine schools in each group for which data were available. Stanines are statistical summary measures for distributions.   The higher stanine figure indicates higher ability or achievement.   Five is the average stanine for the city.   A stanine of 7 indicates that the students score on the average higher than three-quarters of the students in the city, but below the top one-tenth.
† Metropolitan Achievement Test, sixth grade, "Word Knowledge."
# California Test of Mental Maturity, sixth grade.

be expected to achieve.   In the guarded interpretation of the data the Panel felt justified in making the following generalizations:

(1) The median mental ability of students is highest in predominantly white schools, is lowest in predominantly Negro schools, and falls between these two medians in integrated schools.

(2) Achievement test medians in all areas of basic instruction are highest in predominantly white schools, are lowest in predominantly Negro schools, and fall between these two medians in integrated schools.

(3) Achievement test medians of students in each of the three groups of schools, when compared with achievement prediction as reflected by mental ability test data, show that students in predominantly Negro schools do fully as well as those in either predominantly white schools or in integrated schools.   This significant fact suggests that Negro students, as a group, profit from instruction fully as much as other groups. It also suggests that intensified educational opportunities for Negro boys and girls would result in a major closing of the achievement gap between group performance of Negro students and other groups.

(4) It should also be noted that students from low education areas in white, integrated, and Negro schools had lower achievement and mental ability test scores than their counterparts in high education areas; and their achievement test scores, when compared with achievement prediction as reflected by mental ability test data, also show they did fully

as well as students in areas of high educational background (*Report,* 1964, pp. 20–21).

The Panel might have also observed that the children in the white schools did not come as close as did those in Negro schools to achieving what their ability tests predicted. Because the objective of the Panel was to explore possible adverse effects of segregated schooling and to recommend what could be done to improve the education of segregated Negro pupils, emphasis was placed on the finding that Negro children have higher ability and achievement with better neighborhood backgrounds and were able to reach achievement levels predicted by their ability tests. With a different orientation the Panel might also have concluded that white pupils could be induced to achieve more with whatever additional inputs might have been necessary to get them to reach achievement levels equal to the predicted levels given by their ability tests. Moreover, to the extent that the ability tests were also achievement tests it could also have been concluded that since the ability scores were lower for Negro than for white pupils, the former got less rather than more from instruction. But such an interpretation of the data would have ignored the disadvantaged backgrounds of Negro children that would be expected to lower both their mental ability and achievement scores.

In any case, whatever the best interpretation of the data might be, the fact is that the use of the census data in conjunction with the test data made it possible to take into consideration the neighborhood background as well as race in studying the impact of segregated schooling.

ANTICIPATION OF THE FUTURE

The Panel in performing its function felt it desirable to outline to the Board of Education what the population trends, for whites and blacks, indicated for the future. Accordingly, population trends as analyzed from the 1960 and preceding censuses were projected to 1980. This enabled the Panel to state the following:

> Population projections indicate that the non-white population of Chicago, if present trends continue, will increase from 838 thousand to 1.2 million, or by 40 percent between 1960 and 1970. Non-whites 5 to 9 years of age may increase by 62 percent, and those 10 to 14 years old by 106 percent between 1960 and 1970. Non-whites of high school age (15–19) may increase by 135 percent during the decade. (Projections by the Population Research and Training Center of the University of Chicago.)
>
> In contrast, the white population in Chicago by 1970, present trends continuing, will decline by an additional 285 thousand persons to total 2.4 million. White children 5 to 9 years will increase by only 11 percent and those 10 to 14 years by 30 percent. White children of high school age will increase by about 46 percent during the decade.

In consequence, by 1970, non-white children may make up about 44 percent of the elementary school age children and 40 percent of the children of high school age. It is possible with present trends that by 1970 non-white children will make up about 65 percent of all children in the public elementary schools, and approximately 45 percent of those in the public high schools.

Projections to 1980 show the same pattern of change—a more rapid increase in non-white than white school age population and, therefore, in the proportion of non-white children of school age and in the public schools.

Thus, demographic and social trends are exacerbating the problem of *de facto* school segregation and its consequences. The Negro population in the City is increasing rapidly while the white population continues to decrease. Moreover, the Negro school population, because of relatively high Negro birth rates, the enrollment of many white students in parochial schools, and the exodus of whites to the suburbs, is increasing more rapidly than the total Negro population (*Report,* 1964, pp. 6–7).

The projections and conclusions based thereon were presented to the Board to enable it and the school administration to anticipate the course of enrollments and to plan accordingly.

CONCLUDING OBSERVATIONS

It should be mentioned that the Chicago Board of Education approved the Panel's report "in principle" within nine days of its transmittal. It is beside the point, for purposes of this essay, that little has actually been done to implement the recommendations of the Panel in the years that have elapsed since March 1964. The basic policy issues are difficult ones, ethnic and racial hostilities continue to plague the city of Chicago as they do the entire nation, and Chicago, as well as the rest of the nation, is still struggling with the problem of residential and school segregation.

The report of the Advisory Panel, based in part on the use of census data and in part on data acquired locally in Chicago, contains 13 recommendations designed to reduce segregated schooling and to eliminate its adverse consequences. These recommendations include some relating to forms of integration, some to improving the quality of education, some relating to improved school-community relationships, and some relating to meeting the costs of the proposed changes. The report has had wide circulation not only in Chicago but throughout the nation and has probably influenced public school programs in other cities more than in Chicago. There are, of course, varied reactions to the recommendations made and wide divergencies in judgments on educational policy. But the Panel's report, based on census as well as other forms of data, has provided a set of facts that illuminates the situation, and constitutes a sound basis for continued public consideration of this perhaps most difficult of all of our domestic problems.

PROBLEMS

1.   Why did the Panel use the Census statistics to study the problem of school segregation?

2.   Using Figure 1, estimate the proportion of Chicago's land area occupied by the 22% of the city's population which was black.

3.   How does figure 1 indicate the housing patterns in Chicago? Is it clear that Chicago's schools were "neighborhood schools"? In what special sense are the integrated schools "neighborhood schools"? (Restrict your answers to information given in Figure 1.)

4.   In attempting to assess "ability" and "achievement" why did the Panel choose to conduct its analysis separately for areas of different levels of socioeconomic status?

5.   List another intuitively appealing measure of neighborhood "socioeconomic status" in addition to "median years of schooling."

6.   What is a stanine? [Hint: See Table 1.]

7.   How does Table 1 show that "students in predominantly Negro schools do fully as well as those in either predominantly white schools or in integrated schools"?

8.   Did the Panel, in its report, attempt to state all possible findings from the data? Why or why not?

9.   What did the Panel forecast would happen to the extent of *de facto* Chicago school segregation in the future and what was the Panel's explanation for the forecasted trend?

REFERENCE

The Advisory Panel on Integration of the Public Schools.
    Philip M. Hauser, Professor and Chairman, Department of Sociology, University of Chicago; Sterling M. McMurrin, Professor of Philosophy, University of Utah, and former United States Commissioner of Education; James M. Nabrit, Jr., President of Howard University; Lester W. Nelson, formerly principal of Scarsdale High School, and retired as Associate Program Director of the Education Division of the Ford Foundation; and William R. Odell, Professor of Education, Stanford University. Philip M. Hauser was elected Chairman, and Sterling M. McMurrin Vice Chairman of the Panel. 1964. *Report of the Board of Education, City of Chicago.*

MEASURING SOCIOPOLITICAL INEQUALITY

Hayward R. Alker, Jr.   *Massachusetts Institute of Technology*

EQUALITY OR its absence has long been a focal concept in political philosophy. Aristotle spoke against the democratic conception of justice defined as "the enjoyment of arithmetical equality," preferring the enjoyment of proportional equality on the basis of merit or property; in effect he took a conservative, inegalitarian view. In the spirit of the French Revolution and its radical emphasis on "Liberty, Equality, and Fraternity," Thomas Paine argued that "inequality of rights has been the cause of all the disturbances, insurrections, and civil wars, that ever happened . . . ."

More contemporary discussions often implicitly or explicitly rely on definitions of equality. The U.S. Constitution calls for "equal protection" under the law. Citizens of various persuasions argue for or against racial imbalance in their schools, for or against more progressive tax structures. When sufficiently general questions are asked, the great majority of the American people is for "equality of economic opportunity" and against government by or for a privileged few. In an historic 1964 decision against legislative malapportionment (*Reynolds versus Sims,* 1964), Chief Justice Warren enunciated "the fundamental principle of representation for equal numbers of people, without regard to race, sex, economic status or place of residence within the state."[1]

THE CONTRIBUTION OF DESCRIPTIVE STATISTICS

Of course, statistics cannot in and of itself resolve the moral and political issues raised in these quotations, nor, without the explanatory theories and data of social science, can it answer the many rival claims about the causes and consequences of various kinds of inequality.   But I would like to argue that as a set of ideas and practices, statistics has contributed to the clarification of the meanings of sociopolitical *inequality*.   In a sense, this achievement amounts to providing a common notion of sociopolitical inequality, several measures or quantitative descriptions of the extent of inequality in a particular situation, and several criteria for choosing one measure over another as more appropriate to a particular situation or concept.   After a brief review of some conventional aspects of most statistical measures of inequality, we shall present some of these measures and several criteria for choice among them.

Can we think of such modest statistical accomplishments as in any way a political success?   Plato would have considered the formal ideal of justice of ultimate significance, its realization, even imperfectly in the world of politics, a noble human achievement.   But is greater sociopolitical equality always greater justice?   For Tom Paine it often was, for Aristotle it clearly was not.   Evaluations of contemporary progress toward greater political equality or toward multiracial equality of educational opportunity are questions on which citizens disagree.   Many would see such achievements as realizations of ideals of equality whose time has come, others as victories for hollow formalisms.   Statistical assessments of two such controversial achievements will conclude this paper.

STATISTICAL CONVENTIONS IN MEASURING INEQUALITY

The simplest but perhaps the most important choice that statistical treatments of inequality have conventionally made is to focus on the distribution of valued objects, events, and relationships.   These have included income, land, votes, legal treatment, ownership shares, and racially equivalent classmates.   The abstractness of the concept of value distribution means that all these different values can be discussed in the same statistical terms.

Does this focus on value distributions seem an obvious or inevitable choice? To some, familiar with notions of income inequality, the answer may be yes. But some deeper reflection suggests that more subjective, more qualitative, and less comparable aspects of inequality might have been focused on.   What

---

[1] This and subsequent citations of the Supreme Court decisions are from *The New York Times,* June 16, 1964, pp. 28–31.

did it mean, for example, to many of the suburban and black voters who were underrepresented in their state legislatures before the major reapportionment decisions of the Supreme Court? Obviously it meant more to some than to others. Except for the common facts of underrepresentation, these meanings were different for many citizens and would be hard to summarize. The same point could be made about income differences, of course. For some, an extra $5000 income would mean being able to pay urgent medical bills; for others, being able to pay for part of their son's college education, a new car, or a pleasure trip to the Taj Mahal.

Nonetheless, the statistical orientation has focused on quantitative common denominators of meaning—a vote is a vote, a dollar a dollar. This interpretation has increased the possibility of comparisons across individuals, families, cities, states, even if it has lost significant subjective differences. Equality before the law or equality of opportunity means something concrete and comparable: similar treatment in similar circumstances.

Another aspect of most serious statistical measures of inequality is that they are, or they rely on, cumulative measures. Thus, in the interests of comparison, a statistician typically looks at the cumulative effects of inequalities in value distribution over an entire population or measures one person's privileged position in terms of how far it is above the average calculated for the whole population. Comparing the largest and the smallest incomes, for example, might make good newspaper copy, but it would not reflect the cumulative impact of the inequalities involved.

In part because of their reliance on cumulative terms, most descriptive statistical measures of inequality are calculated in generally comparable units. Thus a statistician would want to be able to compare income distributions in rubles with those in francs or dollars; to do so he relies on his knowledge of total incomes in each society. Of the several ways of getting such measures, two cumulative procedures will be used frequently here: working with data in percentage and percentile form, and defining indexes on a 0-to-1 or 0-to-100 scale by dividing the raw numerical value in particular units by its maximum value calculated in these same units. The resulting measures are pure numbers; certainly this property maximizes the possibility of comparison across variables.

SEVERAL MEASURES OF SOCIOPOLITICAL INEQUALITY

In a condensed, but geometrically intuitive, way, Figure 1 suggests at least six different but related cumulative measures of sociopolitical inequality. The data there represent the distribution of 150 seats in the New York State Assembly over a total state population of 17 million around 1960. In drawing Figure 1(a), the various Assembly districts (which have one representative each) were ordered from the largest to the smallest in population size. Because each district, regardless of population, elects one assemblyman, it follows

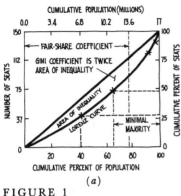
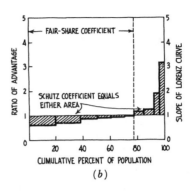

FIGURE 1

Legislative malapportionment in the New York State Assembly, 1960. Source: Alker (1965)

that the most populous districts are the most underrepresented. Plotting the cumulative percentage of seats controlled by the most underrepresented 10, 20, 30, 40, . . . , 100% of the population and then drawing a smooth curve through these points gives us our first cumulative measure of inequality: the *Lorenz curve*. Let us draw a vertical line from 40% on the horizontal axis to the curve, and from there a horizontal line to the right-hand vertical axis. The reading there shows us that the most underrepresented 40% of the population of New York in 1960 controlled only about 28% of the Assembly seats. Note that if representation were equally distributed the Lorenz curve would take the form of a straight line from the lower left corner to the upper right one, a 45° line.

The Lorenz curve also gives us a second measure of inequality: the percentage of the total held by the most "overrepresented" $n\%$ of the population (where n may be 1, 5, 8, 10, etc.). Thus Figure 1(a) shows that the most "overrepresented" 8% of the population elected 20% of the State Assemblymen! Going slightly further down the Lorenz curve gives us evidence that 35% of the state's voting population were potentially, through their representatives, a minimal majority in the Assembly.

Leaving Figure 1(a) for a moment, we realize that it has become convenient to talk about "over-" and "under-" represented citizens. A remarkable property of the Lorenz cumulative curve is that its slope (inclination, or rate of increase) provides us with a measure of this "over-" or "under-" representation. Figure 1(b), a "slope curve," is in fact a plot of the slope of the Lorenz curve shown in (a). To construct the slope curve the data are grouped into ten 15-seat categories. The "ratios of advantage" plotted vertically in (b) show, for example, that the most overrepresented 3.1% of the population, with 15 Assemblymen, elected more than three times as many representatives as the same number of "average" citizens. For this group the value of the slope—their ratio of advantage in the bar graph (b)—is about 3.2. (A little reflection suggests that this is derivable as follows: if 3.1% of the population elects 15 Assemblymen when an "average" 10% should do so in an Assembly of 150 members, their ratio of advantage is $10/3.1 = 3.2$.)

Another quantitative measure of inequality, comparable across states, is immediately suggested by the slope curve. The *equal-share*, or *fair-share, coefficient* measures what percentage of the population gets less than its equal or fair share of representatives. In New York in 1960, it was 77% for the State Assembly. In other words, 77% of the New York citizen population was at least somewhat underrepresented!

Finally, we come to two summary measures most preferred by statistically minded social scientists, the *Gini coefficient* and the *Schutz coefficient*. The Gini coefficient is the area of inequality [between the ideal 45° line and the actual Lorenz curve in Figure 1(a)] divided by the maximum possible area. This area of inequality ranges from 0 in the case of perfect equality (the poorest 50% of the population still get 50% of the values) to ½ (when an infinitesimally small fraction of the population possesses all the values and the resulting area is the right triangle below the 45° line with sides of 100 and 100). Thus the Gini coefficient itself ranges between 0 and $½/½ = 1$. For the data in Figure 1, a Gini coefficient of 0.22 indicates that malapportionment was 22% of its theoretical maximum value.

Another geometrically appealing measure comes from Figure 1(b). Here we sum the area of advantage above the fair-share coefficient or the area of disadvantage below it. The Schutz coefficient is equal to either the total area of advantage or the total area of disadvantage. It is found by summing the areas of the rectangles to the left of the fair-share coefficient or alternatively by summing the areas of the rectangles to the right of the fair-share coefficient. For the Schutz coefficient, the minimum is obviously zero; its maximum is 1.0 (or 100.0 in percentage units). The New York data give a value of 0.156, or 15.6% for the Schutz coefficient.

All of the measures we have mentioned are in some sense aspects of cumulative quantitative value distributions. Typically they measure deviations from an ideal of perfect equality: a 45° Lorenz curve or a constant ratio of advantage equal to 1.0. But Aristotle and others would not accept the notion of perfect numerical equality. For cases such as income distribution, malapportionment, or more controversially, racial imbalance, they might argue that an appropriate ideal was less than complete equality. We might consider as inequitably or unjustly treated only those taxpayers with incomes below one-third of the national average, only those citizens with ratios of voting advantage below 0.90, or only those students with 25% fewer white classmates than the city average. In such cases, we could draw Lorenz curves showing maximum allowable inequality as a standard rather than the traditional 45° line. Slope curves could be derived from such curves, and departures beyond even these lines would then be the basis for a whole class of new cumulative measures of unacceptable inequality.

CRITERIA FOR CHOOSING AMONG MEASURES

Whether defined in egalitarian or inegalitarian terms, two important criteria argue in favor of some version of either the Gini coefficient or the Schutz coefficient. First of all, each is a full information measure in that all the data are used in calculating them, not just the top 8, 10, 20, or 50%. Second, if we assume that each datum is as significant as any other, we want measures that are sensitive to all sorts of differences among them. The minimal majority and fair-share coefficients depend wholly on the location of one point on the Lorenz curve or its slope, but the Gini and Schutz coefficients depend upon location and slope all along the Lorenz curve. More sensational measures such as "the top 1%, who have such and such . . . " can be misleading as to where the rest of the population stands.

A third criterion for choosing among measures, one offered by Yntema for the study of income inequalities, is the stability of a measure's results when different ways of grouping the data are employed. Although this may seem a technical criterion of little substantive import, for those often accustomed to getting incomplete data grouped by deciles or quartiles this criterion is an important one in practice.

Another question raised by social statisticians is whether various measures are highly intercorrelated or not. The implicit criterion is that one needs as many measures of inequality as there are philosophically important and empirically distinct aspects of the phenomenon at hand. It is of particular interest in the malapportionment case, for example, that both the Gini and the Schutz coefficients give very similar, or highly interrelated, results, with the minimal majority measure not far behind. That the fair-share coefficient is much less closely related to any of these coefficients argues for the necessity of thinking in terms of additional important and meaningful aspects of any unequal value distribution.

SOME APPLICATIONS OF STATISTICAL MEASURES

Let us briefly discuss two areas of statistical applications, malapportionment and racial imbalance, and close with some speculations about our initial quotation from Thomas Paine. Although we shall only sketch some of the more interesting findings and the extent to which controversial political achievements have grown from them, it may suffice to convey to the reader the utility of the above conventions, criteria, and measures.

First, a word is in order about New York malapportionment as it was treated by the U.S. Supreme Court. This is perhaps the clearest case of the motivating force of the quantitative equalitarian ideal—the Lorenz line of full equality—stated in "one man, one vote" terms. The close correspondence of the popular minimal majority measures with the more technically adequate Gini and Schutz coefficients in reflecting an underlying reality was seen by the majority of judges in their historic judgment that there was a violation of the Constitutional call for "equal protection" and "the right to vote." Although explicit use of such sophisticated measures is not indicated in the Court's opinion, it is clear that their opinions on *Reynolds versus Sims* reflect statistical perspectives. Even in dissent, Justice Stewart argues that "nobody has been deprived of the right to have his vote counted," while the Warren majority opinion states in quantitative language that "diluting the weight of votes because of place of residence impairs basic right under the 14th Amendment." In language remarkably like the measures we have reviewed, the majority objected to "minority control of state legislative bodies," did not believe that one person ought to "be given twice or ten times the voting power of another," and sympathized when 56% of the citizen population elected only 48% of the Assembly. Realizing that New York in 1960 was less malapportioned than most other American states in terms of the Gini, Schutz, or minimal majority coefficients, we gain some idea of the reenfranchisement that has taken place since then because of the Court's decision. This effort has not been undisputed, however; almost a majority of state legislatures have called for a Constitutional Convention on the reapportionment issue.

Table 1 and the derivative Lorenz curve in Figure 2 show how the Lorenz curve idea can be extended from the cases of income inequality and malapportionment to measure racial imbalance. The cumulative "value" represented in Figure 2 is the percentage of white students available as schoolmates; an analogous figure could have been drawn with percentages of nonwhites. Note that in this case the complete-imbalance picture we described earlier would be impossible unless there were only one white student in all of New Haven. So in order to evaluate the amount of imbalance we have, we must compare the actual Lorenz curve with the one we would have if there were complete imbalance given the existing number of white students. If we were not constrained at all by the sizes of the existing schools, then we could visualize complete imbalance as all of the black students going to all-black schools and all of the white students going to all-white schools. If we then ordered the schools by percentage of white students, we would have a Lorenz curve showing a maximum feasible imbalance that was at first completely flat along the horizontal axis and then slanted up directly to the upper right-hand corner. Because there were four junior high schools in New Haven in 1964 and because these schools had fixed capacities, it is not possible to visualize all the schools as completely segregated; we must have at least one integrated

TABLE 1. The Actual Racial Breakdown of Students in New Haven's Four Junior High Schools, June 1964*

| | NUMBER OF STUDENTS | | | PERCENTAGES | | | | |
SCHOOLS	Whites	Non-whites	Total	Percent of All Whites	Percent of All Students	Cumulated Percent of All Whites	Cumulated Percent of All Students	Ratio: Percent Whites/ Percent Students
Bassett	55	555	610	3	17	3	17	0.18
Troup	419	514	933	19	27	22	44	0.70
Sheridan	741	148	889	34	25	56	69	1.36
Fair Haven	968	140	1108	44	31	100	100	1.42
Total	2183	1357	3540	100	100			1.0 (weighted average)

* These data are taken from a document issued by the New Haven Public Schools, Dr. Laurence G. Paquin, Superintendent, June 8, 1964, entitled *Proposals for Promoting Equality of Educational Opportunity and Dealing with the Problems of Racial Imbalance*, pp. 10–11.

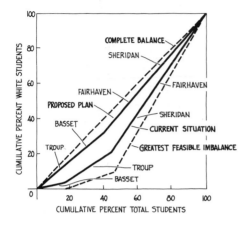

FIGURE 2

Four alternative patterns of racial imbalance in New Haven junior high schools. Source: Alker (1965)

school. Thus the curve of greatest possible imbalance is the lowest one in Figure 2. The actual racial imbalance in the system was measured by a Gini coefficient of 0.25; the maximum possible was 0.40.

Was the cumulative inequality evident in Table 1 and summarized graphically in Figure 2 sufficiently motivating to inspire desirable political action? Of what sort? The New Haven Superintendent proposed, through rezoning and selective busing, to improve racial balance to the considerable extent indicated in Figure 2. The proposed plan would reduce the Gini measure of imbalance from 0.25 to 0.09. After considerable controversy, a somewhat revised plan was evidently put into effect. Many Americans, excluding the majority of the Supreme Court, do not find racial imbalance a compelling issue, so the extent to which such action was a "success" is perhaps more controversial than state legislative reapportionment.

Finally, let us consider some inequalities that have inspired little or no "successful" political action. Recall that Thomas Paine claimed that inequality of rights "has been the cause of all civil insurrections." Statistical analyses do show land and income inequalities within nations to contribute to domestic group violence.

Whether the more extreme and growing inequality between the rich and poor nations of the world has or will occasion similar violence or corrective action remains an open question. Do the members of the poorer, mostly non-white, relatively powerless majority of the world's population have a right to a better life or to more control over their destiny? The answer lies in the success or failure of a revolutionary statistical ideal.

PROBLEMS

1. What is the advantage of a *cumulative* measure of inequality?

2. Consider Figure 1. Suppose the Lorenz curve tells us that the most under-represented 18% of the population elects 15 assemblymen. What is the "ratio of advantage" of this group?

3. Why are the two ways of computing the Schutz coefficient equivalent? (Hint: Refer to Figure 1b.)

4. What is the minimal majority? In Figure 1a, why is it drawn at 49% rather than 51% of the Assembly seats? (Hint: Does minimal majority refer to over- or under- representation?)

5. Study Figures 1a and 1b *together*. What is the name given to the percentage of the population at the point on the Lorenz curve where a tangent of the curve is parallel to the 45° line? (Hint: What is the slope of the 45° line?)

6. What is the major advantage of both the Gini and Schutz coefficients?

7. Explain what is meant by a *summary* (or *full information*) measure of inequality. Of the six measures of inequality the author discusses, which two are summary measures?

8. Consider Table 1.
   (a) What percentage of the students at Fair Haven were non-white?
   (b) What percentage of all non-white students attended Sheridan?

9. (a) Draw the Lorenz curve for *complete* imbalance for Figure 2.
   (b) What is the value for the "area of inequality" in this case?
   (c) Roughly what percentage of the students in a situation of complete imbalance are white?

10. Refer to Figure 2. (Give approximate answers.)
   (a) Using the "proposed plan" curve for Basset and Troup (taken together), what is the cumulative percentage of white students? Of total students?
   (b) Using the "greatest feasible imbalance" curve, what is the cumulative percentage of white students at Basset? At Troup?
   (c) Using the "current situation" curve, what percentage of all white students attend Troup? (Check your answer with Table 1.)

REFERENCES

H. R. Alker, Jr. 1965. *Mathematics and Politics*. New York: Macmillan.

H. R. Alker, Jr. and B. M. Russett. 1966. "On Measuring Inequality." R. Merrit and S. Rokkan, eds., *Comparing Nations*. New Haven, Conn.: Yale. Pp. 349–382.

Part four
OUR PHYSICAL WORLD

The states of nature
Cloud seeding and rainmaking
Looking through rocks
The probability of rain
Statistics, the sun, and the stars

Modern machines
Information, simulation, and production: Some applications of statistics to computing
Striving for reliability
Statistics and probability applied to problems of anti-aircraft fire in World War II

CLOUD SEEDING AND RAINMAKING

Louis J. Battan *University of Arizona*

UNLESS NEW sources of fresh water are found, the world will face serious hardships in the years to come. This is particularly true in the U.S., where the daily bath, the sprinkled lawn and the air-conditioned office have become necessities rather than the luxuries they used to be.

An examination of rainfall and snowfall measurements, going back as far as the records extend, shows wet periods and dry periods, some long, some short. As long as there were few people with small thirsts, there was plenty of water even during dry spells. But the population curve does not rise and fall as does the precipitation curve. The number of people has followed a steady upwards trend and it continues to increase.

We are at that point where several dry years in a row cause great discomfort and concern. This was certainly the case in the northeastern U.S. in the mid-sixties. Fortunately, in the late sixties, wet weather returned and the immediate crisis passed.

Unfortunately, the population continues its relentless climb and each man's use of fresh water does likewise. The next drought will be much more painful to man than the last one, and the one after that still worse. No one can tell when the droughts will come, but you can be absolutely sure they will occur.

What is the solution to this dilemma? It would take very drastic measures to reduce per capita water consumption; therefore, we must find more fresh water. Desalinization of ocean water is a distinct possibility and great efforts are being made in many places to find economical means of removing the salt. Also, progress is being made in the reclamation of sewage water.

The principal source of fresh water today is the atmosphere. On the average, about 40 inches of water falls over the earth every year, but more than half of it lands in the oceans. If we could find an effective way to increase precipitation over the continents, many of our water problems would be reduced greatly. In recent years, meteorologists have begun to feel optimistic about the possibility of attaining such a goal. This essay discusses some statistical aspects of such a search.

FORMATION OF PRECIPITATION

Much of the precipitation reaching the ground originates in the form of ice crystals. On cold days in the winter, the clumping of crystals produces snowflakes that fall to the earth. At other times, when the air is warm, the snow particles melt as they fall and they reach the ground as rain.

Not all rain is formed this way. Some rain develops as small cloud droplets that collide, merge, and grow large enough to fall to the surface of the earth. In this brief essay, it will not be possible to examine this second process any further. We shall concentrate on the precipitation mechanisms involving ice particles because most investigations aimed at rainfall stimulation try to take advantage of the "ice-crystal process."

ICE-CRYSTAL PROCESS

Clouds form as a result of the condensation of water on tiny solid particles in the atmosphere, known as condensation nuclei. They come from blowing soil, swirling smoke, and sea spray. A typical cloud is composed of water droplets ranging in diameter from perhaps 5 to 50 microns. A human hair is about 10 microns in diameter, so it is evident that the droplets are very small, much smaller than typical raindrops, which may have a diameter of 2000 microns (2 mm.).

Water clouds often form in regions where temperatures are below 32°F and still remain in the liquid state. Such clouds are called *supercooled*. The failure of the drops to freeze is attributed to the purity of the water. In extreme cases the clouds can be supercooled to temperatures below −30°F.

Clouds composed of small water droplets, whether supercooled or not, normally cannot produce rain or snow unless some mechanism other than condensation enters the process. The first sound theory for the formation of precipitation was offered in 1933 by the Swedish meteorologist, Tor Bergeron, who theorized that when tiny ice crystals are introduced into a supercooled cloud, precipitation could develop. Because of the difference in the physical properties of water and ice, the ice crystals grow rapidly and the water droplets evaporate. As the crystals enlarge they start falling through the cloud and grow even faster as they collide with other crystals and supercooled droplets. In this way a snowstorm may be produced.

CLOUD SEEDING

In 1946, Vincent J. Schaefer at the General Electric Laboratories at Schenectady, New York, was studying supercooled clouds in a laboratory cold chamber. He discovered that when chips of Dry Ice fell into such a cloud, large numbers of ice crystals formed, grew, and fell to the bottom of the chamber. This was the first clear demonstration of the theory advanced by Bergeron more than a decade earlier.

Schaefer and his colleagues also carried out experiments in the atmosphere. They flew over the tops of thin layers of supercooled clouds (called *stratus clouds*) and dropped Dry Ice pellets from an airplane. In a matter of minutes, the cloud structure in the regions seeded with Dry Ice changed. Ice crystals grew and fell as the water droplets evaporated, leaving a "hole" in the cloud through which the ground could be seen. As the airplane flew below the cloud, light showers of ice crystals and snowflakes were observed to be falling.

This experiment has been repeated in many places by many people. In certain tests, an airplane was flown along prescribed patterns over a uniform cloud deck and seeded along the way. The pattern of cloud dissipation corresponded to the seeding pattern. One of the more spectacular had a shape like a racetrack cut right out of the cloud.

Incidentally, one of Schaefer's colleagues, Bernard Vonnegut, discovered that tiny particles of the chemical silver iodide also are effective ice nuclei. For many cloud-modification operations, silver iodide is more convenient to use than Dry Ice.

Today supercooled fogs and low clouds over certain airports in the U.S. and abroad are cleared on a regular basis by means of cloud-seeding techniques. These operations have allowed airplane movements which would not have been possible if the clouds had not been modified.

RAINMAKING

As noted earlier, when thin, supercooled decks of stratus clouds were dissipated by means of Dry Ice seeding, light snow showers occurred under the seeded regions.   Some meteorologists believed that the precipitation was light because the clouds were shallow.   It was argued that if thicker clouds were seeded, greater quantities of precipitation could be caused to fall.   Skeptics responded that in the case of thick clouds, nature would supply the necessary nuclei and the addition of artificial nuclei would have no effect.

If enough were known about the physical and chemical nature of clouds and precipitations, calculations could be made of how much precipitation would occur with and without seeding.   Unfortunately, we do not have the detailed knowledge required for precise calculations of this kind.

The only satisfactory way to ascertain if cloud seeding can increase precipitation has been by means of experiments in the atmosphere.   The major difficulties faced by scientists engaged in testing the effectiveness of a seeding technique has been the great variability of precipitation in time and space and the inability to make accurate forecasts of precipitation.

Figure 1 shows curves of the August rainfall in Tucson, Arizona, and in New York City for a 39-year period.   The dashed lines show the average rainfalls at the two cities.   The variations from year to year and decade to decade are obvious.   It is evident that August rainfall in New York was substantially below the average during the drought period 1963–66.

In some of the early attempts at rainmaking, very crude schemes were used to evaluate the effects of seeding.   One of them involved the comparison of rainfall during a seeded or several seeded periods with rainfall over the same area during earlier unseeded periods.   In one such test, rainfall during one seeded season exceeded the mean rainfall by 15%, and it was reported that seeding probably had increased the precipitation.   The data in Figure 1 show why such a conclusion is questionable.   It is evident that mean rainfall occurs very infrequently.   Instead, the actual monthly amounts deviate greatly above and below the mean.   Herbert C. S. Thom, a well-known climatologist, has indicated that if you wish to test the hypothesis that seeding does not increase rainfall and you do this by comparing the rainfall over a target area with historical data, extremely large seeding effects would have to occur in order to be detected.   As a matter of fact, with typical rainfall distributions, seeding would have to cause about three times the median precipitaton in order to give a high probability of rejecting the no-increase hypothesis.

In view of the fact that, in 1972, even those convinced of the effects of seeding talk about possible precipitation increases or decreases ranging from about 5 to 50%, it is clear that it is unlikely that seeding effects can be identified simply by examining the rainfall in a target area.   In order to detect seeding effects amounting to about 10% and to do it in a reasonable

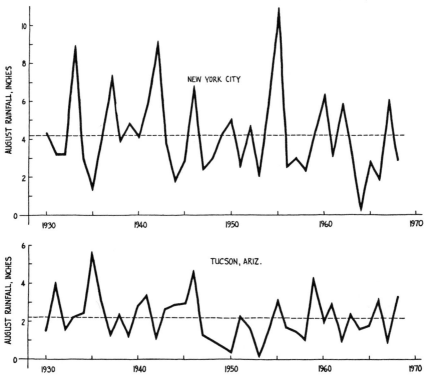

FIGURE 1

Inches of August rainfall in Tucson, Arizona, and New York City, 1930–68. Sources: Institute of Atmospheric Physics, University of Arizona, Tucson. World Weather Records, *U.S. Department of Commerce (1930–60).* Climatological Data: National Summary, *U.S. Department of Commerce (1961–69)*

length of time, it is necessary to use a scheme for estimating how much precipitation would have fallen on the target area if there had been no seeding.   This has brought statistical regression methods into the picture.

REGRESSION METHODS

The character of the weather often is uniform over distances exceeding several hundred miles.   Weather maps show large regions of warm air separated by so-called "fronts" from large regions of cooler air. Within each air mass there are somewhat distinctive atmospheric conditions, clouds, and precipitation. Hence, the average precipitation of two areas some tens of miles in diameter and close to one another are correlated. If care is taken to select areas having similar physical geography and if monthly rainfalls are examined, correlation coefficients exceeding 0.9 are sometimes found. (Correlation coefficients measure similarity between two series; a value of 1 is the highest possible positive value, 0 is low, and —1 means the two series move in exactly opposite directions such as 1, 2, 3 versus 6, 4, 2. See the essay by Whitney for a further explanation of correlation.) This fact has made it possible to obtain results showing that cloud seeding may influence precipitation.

In general the regression methods take into account two areas: a target area and a control area. On the basis of past historical data, a scatter diagram is drawn and a regression line is established as shown in Figure 2. In this illustration the quantities plotted on each axis are not the actual precipitation amounts, but rather the square root of the precipitation. This is called the *transformed precipitation* and is used to give the scatter of the points around the regression line a more symmetrical shape.

It is not necessary to concern ourselves with the details of this analysis. The important point is the following. Having established a regression line such as that in Figure 2, we can, by measuring the precipitation in the control area, estimate the range of rainfall quantities expected over the target area. This knowledge, in turn, makes it possible to estimate the likelihood of the occurrence of results as extreme as those observed during a seeded period (or periods) even if seeding has no effect. For example, the cross in Figure 2 might represent the precipitation amounts measured in the target and control areas during a seeded month. Various statistical techniques exist for calculating the probability that, given the rainfall measured in the control area, the rainfall in the target area occurred by chance.

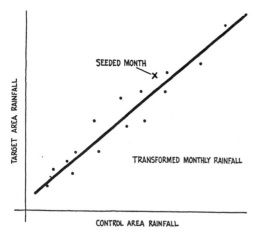

SEEDED MONTH

TARGET AREA RAINFALL

TRANSFORMED MONTHLY RAINFALL

CONTROL AREA RAINFALL

FIGURE 2

Comparison of rainfall in the control area and the target area, plotting square root of rainfall in inches. The dots refer to unseeded months and the cross to a seeded month for the target area. The departure of the cross from the line is similar in size to the departures for the unseeded months

The regression method appears to be sufficiently straightforward, but several pitfalls must be avoided if the conclusions from such an analysis are to be widely accepted. First of all, it is necessary that there be adequate controls over the selection of the seeding periods. An effective way to do this is to use a suitable randomization scheme in deciding when to seed.

Still another danger is the possibility that the historical regression might change significantly from one period to another. For example, would a regression line for the years 1930–50 be the same as the one from 1950–70, or would it be better to use a 40-year regression line? At this time, it is not possible to give a general answer to this question.

CROSSOVER EXPERIMENTS

Researchers in several countries have tested the efficacy of cloud seeding by means of a technique known as the *crossover design*. As in the target-control method discussed above, two areas are involved, but in the crossover design either one of the areas may be seeded, the decision being made according to a suitable randomization scheme.

The higher the correlation between the precipitation in the two test areas, the more efficient is the crossover design in determining whether cloud seeding has increased or decreased the precipitation. If seeding is effective, crossover experiments will detect the effect after fewer tests than would be required if the target-control design were used. The crossover procedure has been employed successfully in Australia where it originated, in Israel, and in the U.S.

SOME RESULTS

Many rainfall modification projects involving randomization have been conducted in various parts of the world. Some of them are still in progress. The results are mixed. In some projects, more precipitation occurred during the seeded periods. For example, experiments conducted near Lake Almanor, California, during the period 1962–67 showed that with cold temperatures and west winds, cloud seeding apparently increased precipitation by an average of 37 percent (Mooney and Lunn 1969). In other tests, such as those conducted by the University of Chicago, in central Missouri between 1960 and 1965, cloud seeding apparently decreased rainfall (Braham and Flueck 1970). In still other programs, it appears that cloud seeding had no effect.

The results of the field tests show convincingly that sometimes cloud seeding influences the quantity of rain or snow reaching the ground. The chances of increasing precipitation from winter clouds over mountainous regions appear to be greater than are the chances of getting more rain from summer showers and thunderstorms, but such a generalization must be treated with caution.

Recent research indicates that by means of ice-nuclei seeding, individual summer clouds having certain specific characteristics were caused to produce more rainfall than they would have produced if they had not been seeded. On the other hand, such special clouds may not contribute substantial fractions of total summer rainfall. For example, scientists in the Soviet Union have reported that rainfall from certain types of summer clouds has been more than doubled by cloud seeding. They also have indicated, however, that such clouds account for no more than one percent of the total summer rainfall.

SUMMARY

The great variability of precipitation and our inability to make accurate quantitative predictions of rainfall have made it necessary to conduct experimental programs in order to establish whether or not particular cloud-seeding techniques can increase precipitation. It has been found that seeding with ice nuclei, that is, Dry Ice or silver iodide particles, may influence precipitation in some cloud systems. We still have not learned how to identify those systems where precipitation can be increased reliably.

Many atmospheric scientists are convinced that it should be possible to increase precipitation by economically significant amounts and hence to help quench the water needs of an increasingly enlarging and thirsty population. An expanded program of carefully designed and executed field experiments is needed to test existing ideas and hopefully generate new ones.

PROBLEMS

1.   Consider Figure 1 and the portion of the article which refers to it.
    (a) What is the ratio of the average 1930-68 August rainfall in New York to the Tucson average?
    (b) What is the ratio of the maximum 1930-60 August rainfall in New York to the Tucson maximum?

2.   Consider Figure 1. Why would it be difficult to test the efficacy of cloud seeding in New York (or Tucson) by comparing the amount of rainfall during a seeded period with the historical average rainfall?

3.   How should the target and control areas in the target-control approach to testing the efficacy of cloud seeding be chosen?

4.   Under what conditions would you expect the rainfall in two areas to be closely correlated?

5.   What is the *transformed precipitation* used in the construction of Figure 2 and why was it used in place of the precipitation itself?

6.   Suppose the seeded month cross of Figure 2 had been *considerably below* the regression line (and the other points). What would you conclude about the impact of cloud seeding on precipitation?

7.   Explain the difference between the target-control and crossover design approaches to testing the efficacy of cloud seeding.

8.   Why do you need a randomization scheme for deciding when to seed?

REFERENCES

R. R. Braham, Jr. and J. A. Flueck.   1970.   "Some Results of the Whitetop Experiment."   Preprints of papers presented at the Second Conference on Weather Modification.   Boston: American Meteorological Society.   Pp. 176–179.

Margaret L. Mooney and George W. Lunn.   1969.   "The Area of Maximum Effect Resulting from the Lake Almanor Randomized Cloud Seeding Experiment."   *Journal of Applied Meteorology* 8:1, pp. 68–74.

LOOKING THROUGH ROCKS

F. Chayes   *Carnegie Institution of Washington*

AIMS AND PURPOSES

Thin enough sections of all crystalline rocks are transparent. They are also among the most beautiful objects studied under the microscope. Most of them consist of aggregates of minerals that can be distinguished from each other by striking differences in optical properties. The science of petrography is devoted to the study of such aggregates.

Although several thousand mineral species are now known, fewer than 15 probably comprise more than 95% of the earth's rocks, or what is called the *lithosphere*. Further, only seven or eight are likely to occur in appreciable amounts in any one rock. Hence, information about the amounts of different minerals it contains usually plays a key role both in the description and classification of a rock and in attempts to decipher its geochemical history.

An estimate of the amounts of the different minerals actually present in a rock is called a *modal analysis,* or *mode.* Although the petrographic use of this once standard philosophical term goes back only to the beginning of our century, what are now called modal analyses began appearing in the technical literature just before the 1850s.

When the geologist uses the term "rock" he is usually thinking of a large mass, or volume of material, underlying an area measurable at least in acres and often in square miles. He can bring home only a few small pieces, or *hand specimens,* of any particular rock, however. The selection of these is the first in a long series of sampling operations that intervene between rocks and scientific knowledge about them.

In the case of modal analysis, for instance, the process consists, in principle, of identifying and estimating the volume of each of a large number of mineral grains, summing the volumes of grains of each kind, and dividing each of these sums by the total measured volume. No one would attempt such a measurement on an entire rock; even the small hand specimen the geologist brings home to his laboratory may contain millions of grains! It is always impractical and usually impossible to separate well consolidated rocks into single grains that can be measured individually, and it is clearly impossible to measure grain volumes directly unless this can be done. So, after all, we do not measure grain volume directly; rather, we measure something statistically related to grain volume, namely, the *area* the section of the grain occupies in a random plane intersecting it. This plane may be a flat surface ground onto a slab or hand specimen of the rock. It is more likely to be the surface of a microscopic preparation called a *thin section,* for many of the mineral identifications required in modern petrography can only be made under the optical microscope or one of its more powerful sisters, the electron microscope and the electron probe.

Our objective is knowledge of the volume composition of a rock body measurable in cubic meters or kilometers. We reach this knowledge by sampling minute amounts of a volume-related characteristic, random cross-sectional area. In the successive stages of sampling, we continually reduce the sample volume, from mountain or hillside to outcrop to hand specimen. In the final stage, volume disappears entirely and its place is taken by area. We shall be concerned here with the product of this final sampling, with the ways of measuring relative areas, and with the reasons we have for supposing that measurements of relative areas contain reliable information about relative volumes.

THE THREE MAJOR FORMS OF MODAL ANALYSIS

In modal analysis we partition a reference area into a set of subareas by direct areal measurement, by measurement of line segments cutting each type of subarea, or by counting points that fall in them. Each technique must satisfy substantially the same basic requirements. We must show that it yields estimates of the relative areas of essential minerals in a plane section of the rock. Then we must show that these relative areas are sound estimates of relative volumes.

If a plane reference surface is sawed and ground on the face of a rock, the measurements will be made at low magnification in reflected light. If, on the other hand, the reference area is the upper surface of a thin section, it will be viewed at considerably higher magnification and in transmitted light. In either case, the result of measurement will be a set of numbers thought to be proportional to the areas occupied by the essential minerals of the rock.

Such numbers were once obtained by tracing the outlines of the various mineral sections onto translucent paper and thence onto tinfoil or cardboard, cutting along the traces of grain or section margins, pooling those of each species, and weighing the pooled fractions. This is the way modes were first obtained, in 1848, by the French mining engineer A. Delesse. Delesse worked on tracings made from rock slabs, but in 1856, Sorby, the great English microscopist, treated tracings of much enlarged projections of microscopic objects in the same way. In 1859, in the *Origin of Species,* Darwin recorded making similar measurements on "Professor H. D. Rogers's beautiful map of Canada," thus discovering that in that country areas of "metamorphic . . . and granitic rocks exceed, in the proportion of 19 to 12.5, the whole of the newer Paleozoic formations." Neither Sorby nor Darwin says a word about Delesse, but it is possible that neither knew of his work.

It is intuitively obvious that if the cardboard or tinfoil is of uniform gauge and the cutting and weighing are exact, the numbers yielded by the Delesse method will indeed be proportional to mineral areas. The "theoretical" basis of the cutout procedure is beyond reproach, but no one has ever attempted a systematic study of the accuracy or precision of the results. In fact, almost no one has used the method for routine modal analysis. Shortly after it was proposed, petrographic interest shifted from the study of hand specimens to the study of microscopic, or *thin,* sections.

Sorby's tracings of enlarged microscopic reference areas were soon substituted for tracings made from polished slabs, but the area of rock represented by one such drawing was a very small portion of the usable area of even a mediocre thin section. Making and measuring a single drawing was a time-consuming chore, and when a prominent petrographer suggested that reliable estimates of the modal composition of a single thin section of a rather common rock would require preparation, tracing, and cutting up of drawings from projections of at least a couple of dozen microscopic fields, the profession seems to have decided to wait for better methods. Attempts to replace the cutting and weighing step by direct areal measurements continued for some time but never attained any real popularity.

The Delesse method cannot be said to have failed; it simply has not been used. But its day may be approaching. The scanning of an image by a TV tube is, for all practical purposes, an area measurement of the Delesse type. It is also virtually instantaneous, and the TV tube by 1971 was beginning to be used for measuring relative areas of materials reflecting (or transmitting) light with different intensities, though the procedure was still far from routine.

Toward the close of the last century direct areal measurement was replaced by linear measurement. The Viennese geologist A. Rosiwal, who was responsible for this change, worked at first on rock slabs, which he ground flat and polished. Inscribing lines on the flat surface, he measured the lengths of segments lying in each of the minerals exposed in it. The proportion of a line lying in a particular mineral species was his estimate of the proportion of that mineral in the rock. It is interesting and rather startling that the lines along which Rosiwal first made his measurements were neither straight nor parallel. His drawings clearly show that the *measurement lines,* or *rock threads,* he inscribed on the surface of a slab were strongly curved and intersected in a complicated, unsystematic fashion.

The petrographic microscope, which had been a novelty when Sorby made his drawings, was standard equipment in Rosiwal's day. He soon attempted to exploit the vastly improved mineral identification made possible by it, and the inadvertent approach to random sampling suggested by his curved and intersecting *rock threads* was ended. Under the microscope it is difficult to use reference lines that are anything but straight, and it is by far simplest to use straight lines that are parallel to each other. *Rosiwal analysis* now refers exclusively to the measurement of mineral intercepts along a set of parallel straight lines. In the original Rosiwal procedure these lines were rulings in the eyepiece of a microscope. In modern equipment, rotation of the calibrated screw of a mechanical stage in which the specimen is clamped moves it past a reference point, usually the intersection of the cross hairs in the microscope eyepiece. Although the measurement technique proposed by Rosiwal was hopelessly time-consuming, the measurement itself seemed so proper on the basis of geometric intuition that he did not bother to justify it. The first attempt at formal justification did not appear until 1913. Few persons applied the Rosiwal technique in routine work, for by his method the analysis of a single thin-section area of less than a square inch required a number of hours. Subsequent mechanical improvements finally reduced the time requirement substantially, and the method was just beginning to find wide application when the whole basis of measurement shifted once more.

Just as it had previously been discovered that it is much easier to measure lines than areas, it was now discovered that it is much easier to count points than to measure lines, as the Russian mineralogist A. A. Glagolev suggested in 1933. Like most Russian work of that period, this suggestion was ignored in the West, where the advantages of point counting were discovered independently 16 years later. A short English note on the subject by Glagolev in an American mining journal seems to have escaped almost unnoticed, probably for the reason that the journal was not one widely read by either petrographers or metallurgists. Except for elaboration of the instrumentation, the principal effect of which seems to have been to increase the cost of the operation, the replacement of line measurement by point counting brings to a close the preelectronic development of the art of point counting.

ERRORS IN MEASURING AND IMPRECISION IN THINKING

The analogy between the areal method of Delesse and the linear method of Rosiwal is evident. As the distance between lines decreases, the precision of estimates of area yielded by the latter increases. In the limit, as the traverse interval becomes infinitesimal, the distinction between a Rosiwal analysis and an (errorless) Delesse analysis vanishes. For what now seems a rather bizarre reason, no systematic comparison of the precision and accuracy of the two procedures seems to have been made.

The reason is just that until long after the demise of direct areal measurement, traverses were, in fact, neither uniformly nor randomly spaced over the reference area in Rosiwal analysis. Rosiwal always regarded his *rock thread*—the line along which measurements were to be made—as a sample of the rock rather than of the surface on which it was drawn or imagined. When he abandoned the curvilinear, intersecting "threads" of his earlier work in favor of straight, parallel traverses, he prescribed only that the distance between traverses should be adjusted to insure that no grain was cut by more than one traverse. Before 1923, no one suggested in print that a uniform traverse interval ought to be used in any particular analysis or set of analyses. Even now, the supposed undesirability of traversing any grain more than once is rediscovered every few years. Failure to insist on a traverse interval that either varied randomly or was uniform—a failure that makes it difficult if not impossible to evaluate precision and accuracy—stems from something very like what is nowadays called the confusion of target and sample populations. The target is the rock; the sample is the surface of the thin section. The ultimate aim may be knowledge of the composition of a hand specimen, an outcrop, a hillside, or a continent. But the sample always intervenes between the observer and the target, and if we wish to reach useful inferences about the target our first business is to obtain reliable statistics from the sample. In an individual Rosiwal, Delesse, or other modal analysis, we are not directly estimating the composition of a rock; rather, we are estimating the proportion of a reference area occupied by each of the essential minerals.

How well can we do this? However unsatisfactory the situation as regards the earlier techniques, the theoretical precision, or reproducibility, of modal

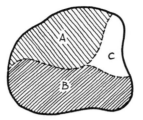

FIGURE 1

*Schematic drawing of reference
area showing regions occupied
by each of three minerals. The
relative areas of regions A, B,
and C can be estimated by
point counting*

analysis by point counting is fairly well understood. Consider the heavy
line of Figure 1 as the boundary of a reference area, whether thin section
or polished slab, and the blank, stippled, and striped regions within the
boundary as areas occupied by each of three minerals. Suppose that we
select points within the boundary of the reference area and record the subfield,
or mineral, in which each point falls. If the points are chosen simply at
random in every sampling, the probability that any point selected will lie
in, say, area *A*, is precisely the ratio of area *A* to the reference area, the
very thing we want to know. If we select *N* points at random, the number
falling in *A* should be approximately $N_a = Np_a$, where p_a is the (unknown)
proportion of the reference area occupied by *A*. In practice, we use the
observed ratio, N_a/N, as an estimate of p_a. Now N_a/N will nearly always
differ somewhat from p_a, but the error involved in estimating p_a in this
fashion is well known and can again be estimated directly from *N* and
N_a.[1] Exactly the same situation holds for minerals *B* and *C*, and however
many more there may be. Estimating areal proportions by selecting points
simply at random within a reference area is one way of generating what
some readers of this book will recall from high school mathematics as the
binomial distribution.

We may draw one valuable lesson from this picture of the point-counting
process even though, as we shall see, it does not apply exactly to our problem.
Clearly it does not matter whether mineral *A* occurs in one large continuous
subarea or is scattered over the reference area in a large number of discrete,
isolated subareas. In a properly random sampling the probability that a point
will fall on mineral *A* depends *only* on the proportion of the reference area
that is occupied by *A*.

As with Rosiwal's technique, however, and for the same reason, the selection process actually used is systematic, not random; this is why we said
the precision or reproducibility theory of modal analysis by point-counting
is fairly well—instead of exactly—understood. Nowadays we certainly could

---

[1] For instance, if p_a is really 0.2 and $N = 100$, the observed values of N_a/N will almost
all lie between 0.1 and 0.3, and most of them will be much closer to 0.2 than to 0.1
or 0.3.

arrange a suitably random scheme of sampling, but the convenient way to collect the data from a single reference area is to locate the points systematically, at regular intervals along uniformly spaced traverses. In fact, we use a rectangular point grid so the selection of points is systematic, not simply random. Once the position of the grid is chosen, the location of all the points is fixed in principle and varies in practice only because of unavoidable, but variable, slippage in the traversing mechanism of the mechanical stage. Although a priori evaluation of the random error of counts made on a rectangular grid is difficult, this error is readily estimated experimentally. A rather sizable body of data now indicates that the reproducibility of modal analysis based on points distributed on a rectangular grid is not substantially different from that associated with the same number of points distributed simply at random over the same area. (This, in fact, is why we discussed random point-counting error in such detail. As is often the case in the natural sciences, much may be learned from a simple statistical model even though the model is only approximate and may ultimately be modified or abandoned.) There are limitations, to be sure (for instance, the relation of distance between traverses to distance between points along a traverse must be such as to insure good coverage of the reference area by the systematic count), but for all practical purposes, the theoretical reproducibility of means determined by simple random counts of size N provides satisfactory estimates of the observed reproducibility of modal analyses based on N points distributed over a rectangular grid.

FROM AREAS TO VOLUMES

The sample statistics are thus in good order. We have a convenient way of estimating the "composition" of our reference areas, or thin sections, and we also have satisfactory knowledge and control of the precision of these estimates. Now we must examine briefly the relation between sample and target, the reason for supposing that measurements of areal proportions are sound estimates of volumetric proportions, so that observations on thin sections can tell us about the composition of rocks.

Delesse was acutely aware that it was a big step from areal to volumetric proportions, and seemed to sense it was a step he could not take. Lacking the modern vocabulary of sampling and estimation, which makes possible the discussion, and sometimes the solution, of such problems, he was obliged to rely heavily on geometrical intuition. In the end, he did not get much past the simple announcement that it seemed reasonable to suppose the proportion of the measurement area occupied by a mineral was the same as the volume percentage of that mineral in the rock from which the measurement areas had been obtained.

No one objected—indeed, the question was not raised again in print for nearly a century—but in retrospect, it seems quite possible that doubts about the validity of areal measurements as estimates of volumetric proportions were responsible for the sluggish development of the subject. The information was of a sort that every petrologist badly wanted. The period was one of considerable advancement in the instrumentation of other forms of microscopic observation. It is true that the instrumentation proposed by Delesse, Sorby, and Rosiwal was crude and inefficient, making the process of modal analysis painfully slow and difficult. But mechanical devices which vastly reduced the time requirements were in fact developed in the first quarter of the present century, and there was nothing about these instruments that would have taxed the ingenuity of the machinists and gadgeteers associated with the remarkable development of instrumentation for the sister sciences of optical mineralogy and crystallography in the last half of the nineteenth century. Had there been a demand for suitable measuring devices, it almost certainly would have been met. But there was no point in measuring relative areas unless they were sound estimators of relative volumes, and there was no assurance that this was so.

The situation today is very different, largely because of the development of a vocabulary specifically designed for analysis of problems such as this one. Basically, we are concerned with the properties of relative areas as estimators of relative volumes. It turns out that relative area is always a *consistent* estimator of relative volume, in the sense that the average of results for an increasingly large number of randomly chosen reference areas is less and less likely to differ from the true relative volume by any given amount, however small.

This is good, but it is not good enough. No one is going to measure an infinite or even a very large number of reference areas of the same sample volume. What we would like to be told is that relative area is a good estimator of relative volume, in the sense that the average *relative* area obtained from any number, however small, of reference areas is likely to be close to the true *relative* volume. Now this, unfortunately, is untrue. It is stretching things very little to say that during the century following Delesse's discovery petrologists refrained from exploiting modal analysis because they suspected some such defect in the area-volume relation, even though for most of that time language which would permit a concise description of their suspicions was either not available to or not known by them.

The remedy, however, is surprisingly simple. Areal proportions measured on sections parallel to any face of a parallelopiped do give good estimates of volumetric proportions in the parallelopiped. If all reference areas used in a particular study are of the same size and shape, the relative areas do average close to the relative volumes. As Delesse suspected, but could not prove, the possibility that this area-volume relation fails is by no means an insuperable obstacle. Instead, it is merely an altogether unexpected but sound reason for using, in any particular study, reference areas which are parallelograms of the same size and shape.

Given experimental evidence that the error of point counting follows a simple, well-known statistical rule and the a priori assurance that relative areal proportions may indeed be good estimators of volumetric proportions, we ought to be prepared to evaluate systematically the errors attaching to the remaining sampling steps required if, as Sorby insisted more than a century ago, "the mountains must indeed be examined under the microscope." There has been considerable progress in this direction, largely stimulated by the substitution of point counting for linear analysis in the decade following World War II.

But it is one of the rules of natural science that a new or improved analytical technique creates a demand for far more data than can be provided. The modal analyst of 1970 can do in 15 minutes what his scientific forebearers of 1920 probably could not do in less than two hours. This striking improvement has created interest in sampling problems whose successful solutions require far more than eight times the number of analyses that the same amount of work would have generated in 1920. The petrologist works with a number of closely interrelated sets of variables characterizing rock composition, and if he cannot get enough information from one set, he turns to another. So interest has shifted recently to rapid methods of chemical analysis, and after a flurry of productive activity extending from 1945 to about 1960, modal analysis seems to be caught in another of the standstills that have characterized its history. Probably the next great revival of interest in it will be prompted by successful electronic automation of the analytical process, as suggested above.

This brief review of what is obviously a highly specialized scientific activity may perhaps best conclude with a reminder that science itself is something more than a collection of scientific specializations (or, as they often seem, overspecializations). The practical day-to-day activity of all natural science does consist, for the most part, of learning more and more about less and less. It often happens, however, that the means by which we seek knowledge in one field are independently developed by, find application in, or are borrowed from, another. The potential yield of a forest, for instance, may be estimated by a kind of traverse sampling not unlike that used on microscope slides in petrographic point counting. What petrographers call modal analysis is probably more widely used in the study of metals and alloys than in the study of rocks. Application of the electron microscope to biological tissues has revealed an enormous amount of detail in the previously featureless cytoplasm, or nonnuclear portion, of the cell; here, too, modal analysis is beginning to find application. The objects being examined and the methods of observation differ widely from field to field, so widely, in fact, that a specialist in any one usually knows little or nothing about the others. But the *statistical* concepts on which modal analysis rests are the same whatever the nature of the material being analyzed. A derivation of the area-volume relation developed for one field is applicable to all, as are the results of a properly designed study of precision or reproducibility. It has been said that although the nouns of the technical languages of various fields of specialized scientific activity are very different, the verbs are pretty much the same. To this we may add, extending the analogy a bit, that the roots of many of the verbs common to the diverse languages of the natural sciences are basically statistical.

PROBLEMS

1. What is *modal analysis?*

2. What is the first sampling operation in studying a rock?

3. What is *Rosiwal analysis?* What was peculiar about Rosiwal's initial rock thread?

4. Explain the differences between the Delesse, Rosiwal, and Glagolev methods of modal analysis.

5. At several points the author suggests that a certain worry may have been operating in the unconscious minds of petrographers and slowing down the development of the science. Explain the worry.

6. Consider Figure 1. How would you locate the points for point counting modal analysis? Give reasons for your choice.

7. You are doing modal analysis by point counting, as shown in Figure 1. What is the actual (not estimated) probability that a given point will fall in area B?

8. Suppose 1000 points are selected at random in each of 4 equal size and shape parallelogram reference areas on a given rock containing three minerals with the following results:

	# Mineral A Points	# Mineral B Points
Reference Area 1	300	200
Reference Area 2	311	196
Reference Area 3	282	211
Reference Area 4	291	204

(a) Estimate the relative volume of mineral B in the rock.
(b) Estimate the relative volume of mineral C in the rock.

9. Suppose the measurements of question 8 are conducted again, only the four reference areas this time are *not* equal size and shape parallelograms. Would estimates of the relative volume of mineral B and the relative volume of mineral C based on the new results be as useful as those made in question 8? Explain.

10. What is a *consistent* estimator? Why is it "not good enough" in this situation?

11. Suggest another field, besides petrography, where the volume-estimating techniques described in the article might be useful.

THE PROBABILITY OF RAIN

Robert G. Miller   *Life Insurance Agency Management Association*

THE NOTED mathematician Norbert Wiener has referred to meteorology as one of the semi-exact sciences.   In his book *Cybernetics* he states:

> . . . in meteorology, the number of particles concerned is so enormous that . . . if all the readings of all the meteorological stations on earth were simultaneously taken, they would not give a billionth part of the data necessary to characterize the actual state. . . . [A]ll that we can predict at any future time is a probability distribution . . . and even this predictability fades out with the increase of time.

An example of one of Professor Wiener's probability distributions might be the 70% issued by a weather forecaster on radio or television as his estimate of the probability of rain.

Some people feel lost in the face of a probability forecast.   "Just tell me the answer!" they say.   But, like many other forecasts of events in life, the future state of the weather is uncertain; no one knows for sure what will happen. Consequently, a forecast that admits to uncertainty seems appropriate. Such a forecast gives extra information to people who may want to take action in the face of weather threatening their pocketbooks. For example, suppose that it costs $100 to take preventive measures against a threatening storm and that the loss if the storm occurs is likely to be about $300. If the chance of the storm is only 0.1, then the expected loss is $30, which is less than the cost of the preventive measure; it would be uneconomical to take steps. But if the chance of the storm were 0.9, the expected loss is $270, and preventive measures look worthwhile.

The ordinary citizen, deciding whether to go out equipped for rain, will consider inconvenience versus risk of drenching and find a probability to use as a cutoff point. For example, he may decide to equip for rain when the probability of rain is 0.5 or higher, otherwise not.

Probability distributions may be arrived at in various ways. The most common method is to use the human judgment of an experienced weather expert. He considers all the evidence and on the basis of his experience chooses a number that he thinks expresses the chance of rain. Another way of generating a probability distribution is to apply statistical methods to weather data stored in government archives. This essay describes a method for arriving at such a probability distribution based on the statistical evidence of past years.

POSSIBLE FORECASTS

As an example, we may want to estimate at 7:00 AM (0700 EST) each day the probabilities for each of five possible precipitation conditions at Hartford, Connecticut, during the next six hours. The five conditions listed in Table 1 are: dry, a little rain, a little snow, rain, and snow. The prediction consists of five numbers, adding to 1.0 and representing the probabilities of each of the five possible outcomes. For example, the numbers 0.4, 0.2, 0.2, 0.1, 0.1 mean a 40% chance of dry weather, a 20% chance each of a little rain or snow, and a 10% chance each of substantial rain or snow. The prediction might group the last four numbers together and report a 60% precipitation probability. Of course, one or the other of the five weather possibilities *must* occur, and we do not report more than one for the period, even though we sometimes have "snow changing to rain."

LEANING ON DATA FROM OTHER PLACES

Meteorologists have found from experience that observing present weather conditions over a fairly large region enables them to predict future weather conditions at points within the region. They have not ordinarily made forecasts with quantitative probabilities, though "a slight chance of rain" was a common way of describing a small probability without quantifying it. The

TABLE 1.   Detailed Definition of Five Precipitation Categories for Hartford, Connecticut

CATEGORY	CONDITIONS
(1) Dry	No precipitation of any kind over the period 0701 EST–1300 EST.
(2) Little rain	Rain or freezing rain reported at some time over the period 0701 EST–1300 EST in the amount of at least a trace but not more than 0.05 inch. No snow or sleet reported at any time over this six-hour interval of time.
(3) Little snow	Snow or sleet reported at some time over the period 0701 EST–1300 EST in the amount of at least a trace but not more than 0.05 inch of melted water equivalent.
(4) Rain	Rain or freezing rain reported at some time over the period 0701 EST–1300 EST in the amount of greater than 0.05 inch. No snow or sleet reported at any time over this six-hour interval of time.
(5) Snow	Snow or sleet reported at some time over the period 0701 EST–1300 EST in the amount of greater than 0.05 inch of melted water equivalent.

statistical approach, in attempting to refine the meteorologist's forecast, will also use data from a fairly large region around the point of interest; in the case of our example, from weather stations with Hartford, Connecticut, considerably east of center because of the weather's general movement from West to East in the Northern Hemisphere.   The dots on the map in Figure 1

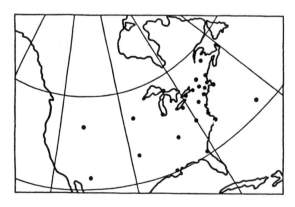

FIGURE 1
25-station network

TABLE 2.   25 Network Locations

Hartford, Conn.	Hatteras, N. C.
Caribou, Me.	Jacksonville, Fla.
Portland, Me.	Sault Ste. Marie, Mich.
Boston, Mass.	Chicago, Ill.
Nantucket, Mass.	Cleveland, O.
Burlington, Vt.	Knoxville, Tenn.
Albany, N. Y.	New Orleans, La.
Buffalo, N. Y.	Sioux Falls, S. D.
New York, N. Y.	Oklahoma City, Okla.
Syracuse, N. Y.	Pocatello, Ida.
Harrisburg, Pa.	Tucson, Ariz.
Norfolk, Va.	Saint George, Bermuda
Roanoke, Va.	

indicate the locations in the network; Table 2 lists the stations in the network.

VARIABLES USED IN THE FORECAST

Out of the many weather elements observed hourly at each of these locations, seven were used to characterize the state of the weather at forecast time, 0700 EST. These seven, listed in Table 3, are: barometer reading and its three-hour change, temperature, moisture, wind direction and velocity, and cloud cover. They were chosen because as a group they seemed to predict the weather at least as well as others that were available. The inclusion of 0700 EST rain or snow conditions would have enhanced the predictions but they were not among the elements available in this sample study.

As data for developing the procedure we used the 1096 daily 0700 EST observations of each weather element for the years 1951–53 for all locations. (These are stored in the U.S. Weather Bureau files in Asheville, North Caro-

TABLE 3.   Meteorological Elements

Sea level pressure (millibars)
Past three-hour change in sea level pressure (millibars)
Dry bulb temperature (degrees Fahrenheit)
Temperature-dew point depression (degrees Fahrenheit)
East-West wind component (knots)
North-South wind component (knots)
Total cloud cover (tenths of the sky covered)

lina.)   An additional sample of 221 observations from 1954 are used in an independent test of the chosen forecasting procedure, described later.

A typical 0700 EST observation at one of the 25 locations, say, Hatteras, North Carolina, for one of the 1317 days in the two samples might have been:

Element	Measurement
Sea level pressure	1000.2 millibars
Three-hour pressure change	−1.0 millibars
Dry-bulb temperature	55° Fahrenheit
Dew-point depression	0° Fahrenheit
Wind	Northeast at 25 knots
Cloud cover	10/10

The pressure measurement of 1000.2 millibars (1000.2 × 1000 dynes per square centimeter) indicates low pressure conditions because the normal value is 1017 millibars.

A drop in pressure of one millibar was observed over the last three hours, suggesting the approach of even lower pressure and a worsening of weather conditions.

The air temperature was observed to be 55° Fahrenheit.   The dew-point temperature (the temperature at which the air becomes saturated) and the air temperature were precisely the same (a zero depression), so that it was a humid morning.

The wind blew from the Northeast (45°, at 25 knots. We can, with a little trigonometry, express the wind in terms of its East-West and North-South components:

u (East-West component) = 17.68 knots
v (North-South component) = 17.68 knots

On this morning, moist air from over the ocean was carried over Hatteras generally making for precipitation.   A sky cover of ten-tenths of clouds was also observed, further enhancing the chance of precipitation.

THE STATISTICAL TECHNIQUE

The statistical method used to predict the probability distribution of precipitation at Hartford, Connecticut, has the technical name of *multiple discriminant analysis*.   Elsewhere in this book (see the essay by Howells) the general idea of discriminant analysis is described.   Ordinarily we find several variables, or measurements, each of which is related to the presence or absence of the categories we are predicting—here dry, a little rain, a little snow, rain, and snow. Then it may be possible to make, out of the several measurements, a single measure that is a better predictor than any individual one.   Often the better predictor is a weighted sum of the individual variables.   Two important steps are required: first, to select a few from among all the variables and, second, to determine the weights to get a good predictor.

Howells wanted to decide or "predict" whether a bone was human or not and he used a single weighted sum of properties of the bone.   We have more categories and our forecasts will be improved by using more than one weighted sum of the same variables to summarize the predictive data.

Let us describe our steps in more detail.

Step 1: Selecting the Variables.   Using statistical procedures, we selected out of the original 175 weather variables (7 elements at each of 25 locations) those 16 that contained most of the predictive information about the subsequent six-hour precipitation conditions at Hartford, Connecticut (see Table 4).   One way to do this is to provide a high-speed computer with the data and give it a rule for deciding which among several collections of variables makes the best predictions.   In one approach, the computer tries all 175 variables one at a time and chooses the best of these.   Next it tries combining each of the remaining 174 variables with the first one selected, and chooses the second variable that goes best with the first one chosen, and so on until adding further variables doesn't help much.   If we had used data from years different than the ones we chose, no doubt a somewhat different set of variables would have arisen, and the order would probably have been different.

TABLE 4.   Selected Predictors in Order of Selection

STATION	ELEMENT
Boston, Mass.	Total cloud cover
Portland, Me.	Past three-hour pressure change
Sault Ste. Marie, Mich.	Dry-bulb temperature
Hartford, Conn.	Temperature–dew-point depression
Buffalo, N. Y.	Dry-bulb temperature
Boston, Mass.	East-West wind component
Hatteras, N. C.	North-South wind component
Norfolk, Va.	Past three-hour pressure change
New York, N. Y.	Dry-bulb temperature
Portland, Me.	North-South wind component
Nantucket, Mass.	North-South wind component
Norfolk, Va.	Dry-bulb temperature
Oklahoma City, Okla.	North-South wind component
Caribou, Me.	Dry-bulb temperature
Boston, Mass.	Dry-bulb temperature
Albany, N. Y.	North-South wind component

Because Hartford is only 100 miles from Boston, cloud cover at the latter is a reasonable variable. That temperatures appear as six variables is not remarkable, though most of us would not have expected a Michigan temperature to appear so early on the list, even though weather usually moves from West to East. Having temperatures at a variety of places gives the discriminant a way of measuring the changing temperatures from one station to another. That changes in the barometer matter is to be expected, but that the computer would choose stations to the East and to the South is not at all obvious. Lacking a direct measure of 0700 EST rain or snow, the selection of a moisture variable (dew-point depression) at Hartford seems a very reasonable choice. The reader will not be surprised to see that among these 16 variables, six measure wind direction and velocity, four at stations near Hartford. The selection of these wind measurements indicates an attempt to include circulation characteristics as well as the location of weather fronts.

While selecting the 16 variables to be used, the computer also computed two sets of 16 weights to be attached to them. They are used to give two weighted sums, or *discriminant functions,* to be used as predictors. In principle, more than two sets of weights could be used, but we stopped with two because they seemed to wrap up nearly all the information.

Step 2: Finding the Weights. The discriminants are scores that predict the weather. The first discriminant forecasts *dryness*—a high score forecasts very dry, a low score considerable precipitation—rain or snow. The second discriminant helps forecast the kind of precipitation—high scores implying snow, low scores rain, middle scores dry; let us call this the *snowiness* score.

For each historical observation, the numerical score for dryness and the one for snowiness were plotted on a graph (see Figure 2) and each point was labeled to show the actual weather that occurred. Once this was done, we found that the (dryness, snowiness) scores for a given weather, say snow, clustered around a point and that larger and larger ellipses drawn around such a central point included more and more of the observations of that weather. Figure 2 shows all the points corresponding to historical instances where a little rain fell. It also shows the ellipse that includes 50% of these points. Each category has such a scatter of points and has a corresponding ellipse. The choice of this particular ellipse and its orientation are guided by some calculational considerations we need not go into here.

Figure 3 shows the placement and shape of five ellipses, each of which contains 50% of the points corresponding to its kind of weather. These 50% ellipses overlap considerably. The overlap illustrates why one may not be able to give a simple "it will rain" as a forecast. Only the ellipse for snow stands a bit off from the others. Remember, these are only 50% ellipses, so the snow points fall among the others more than Figure 3 may suggest; nevertheless, "heavy snow" observations are found mostly in the upper left-hand region of the graph (in and around the ellipse labeled 5). Most of

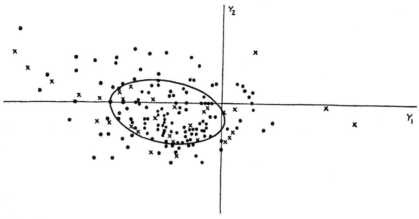

FIGURE 2

Empirical distribution of sample points (dots) from the original 1096 observations and sample points (crosses) from the later 221 observations for those cases of a little rain, group 2. About half the points fall inside the ellipse

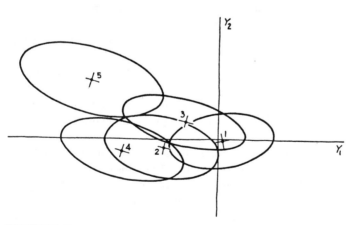

FIGURE 3

The ellipses shown encompass about 50% of the observations of the corresponding precipitation categories: (1) dry, (2) a little rain, (3) a little snow, (4) rain, (5) snow the observations located near the intersection of the axes (inside ellipse 1) are for "no precipitation."

Step 3: Calculating the Probabilities. The next step is to calculate the probabilities of the precipitation categories for each of the 1096 observations and for the 221 independent sample observations to be used to test the method. This is accomplished as follows.

As we have seen above, the set of observed conditions of the 16 selected weather elements at 0700 EST on any particular day corresponds to a point in Figure 3. The probability of each kind of weather is determined by where the point lies. If it lies inside the ellipse labeled 1, then the probability that there will be no precipitation within the next six hours will be high, say, 0.8 or 0.9; whereas if the point lies inside the ellipse labeled 5, the chances for snow will be high. Naturally, computing the probabilities requires a complicated procedure, but we can give the idea if not the exact method.

Suppose, for a moment, our 16 variables produced a point somewhere in Figure 3. You could swing a small circle around that point just large enough to include, say, 100 of the 1096 historical points. Suppose 60 were in category 1, 20 in category 2, 10 in category 3, 9 in category 4, and 1 in category 5. Then we could forecast 0.6 chance of no precipitation, 0.2 chance of a little rain, and so on, and practically 0.0 chance of heavy snow. The actual method uses more mathematics, but this idea is adequate for understanding.

In place of providing all of these predicted probabilities, we give two tables that summarize the predictions of the 1096-case sample (see Table 5) and the 221-case sample (see Table 6). These tables, for particular ranges of predicted probabilities, include: the number of forecasts F made within each group of weather conditions and within each range of probability (notice that across each row the Fs sum to 1096, because in each case *some* probability was assigned to each group), the number of actual occurrences U of the designated group when the predicted probability was in that range, and the sum of the probability over all F forecasts, ΣP, gives the expected or average number of occurrences.

Let us give a few examples of reading Table 5. In row F of the "dry" group, the first entry tells us that 50 of the 1096 historical points would have forecast "dry" with probabilities at least 0 but less than 0.1; the next entry tells us that 80 points would have forecast "dry" with probability at least 0.1 but less than 0.2, and so on. In row U of the dry group, the first entry tells us that in 2 of the 50 cases in which we predicted "dry" with probability between 0 and 0.1, "dry" actually occurred. If we look at the total U for the "dry" group, we see that, in all, "dry" occurred in 817 of the 1096 observations. Comparing this line with the third gives us an idea of the comparison between occurrences and probabilities as we pointed

TABLE 5. Summary of Probability Predictions (1096-Case Sample)

Group		$0.0 \leq P < 0.1$	$0.1 \leq P < 0.2$	$0.2 \leq P < 0.3$	$0.3 \leq P < 0.4$	$0.4 \leq P < 0.5$	$0.5 \leq P < 0.6$	$0.6 \leq P < 0.7$	$0.7 \leq P < 0.8$	$0.8 \leq P < 0.9$	$0.9 \leq P \leq 1.0$	Total
Dry	1 F	50	80	37	58	37	42	22	82	145	543	1096
	U	2	10	8	25	16	23	14	68	122	529	817
	ΣP	2.36	12.88	9.60	21.36	16.84	23.12	14.60	63.48	125.76	528.28	818.28
Little rain	2 F	616	212	89	130	30	18	1	0	0	0	1096
	U	11	30	22	45	17	10	0	—	—	—	135
	ΣP	12.88	33.24	23.00	46.84	13.64	9.76	0.64	—	—	—	140.00
Little snow	3 F	1034	46	15	1	0	0	0	0	0	0	1096
	U	19	6	4	0	—	—	—	—	—	—	29
	ΣP	17.56	7.04	3.92	0.32	—	—	—	—	—	—	28.84
Rain	4 F	821	100	54	63	20	20	12	6	0	0	1096
	U	8	11	16	20	11	13	10	3	—	—	92
	ΣP	6.84	15.52	13.88	22.16	9.20	11.32	8.00	4.32	—	—	91.24
Snow	5 F	1031	38	12	12	3	0	0	0	0	0	1096
	U	4	5	2	9	3	—	—	—	—	—	23
	ΣP	3.28	5.56	3.08	4.32	1.40	—	—	—	—	—	17.64

Table 6. Summary of Probability Predictions (221-Case Sample)

	Group		$0.0 \leq P < 0.1$	$0.1 \leq P < 0.2$	$0.2 \leq P < 0.3$	$0.3 \leq P < 0.4$	$0.4 \leq P < 0.5$	$0.5 \leq P < 0.6$	$0.6 \leq P < 0.7$	$0.7 \leq P < 0.8$	$0.8 \leq P < 0.9$	$0.9 \leq P \leq 1.0$	Total
Dry	1	F	12	17	7	11	10	10	4	22	27	101	221
		U	—	1	5	4	1	7	3	17	23	94	155
		ΣP	0.56	2.52	1.76	3.92	4.60	5.60	2.60	17.00	23.28	98.20	160.04
Little rain	2	F	114	48	22	21	11	5	—	—	—	—	221
		U	6	8	4	8	4	3	—	—	—	—	33
		ΣP	2.52	7.24	5.68	7.28	5.00	2.72	—	—	—	—	30.44
Little snow	3	F	211	6	4	—	—	—	—	—	—	—	221
		U	4	2	2	—	—	—	—	—	—	—	8
		ΣP	3.68	0.96	1.04	—	—	—	—	—	—	—	5.68
Rain	4	F	159	24	12	11	3	9	2	1	—	—	221
		U	3	5	1	5	1	3	1	1	—	—	20
		ΣP	1.48	3.64	3.08	3.76	1.36	4.92	1.28	0.72	—	—	20.24
Snow	5	F	203	10	3	4	1	—	—	—	—	—	221
		U	1	1	—	2	1	—	—	—	—	—	5
		ΣP	0.52	1.40	0.76	1.44	0.48	—	—	—	—	—	4.60

out in Table 5. On 2 occasions it was actually "dry" when the probability of "dry" was only between 0 and 0.1. The sum of the computed probabilities for the 50 cases falling into this category was 2.36. That means the method forecasts 2.36 instances of dry weather even though it forecasts none of the 50 as likely. The number observed, 2, is very close to the computed count 2.36. The largest discrepancy is in category 5, snow, for probability between 0.3 and 0.4, where there were 9 occurrences versus 4.3 predicted. By and large the agreement is good. This shows that we can find a way to forecast the past based on that same past. But will the same method work for another period of time?

Step 4: Validating the Approach. In Table 6, we show the result of applying the same technique to the new 221 observations which were not used to construct the discriminant functions. Again the agreement between the second and third lines is usually close, and this validates the method. Thus, Tables 5 and 6 show us a lot about the weather in Hartford and the forecasts. Seventy-five percent (817/1096) of the time it is dry, 12% (135/1096) of the time there is some rain, 2% of the time it snows a little, 8% it rains heavily, and 2% it snows heavily. As for the forecasts, often "dry" can be predicted with high probability; indeed about half the forecasts are "dry" with probability greater than 0.9. Because the sum of the probabilities must be 1.0 for each forecast, it follows that forecasts of low probability can be given frequently for the other four categories. But it is a general feature of both tables that a specific category of precipitation is never forecast with probability even as great as 0.8. One requirement for obtaining higher probabilities of these rare events is a much larger data sample. A smaller, more concentrated ellipse would then be capable of encompassing the 100 or so sample points needed to estimate the probabilities.

CONCLUDING REMARKS

From the results obtained the following observations can be made:

(1) From Figure 3, it can be seen that the horizontal axis orients the precipitation categories such that larger amounts of precipitation, irrespective of type, lie to the left, and lower amounts to the right. This is contrasted with the vertical axis, which appears to "discriminate" the snow groups (categories 3 and 5) from the rain groups (categories 2 and 4).

(2) It may be of interest to know how many correct predictions were made of the conditions—precipitation or no precipitation—in the test sample. Taking those situations in which the probability of no precipitation was 0.5 or greater as an arbitrary criterion for categorically predicting no precipitation, the number of forecasts of "no precipitation" was 164

out of 221, with 144 correct. The number of forecasts of "precipitation" was 57 out of 221 with 46 correct. Altogether there were 190 out of 221 correct, or 86% accurate. This compares favorably with the 957 out of 1096, or 87% for the sample used to develop the relationships.

(3) There appears to be good agreement between U, the number of observed precipitation events, and ΣP, the sum of the probabilities in each of the five categories in each of the ten columns of Tables 5 and 6. For example, Table 6 shows that for the 21 times that category 2 (light rain) was predicted with a probability from 0.3 to 0.4 the light rain was observed to happen eight times while its expected number of occurrences was 7.28 times. This good agreement can be seen in all of the independent data results of Table 6. The discrepancies that do exist can be largely attributed to sampling fluctuations.[1]

The method of discriminant analysis has been instrumental in enabling the prediction of weather probabilities objectively. However, operational weather prediction utilizing the method requires a computer to do the calculations. Alternative discriminant methods have been developed which do not have such requirements. These alternative methods possess the following features:

(1) Qualitative as well as quantitative predictors can be used.

(2) Probabilities can be predicted for categories which are not necessarily mutually exclusive.

(3) More varied shapes of clustered points than those characterized by simple ellipses can be dealt with.

(4) Operational probabilities can be obtained directly by merely adding a small set of numbers together.

(5) Large numbers of variables with many sample cases may be processed with ease.

(6) Results are more easily interpreted.

(7) Missing and erroneous or incomplete data are handled systematically.

Prospects for the future are that discriminant type methods will make it possible for users to request weather probabilities by telephone and to receive a voice response directly from a computer.

---

[1] For a detailed exposition of this and one other weather forecasting example using statistical methods see R. G. Miller. 1962. *Statistical Prediction by Discriminant Analysis*. Meteorological Monographs, No. 25. Boston: American Meteorological Society.

PROBLEMS

1.   Refer to Figure 1. Why are so many stations in the network in the north-eastern U.S.?

2.   Notice in Figure 1 that Hartford, Connecticut is far from the geographic center of the network of stations used to predict its weather. Why?

3.   What are the two stages of multiple discriminant analysis as it was used in this study? Can you say that the result was a single good predictor? Explain your answer.

4.   What are the Y_1 and Y_2 axes in Figures 2 and 3?

5.   Refer to Table 5, "total" column, category "little snow," $U = 29$ and $\Sigma P = 28.84$. What do these numbers mean?

6.   Consider Table 5. For which group of weather conditions was

$$\frac{\text{(expected number of occurrences-actual number of occurrences)}}{\text{actual number of occurrences}}$$

the greatest?

7.   What is the sum of all the predicted probabilities involved in the 1096-case sample? Why? (You may wish to refer to Table 5.)

8.   Consider Table 6.
   (a) How many of the 221 points forecast "rain" with probability $\geq .5$?
   (b) In how many of the cases in (a) did "rain" actually occur?

9.   In Table 6, category "snow," it is indicated that for $0.2 \leq p < 0.3$, $F = 3$ and $U = 0$. Does this mean that it did not snow during the study period on those three days? Explain your answer.

10.   In what way does the sample of 221 observations serve as an independent test of the forecasting procedure?

STATISTICS, THE SUN, AND THE STARS

C. A. Whitney   *Harvard University*

THE VISIBLE stars appear to be scattered at random in space, so it is natural that astronomers should have turned to statistical methods and probabilistic arguments. But this is a relatively recent development. Babylonian and Greek astronomy was based on cycles, and the concepts of probability were rarely used in modern astronomy before the twentieth century.

An early use of the idea of randomness and probability in astronomy was John Michell's mathematical demonstration, in the eighteenth century, that most stars which appear to be very close together in the sky actually are close together in space. William Herschel, a contemporary of Michell, discovered an abundance of very close pairs of stars—*double stars* they were called—but he had no way of telling whether the stars really were paired or merely appeared to be. Herschel hoped that the individual stars in a pair were quite far apart and that they accidently lay along the same line of sight; from observing such pairs he hoped to detect the motion of the earth about the sun and, ultimately, to determine the distances of the stars from earth. But Michell computed the likelihood that such apparent pairing could arise in the numbers found by Herschel if the stars were scattered at random through space; the probability was so minute that Michell believed the stars to be physically coupled. He was later proven correct, by Herschel himself. Herschel found many pairs in which the members were rotating about a common center of gravity, and he was thus led to the first demonstration of the application of Newton's concept of universal gravitation outside the solar system.

Similar calculations of improbability have been applied recently to argue that chains of stars seen occasionally in the sky (four or five stars strung in a short necklace) must actually result from recent, and presumably spectacular, multiple star births. Convincing explanations, however, have not been advanced for these peculiar collections.

On a grander scale, statistical studies of the clustering of galaxies and loose groups of stars within galaxies have provided some clues to the early history of the universe, and they have helped to clarify the tests that a satisfactory theory of the origin of these clusters must meet.

Generally speaking, the statistical arguments now used by astronomers in their attempt to unravel the causal connections woven into the sky are of two classes: first, statistical analyses of data and, second, physical theories based on statistical or probabilistic concepts. An example of each will be presented.

ANALYSES OF DISTURBANCES IN THE SOLAR ATMOSPHERE

The solar atmosphere seethes with activity, and some disturbances, the gentler ones, are especially amenable to statistical analysis. The solar weather, so to speak, is not altogether chaotic, and astronomers are anxious to glean what they can about the regularities of the pattern because such regularities invariably assist the construction of theoretical explanations.

The vapors of the solar atmosphere are intensely hot, so hot, in fact, that they radiate visible light. Yet, their heat is far from uniform; a telescope reveals a welter of evanescent detail that surges and disappears from place to place within brief minutes. Disturbances are strewn in an irregular pattern—the hot and cool areas cover hundreds of miles and their outlines are roughly hexagonal. This pattern is called *granulation* and the hexagon-shaped elements, called *granules,* are evidently bubbles of hot gas welling up from the interior, carrying heat from the center and disturbing the delicate outer layers. Even the most powerful telescope cannot penetrate beneath the solar atmosphere, so astronomers rely on mathematical analysis to assess the solar interior. But this analysis requires an observational check, and the detailed nature of the atmospheric fluctuations provides one such check.

FIGURE 1

Photograph of the sun's surface showing fine details that have been subjected to statistical analyses of various sorts. The edge of the sun appears at the top; the large dark spot is a cool region of intense magnetism; the bright area is a magnetic storm, known as a flare; the dark filamentary structure is a solar prominence seen in projection against the face of the sun. Source: Harvard College Observatory

BRIGHTNESS

Two types of measurements can be made to infer the structure: the brightness and the velocity of the gas. The earliest studies showed that the pattern of brightness changed drastically every five minutes or so. Quantitative data have been obtained since from series of photographs exposed briefly every 10 or 20 seconds. The exposures are typically one-hundredth of a second, and the best series are taken immediately after sunrise, when the ground is cool and the air is steady. These series can sometimes cover as much as half an hour.

Comparisons of individual photographs (Figure 1 is an example) separated by longer and longer intervals reveal changes in brightness associated with time lapses. The cross-correlation of the patterns on different films is determined in the following way: a line is specified on the sun's surface, and the intensity of the light at points along the line is determined on each photograph. (Those who are not interested in the measure of similarity or who are familiar with the correlation coefficient may want to skip past the following material to the heading "A Possible Model.")

Table 1 represents five points along a line on the sun's surface. The first row of numbers gives initial brightnesses. These are measured in a standardized manner so that brightness at a point is measured as a departure from the mean in units of the original variability. This standardizing gives these numbers the properties that they average to zero and their squares average to 1. (In this table, these averages do not work out exactly because of rounding off.) The second row gives standardized brightnesses a little later in time, but at the same positions. The third row gives standardized brightnesses a few minutes later.

Note that the second row is almost the same as the first, but the third row differs a great deal from the first. One way to measure similarity between two sets of five numbers is to multiply the two standardized numbers of each pair, add the five products, and divide by five to get the average. The underlying idea is that this average will be near 1 if there is high similarity, near 0 if there is little connection, and near −1 if there is high dissimilarity. The average product is called the *correlation coefficient*.

TABLE 1.   The Idea of Correlation

	POINTS				
TIME	1	2	3	4	5
1	1.5	0.5	0	−0.5	−1.5
2	1.4	0.7	0	−0.7	−1.4
3	1.0	−1.4	0.4	−1.0	1.0

Let's first look at an extreme case of similarity; when we compare the first set of 5 numbers with itself, we should get perfect similarity. The correlation coefficient is

$$\tfrac{1}{5}[1.5 \times 1.5 + 0.5 \times 0.5 + 0 \times 0 + (-0.5) \times (-0.5)$$
$$+ (-1.5) \times (-1.5)] = 1,$$

as we said it would be. The similarity of a set of such standardized numbers with itself when measured this way always gives a value of 1, just as in this example. The correlation coefficient can range from $+1$ to -1. If we can't predict the values at one time from those at another any better than we would by guessing, the correlation coefficient gives the value 0.

Correlating the first set of brightnesses with the second, we get

$$\tfrac{1}{5}[1.5(1.4) + 0.5(0.7) + 0(0) + (-0.5)(-0.7) + (-1.5)(-1.4)] = 0.98.$$

This pair is very highly correlated, though, of course, slightly less than the original numbers with themselves.

Correlating the first set with the third gives us

$$\tfrac{1}{5}[1.5(1.0) + 0.5(-1.4) + 0(0.4) + (-0.5)(-1.0)$$
$$+ (-1.5)(1.0)] = -0.04,$$

a slightly negative value, but not far from 0. The example is primarily for illustrative purposes, as the correlation ordinarily stays positive.

If every standardized brightness were replaced by its negative, we would find a coefficient of -1 between the original and the new values. Let's try it with the first set of numbers:

$$\tfrac{1}{5}[1.5(-1.5) + 0.5(-0.5) + 0(0) + (-0.5)(+0.5)$$
$$+ (-1.5)(1.5)] = -1.$$

Although the correlation coefficient ranges between -1 and 1, other ways of standardizing are possible. But the -1, 1 interval is conventional and convenient.

A Possible Model. This correlation will be positive if bright points on one film correspond with bright points on the other film—as will be the case if the films were separated by a very short interval of time. The correlation will be negative if bright points on one correspond to dark points on the other, and if we cannot forecast one set from another better than guessing will do, the correlation is zero.

As the time interval increases, the correlation has been found to decrease steadily toward 0 without actually going negative. The value of the correlation is reduced to about $\frac{1}{2}$ in an interval of 5 minutes and to somewhat less than $\frac{1}{4}$ in 10 minutes.

A model for this process has been suggested. It assumes that at random times, averaging about 5 minutes apart, new granules appear and gradually cool. Each is replaced by another at the next random time. If the cooling is slow and the replacement is slow in coming, the correlation is slow in going to 0. If a replacement comes rapidly, the correlation goes to 0 rapidly. Theoretical work not given here shows that this model will create correlations that reduce from 1 to $\frac{1}{2}$ in about 5 minutes on the average. This agreement with the facts lends some support to the model.

VELOCITY

The other type of measurement, velocity, has been more exciting in its consequences; it is also more difficult to obtain, but during the past ten years a vast amount of data has accumulated, and these data show a behavior markedly different from that of brightness changes. Correlations can be measured for the velocity in much the same way they are for the brightness; they also decline with time, but they dip down and become negative before they rise again to 0. In fact, a detailed plot of the velocity at a particular point on the sun's surface shows an oscillation that is quite striking in its apparent regularity. Figure 2 shows one example; it is not perfectly regular, but there is no doubt of an actual oscillation. Why are oscillations seen in the velocity, while the brightness pattern is irregular in both space and time?

The answer to this question remains incomplete, but studies of the detailed frequency composition of these oscillations are beginning to clarify the phenomenon—and to show its real complexity. These studies have shown that the oscillations are amazingly similar to waves of musical notes emitted by a violin string. Overtones are present, but they are relatively weak. In fact, over half the energy is contained in oscillations whose periods are confined between 2 and 6 minutes. As a terrestrial example, we might say that a slow motion movie of surf breaking on a beach would show about the same degree of periodicity.

Astronomers were astonished by the discovery of this regularity, because they had come to think of the sun's atmosphere as the seat of mere chaos; also, the brightness variations had shown no such periodic oscillations, and astronomers had assumed that the pattern of the upward and downward motions would closely mimic the irregularity of the brightness changes.

The periodicity of the motions, and its sharp contrast with the randomness of the brightness, showed at once that astronomers were observing two different parts of the solar atmosphere; it has since been proven that the brightness variations are produced low in the visible atmosphere while the observed motions take place in the upper layers of the atmosphere. And, what is

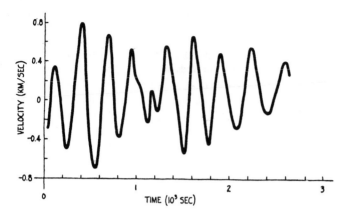

FIGURE 2

Measurements of the vertical component of the velocity of gas in the sun's atmosphere are plotted here for a single point on the sun. The marked periodicity of the motions is typical of the solar photosphere, and it has been the object of many correlation studies

more, the nature of the motions alters with increasing height in the atmosphere —the average period shortens by a factor or two, and at the greatest observable height (several thousand miles above the "surface") the velocity fluctuations become quite chaotic, resembling noisy static without any pronounced periodicity.

Why? Astronomers assume that we are witnessing the upward flight of very long "sound waves" in the solar atmosphere—waves that are generated deep in the solar atmosphere, perhaps by the rising granules. Some waves are trapped in that atmosphere, predominantly those with periods of about 5 minutes; waves of shorter period escape quickly to the upper levels, where they predominate. Waves of very long period die out quickly, and in fact, they are not easily excited by the granules, so they are very weak at all levels.

Even this brief explanation makes it clear that a study of the frequencies of these oscillations may reveal several features of the solar envelope: the nature of the deeper disturbances that generate these sound waves, the rate at which waves of different periods are dissipated as they propagate, and the extent to which the solar atmosphere is capable of trapping waves of different periods.

PROBLEMS

1.   What role have "calculations of improbability" played in the theory of double stars?

2.   What are the two types of statistical arguments used by astronomers?

3.   Refer to Table 1. We are interested only in the measurements taken at time 2. How does the zero entry at point 3 compare to the average brightness at all points?

4.   Using Table 1, calculate the correlation coefficient for the second set of measurements with the third.

5.   From the model described in the text, would you expect the correlation coefficient for times 1 and 3 to be larger than that for times 2 and 3? Did your calculation above meet your expectation? Explain.

6.   Refer to Figure 2. How many oscillations fail to dip below 0? below -0.4?

7.   Draw an analogue of Figure 2 for the correlation coefficient of measurements of brightness of the sun's atmosphere made at several points on the sun.

8.   The velocities presented in Figure 2 show marked periodicity. Would you expect similar graphs from other measuring points in the solar atmosphere? Explain your answer.

9.   (a) Describe the differences in the observed brightness and velocity patterns of gas in the sun's atmosphere.
     (b) Give a partial explanation of this difference.

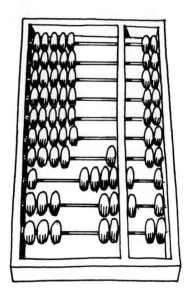

INFORMATION, SIMULATION, AND PRODUCTION: Some Applications of Statistics to Computing*

Mervin E. Muller   *University of Wisconsin*†

STATISTICS CAN play a central role in fully utilizing the computer's power, economy, speed, and precision. The use of computers often depends upon the availability of statistical methodology to provide insight and direction in the specification of problems and in methods of solution. There are problems that, in spite of their social importance and in spite of the speed and power of computers, would be prohibitively expensive or time-consuming to solve without the use of statistical techniques to guide in the collection and analysis of data. We shall deal here with some of the ways statistics helps in the use of computers. Many other articles could be written to show how

* The author is grateful to Alan G. Merten, Wayne Rayfield, and Peter J. Wolfe for their careful reading of the manuscript and helpful suggestions.
† Currently with the World Bank.

the availability of computers has made it possible to perform many kinds of statistical analyses and to use many kinds of statistical techniques that would have been impossible without a computer, because they would require excessive amounts of time, effort, or cost, or would be subject to excessive computational errors, but in general, we shall not treat these in this essay.   Several essays in this volume base their analyses heavily on the use of the computer.

INFORMATION STORAGE AND RETRIEVAL

One important area of actual and potential use of statistics and computers is in the storage and retrieval of information.   For example, with the increased mobility and size of our population, it is increasingly apparent how valuable it is to have fast access to one's medical history.   One dimension of this problem is that patients and physicians are on the go, another is the explosion in the amount of medical literature and information now available for physicians; still another is the potential total size of medical files resulting from the size of our population.   In some medical situations (arising from accidents, for example) it is exceedingly desirable to be able to locate rapidly a patient's medical history to determine possible dangers or side effects to the injured person if certain treatments are applied.   Also, planned within the past few years is a pilot computer-based information storage and retrieval system for a National Poison Center.   When this Center exists, any physician may call in, describe symptoms, and then be advised rapidly about the best possible treatments.   The trouble is that as any of these medical files (or any other files of information) become large, the cost of storing and retrieving information can become large and the time required to find a desired item of information long.   We need to arrange the information in storage so as to minimize these costs.   Here is where statistical techniques can aid in providing criteria and methods of analysis for designing, implementing, and maintaining large computer-based information storage and retrieval systems which are both effective and economical.

Because we cannot usually afford to store all information items so that they can be retrieved in a short time, we are forced to take into account the demands for use of the items of information so that, in general, the most used items can be retrieved most quickly.   But items vary over time in their frequency of use, so statisticians have developed theories for characterizing demands and estimating their fluctuations.   This type of analysis can aid in determining where to store items of information because there will usually be varying demands for the use of information within a system.   Hence we need ways to characterize such variations and then ways to use such analyses to design and evaluate the performance of an information system.   Given such a design, other considerations, basically economic, control the type of computer hardware used and the actual speed of retrieval.

We shall now consider a very simplified problem in the organization of files to indicate some of the roles of statistics. Imagine that we must design an information retrieval system for information on symptoms and treatments of poisons. Suppose that the poison information file is to be kept in a sequential file. (To simplify the presentation, suppose the file contains information on only five types of poison.) We refer to the information on a particular poison as an information item, or more succinctly, as an item. Suppose that the five items (the poison information) are identified by labels A, B, C, D, and E. Usually there will be a different amount of information on each poison. For illustrative purposes, suppose also that the 5 items respectively contain 1, 2, 3, 4, and 5 information records so that the time it takes to scan item C is 3 time units because item C contains 3 records.

Once we decide on the order in which the items are to be placed in the file, we may assume that each time we have a request for some information on a particular poison we will start at the beginning of the information file and look at the label of each item until the desired one is found. (In place of a sequential file, we could consider other kinds of computer-information storage schemes so that we would not always have to start searching from the beginning of the file. These schemes present their own problems for determining the best file organization, but they will not be considered here. Such schemes offer attractive access speed but are generally more expensive to implement.) Suppose it has been possible to measure the relative frequency of demand for the five items as $\frac{1}{10}$, $\frac{2}{10}$, $\frac{4}{10}$, $\frac{2}{10}$, $\frac{1}{10}$. Thus, items A and E are requested equally and least often, while item C is requested most often, 4 out of every 10 times a request is received.

What is a good way to arrange the file? We can call on the statistician to help us select a criterion of best arrangement. He might ask us to say whether or not in this case "best" implies that the average time to locate an item should be as small as possible or that the longest necessary search time should be as small as possible or maybe that the variability in search time should be as small as possible. Depending on the particular circumstances, each of these performance criteria could be the one of primary importance.

One criterion that is often selected is the minimum average access time, since it appears desirable to reduce the average time that someone must wait to obtain a desired answer to an inquiry. However, a file organized to minimize the average access time might cause some inquiries to the file to wait excessively long for a response. In some applications, minimizing the variability of waiting time for responses is considered to be most important. One measure of variability is the *range*, which measures the difference or spread between the longest and shortest wait. A different measure of variability is the *variance*, which measures the average squared deviation from the average access time.

In most cases, the file organization that is best with respect to one criterion will probably not be best with respect to any other criterion, as will be illustrated in an example. Therefore, the relative importance of the performance criteria must be taken into account when selecting one; this selection, in turn, will determine the file organization to be selected. For our example of five items, we might be prepared to use brute force, that is, to examine *all* the 120 possible arrangements of five items. If there were ten items in the file, a brute-force approach would require examination of over three million possible file arrangements, in fact, 3,628,800. With 10 items, brute force appears to be out of the question. However, such problems involving large numbers of possible arrangements are the sort that can be solved with the aid of statistical techniques.

The columns of Table 1 contain 10 illustrations of the 120 possible arrangements of the five items, and the last five rows contain some numerical values that indicate the statistical performance of the ten arrangements according to five criteria: average (mean) access time, variance of access time, minimum access time, maximum access time, range of access times. We see that orderings 1 and 2 possess average access time of 6.3 time units, the minimum for the ten chosen orderings and for all 120 possible arrangements as well. But the variance of the access time for arrangement 1 (15.21) is greater than for arrangement 2 (15.01). Arrangement 7 provides the ordering with the minimum variance (8.04) in the table and of all 120 possible arrangements, but a mean access time of 11.4 time units. Maybe we would prefer arrange-

TABLE 1. Ten of 120 Arrangements of Items A, B, C, D, E, Showing the Average Access Time, Variance, and Other Statistics

	ARRANGEMENT									
	1	2	3	4	5	6	7	8	9	10
First item	C	C	A	A	D	C	E	E	E	A
Second item	A	B	B	B	C	D	D	D	D	B
Third item	B	A	C	E	B	E	C	B	A	D
Fourth item	D	D	D	D	A	B	B	A	B	E
Fifth item	E	E	E	C	E	A	A	C	C	C
Statistics on access time										
Mean	6.3	6.3	6.6	9.9	7.9	8.1	11.4	11.7	11.7	9.3
Variance	15.21	15.01	15.24	29.09	9.09	23.89	8.04	10.41	10.61	29.61
Minimum	3	3	1	1	4	3	5	5	5	1
Maximum	15	15	15	15	15	15	15	15	15	15
Range	12	12	14	14	11	12	10	10	10	14

ment 5 because it has a reasonable (smaller than average) average access time and a relatively small variance.

For this example, the average of all 120 average access times is 9.0 time units. This average of all averages provides a basis for comparing how much better one can do in terms of average access by using the best organization instead of just leaving the file organization to chance and accepting the access performance associated with a randomly selected arrangement. Orderings 8 and 9 have the maximum average access time (both in the table and of the 120 orderings) of 11.7 time units and correspond to the reverse of orderings 1 and 2, respectively. Finally, ordering 10 has the maximum variance of access time of all 120 orderings, 29.61.

The complexity of analysis of data file organizations is much greater than the above example conveys. Statistical techniques used to design computer-based file organizations to have desirable storage and retrieval characteristics have many areas of applications, for example, keeping track of air-traffic patterns, voting records of elected public officers, dangerous drivers, and inventories of natural resources. One such technique, called the Monte Carlo method, will be described briefly later.

One of the oldest and most active uses of statistics in computer-based storage and retrieval of information involves dairy herds. The various dairy-herd improvement associations have record systems that contain extensive details about bulls, their offspring, and the associated milk production of the cows. Statistical techniques are used to help in determining what information should be kept in a file of this type so as to have an effective store of information.

MONTE CARLO APPLICATIONS

A class of statistical applications that was stimulated during World War II is identified as *Monte Carlo applications*. Some of the problems that needed solutions using these methods arose in the design of atomic reactors. Scientists had to estimate the behavior and interaction of elementary atomic particles so as to predict the expected termination site of the particles confined to the protective lead shieldings of the walls of the reactor. The term "Monte Carlo method" was selected because such methods use "computational statistical games" involving sampling and probability theory to provide understanding of the behavior of chance events involving atomic particles.

Insight into probability and statistics has been obtained throughout history by playing a particular game of chance many times. One use of the Monte Carlo approach would be to roll repeatedly a pair of dice, record the outcome of each roll, and then analyze all the outcomes in an attempt to determine if the dice were loaded. More generally, in the Monte Carlo method, we formulate and build statistical models and perform the necessary computations to generate samples from the models. When the results of these samples are analyzed, they provide useful estimates of the solution to a mathematical equation or equations.

There are several reasons why the Monte Carlo method is used. In some situations the real life process being studied has such complexities and uncertainties that even if a set of completely determined mathematical equations could be obtained to describe the process, no direct solution could be found, even if a computer with all its speed and accuracy were used. In other situations, no set of completely determined equations can be obtained at all.

The computer is made an effective tool in some of these situations by the application of statistical sampling techniques and mathematical models executed on computers. In this class of applications, some of the problems to be solved do not themselves directly involve statistics or probability. However, by use of statistical techniques involving sampling we can carry out on a computer a process whose statistical properties yield an estimate of the answer to the original nonstatistical problem. Thus the Monte Carlo method is an example of an interesting and important interaction between statistics and computers.

The Monte Carlo method requires developing statistical and computational procedures (called computing algorithms) that can be performed on computers to simulate various kinds of statistical or random behavior; for example, the computing algorithms provide the simulated probabilistic interaction of colliding atomic particles for a physics application, or a distribution of poker or bridge hands for one interested in understanding games of chance. It was partly through his desire to obtain the computational power needed for Monte Carlo–type analysis during World War II that one of the pioneers of the Monte Carlo method, John von Neumann, also became one of the most important pioneers of the modern digital computer. The term *simulation* is now often used in place of the term Monte Carlo method. Both terms are used to describe applications where statistical techniques directly influence the control and use of computers. In the file organization problem, we could obtain useful insight by performing Monte Carlo analysis or simulation to sample various orderings of the file and thus estimate the performance and characteristics of a near-optimal ordering.

COMPUTERS IN INDUSTRIAL PROCESSES

There are important industrial processes, in the chemical industry, for example, where the computer is recognized as a vital contributor. As an example, consider an industrial process from which some intermediate chemical B is desired. The process starts with product A, then yields B, and then yields C.

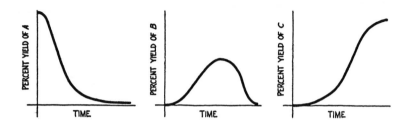

FIGURE 1

A chemical process that starts
with chemical A has an inter-
mediate yield of B and a final
yield of C

These three parts of the process are shown separately in Figure 1 and
then superimposed in Figure 2, so that the state of affairs at any time can
be more easily seen.

The chemical engineer can be faced with considerable computational complexities in estimating the optimum time at which to begin extracting product B.   Often he must use statistical techniques to determine a useful set of
equations describing the chemical process.   Of course, if B is extracted too early
or too late in the process, only a small yield will be obtained.   The engineer
wants to find an interval around time b when the amount of B available
is greatest.   To characterize the process, he must collect data to verify the
appropriateness of the equations describing the process and then collect data

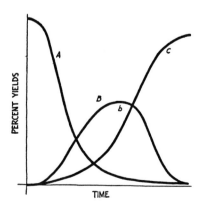

FIGURE 2

The process of Figure 1 on one
time axis

to estimate the parameters of the equations that are important in controlling the process. Without careful statistical analysis to determine *when* and *how* to collect data, not only will the computer analysis required to solve the complex and lengthy arithmetic be of questionable accuracy, but also the yield of the process may be substantially below what is required to have a useful and profitable product.

PROBLEMS

1.   What is meant by a *sequential* file?

2.   Two measures of variability are *range* and *variance*. Define them.

3.   If there were only four items in a file, how many possible file arrangements would there be?

4.   Refer to Table 1. For each arrangement the entry for maximum access time is 15. Why?

5.   How does the average of the average access time of the 10 selected arrangements of Table 1 compare with the average of the average access time of all 120 possible arrangements?

6.   How are samples generated in the *simulation* method? Why is it also known as the *Monte Carlo* method? Describe when this method is used.

7.   Suppose one were interested in extracting an equal amount of both B and C from the process depicted in Figures 1 and 2. Around what point in Figure 2 would you center your study of the process?

REFERENCES

Alan G. Merten. 1970. "Some Quantitative Techniques for File Organization." University of Wisconsin Computing Center Technical Report No. 15.

Mervin E. Muller. 1969. "Statistics and Computers in Relation to Large Data Bases." R. C. Milton and J. A. Nelder, eds., *Statistical Computations*, New York: Academic. Press. Pg. 87-176

STRIVING FOR RELIABILITY*

Gerald J. Lieberman   *Stanford University*

THE APOLLO mission to land men on the moon required the creation of an artificial world to support the needs of three men and complex electronic gear in a hostile environment for approximately two weeks. For a successful mission all of the components of the Apollo system had to function in a satisfactory manner. Failure in any one of innumerable areas could lead to a failure of the mission.

If reliability is defined as the probability that a device performs adequately for a specified period in a given environment, it is clear that the reliability problem was one of the greatest challenges facing the Apollo program. What is the role that probability and statistics play in reliability? Reliability is just a "probability," and hence the mathematical structure of probability the-

* This paper was prepared while the author was visiting Imperial College of Science and Technology. It was supported in part by the Army, Navy, and Air Force under contract Nonr 225(53) (NR-042-002) with the Office of Naval Research.

ory is important in its evaluation.   There are two types of general problems that arise in reliability.   First, we must develop a mathematical model to represent the system, and second, we must be able to provide estimates of the numbers that enter in such a model to provide a numerical value for the reliability.

In the context of the Apollo, the command and service modules have approximately two million functional parts, miles of wiring, and thousands of joints.   These parts, wiring, and joints may be broken into subsystems in some fashion.   Each subsystem may be assumed to have a known reliability associated with it.   A mathematical model of the Apollo system can then be abstracted from the physical processes, and the theory of combinatorial probability utilized to predict the reliability of the Apollo system.   Simplified models will be discussed subsequently.

The second general problem is the statistical problem associated with reliability.   Again in the context of Apollo, the system may be broken into subsystems in some fashion, with each subsystem having an assigned reliability.   In practice, estimates of these reliabilities must be obtained from experimental data.   Ultimately, an estimate of the overall reliability of the Apollo system based upon these subsystem estimates is desired.   Alternatively, performance data may be obtained directly on the complete Apollo system, and these experimental results used to obtain an estimate of the reliability.   These are some of the statistical problems associated with reliability.

RELIABILITY MODEL

In order to describe the basic problem of model development encountered in reliability theory, a simplified version of the basic Apollo module is presented in Figure 1.   Suppose that the Apollo system works if, and only if, all five components shown operate properly; that is, the system will fail if any component of the system fails.   Such a system is called a *series system* and a

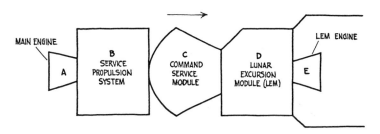

FIGURE 1
Simplified model of the Apollo module

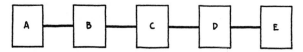

FIGURE 2
Black-box representation of series system

typical example is a string of Christmas tree lights of 1940 vintage (if one
bulb fails, the entire string is darkened).

If Figure 1 is a series system it may be replaced by the slightly more
abstract (*black box*) description shown in Figure 2, where each black box
is in one of two states, operating or failed.   The Apollo system, then, will
fail if any of the five components fails during its mission.   If it can be assumed
that the five components of Apollo depicted in Figure 1 operate independently
and are connected in series, the system reliability (the probability that Apollo
performs adequately for a specified period in a given environment) is easily
calculated in terms of the component reliabilities.   If the component reliabilities are respectively R_A, R_B, R_C, R_D, and R_E, then probability theory indicates
that the system reliability is given by R, where

$$R = R_A R_B R_C R_D R_E.$$

Thus, for example, if each component has reliability 0.999 for a mission.
the reliability of the Apollo system is given by

$$R = (0.999)^5 = 0.995,$$

to three decimals.

A closer examination of these components must be made before calculating
system reliability.   Do the performance characteristics of these components
interact with each other?   Does a successful main engine performance portend
that the LEM engine will also behave properly?   Does degradation of one
subsystem put a high load on other subsystems?   If the answers to these
and similar questions are positive, then the performances of the subsystems
may not be assumed to be independent of each other.   Unfortunately, if the
components are not independent, a more complex expression is required for
calculating the system reliability.

Another simple type of configuration for components is a *parallel configuration*.   Suppose that two (possibly smaller) engines A_1 and A_2 replace the
main engine A.   Suppose further that they both are turned on during the
mission and that the engine function is satisfactorily performed if either A_1
or A_2 is operative, i.e., a failure occurs if and only if both engines A_1 and
A_2 fail.   Such a parallel configuration can be represented as black boxes
as shown in Figure 3.

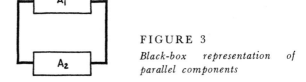

FIGURE 3

Black-box   representation   of   parallel components

A representation of the entire system is shown in Figure 4. An actual example of a parallel system is the second stage of the Saturn rocket used in the Apollo program. This has five J.2 rocket motors, and even though, on the Apollo 13 launch, the center engine failed early, a satisfactory earth orbit was achieved by extending the burn of the remaining engines.

Again, if components connected in parallel can be assumed to be independent, the system reliability can easily be obtained from probability theory. If the reliability of the engine function is denoted by R_A and the engines have reliabilities R_{A_1} and R_{A_2} respectively, then

$$R_A = 1 - (1 - R_{A_1})(1 - R_{A_2})$$

that is, the probability that the engine power function fails, $(1 - R_A)$, is just the probability that both engines A_1 and A_2 fail. Suppose that both engines A_1 and A_2 have reliability 0.980 (considerably less than the reliability of the single main engine described earlier) ; then

$$R_A = 1 - (1 - 0.980)^2 = 1 - (0.020)^2 = 0.9996,$$

to four decimals, which is higher than the reliability of the single main engine described earlier. In fact, the black box representation of the Apollo system shown in Figure 4 can be replaced by that shown in Figure 2, provided the reliability of the main engine is equal to 0.9996.

A third type of configuration is a standby parallel system. If engines A_1 and A_2 were connected as in Figure 4 but so that engine A_2 was switched

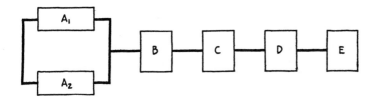

FIGURE 4

Black-box   representation   of Apollo   system   with   parallel main   engine   units   and   other components   connected   in   series

on only after engine A_1 failed, engines A_1 and A_2 would be considered as a parallel standby system. An example of a parallel standby system appeared in the ill-fated Apollo 13 mission when the astronauts actually used the oxygen supply in the LEM after the oxygen supply in the command service module failed. In both the regular parallel configuration and the standby parallel configuration, the additional units are called *redundant units,* and their existence tends to increase the reliability of the overall system (as shown in the parallel engine example) usually at the expense of cost and/or weight. In fact, some of the most interesting and important reliability design problems are those which seek to maximize the reliability of a system subject to constraints on weight or cost or both.

If all complex systems could be decomposed into independently behaving series and parallel systems, reliability calculations would become relatively simple. This is seldom the case, unfortunately, and systems such as Apollo are so complex that high-speed digital electronic computing machines are required to calculate the system reliability (assuming the reliability of individual components is known). The existence of such computing programs is also useful to assess the effect of design changes on the overall reliability of the system. In some situations, systems are so complex that detailed reliability computations are too cumbersome, and other techniques are required to assess the reliability of the system and to determine the effect of design changes.

STATISTICAL ANALYSIS OF RELIABILITY

The previous description of the assessment of the reliability of a system assumed that the reliability of the individual components was known, but nothing was said about how these numerical values were to be determined. As indicated previously this is the statistical aspect of the reliability problem. There are two ways in which estimates of the system reliability can be obtained. The simplest from a statistical point of view, but generally the most costly (and therefore most impractical), is to test by performing the mission using the final complex system. This testing is often destructive in that the system cannot be used again (e.g., launching an Apollo system) so that clearly this mode of testing is used infrequently. The other method is to test the subsystems in an environment, possibly simulated, but presumably similar to that encountered during the mission. Based upon test data, estimates of the subsystem reliabilities can be obtained, and a subsequent estimate of the system reliability then can be found.

Much of the remaining discussion will be concerned with estimating the reliability of a component because a subsystem (and even a system) can be thought of as a single component. Two important ways of testing a component directly are as follows: the first method, called *testing by attributes,* places the unit on test in the appropriate environment and for the appropriate time and determines whether the unit fails or operates successfully; the second method, called *life testing,* places the unit on test in the appropriate environment, and the time to failure is recorded.

The statistical model for testing by attributes is relatively simple to describe. A random sample of n (independent) units are placed on test in a specified environment for a predetermined time t. It is assumed that each unit has the same reliability R, which is now assumed to be unknown. The total number of "successes" is recorded. Based upon these data, the reliability is to be estimated. In technical terms, the number of successes in the n trials is said to have a binomial distribution with parameters n and R, and standard techniques can be used to estimate the reliability. A general comment is in order. Usually, a large number of items must be tested in order that the estimate of the reliability be "good." For example, 230 consecutive units must be tested successfully in order to demonstrate (with probability 0.90) that the reliability of a system exceeds 0.99.

We will now discuss the statistical problems in life testing. Recall that units are placed on test in the appropriate environment, and the time to failure is recorded. In order to use these data to estimate the component reliability, some assumptions are usually required about the underlying failure mechanism. A simple characterization of the failure mechanism is in terms of the instantaneous failure rate, commonly called the hazard function (i.e., the probability of the unit failing in a small additional increment of time given that it has survived to the present). This is analogous to those mortality tables in life insurance, which are concerned with the probability of an individual surviving an additional year given that he has survived to the present. A reasonable hazard function that a component may possess is referred to as the bathtub instantaneous-failure function and is shown in Figure 5.

Initially, the hazard function tends to decrease, reflecting the "break-in" of marginal parts which, though operative, are operating improperly and hence may cause premature failures. The longer these parts function, the better

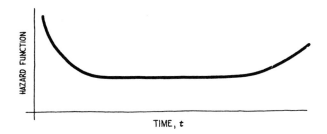

FIGURE 5
Bathtub hazard function

they perform.   After this break-in period, the hazard function tends to remain constant for what is believed the normal operating period.   This constant failure rate implies that no wear and deterioration are present; that is, during this period, the probability of a still working unit failing in a small additional time interval is independent of the age of the item.   (It is the same for a unit that has lasted a thousand hours as for a unit that has lasted one hour.) Finally, the last part of the hazard function reflects an increasing segment which implies that wear or deterioration is present.   Unfortunately, the locations of these break points are not easily determined, so erroneous assumptions about the location of the normal operating interval frequently occur.

OTHER USES OF RELIABILITY CONCEPTS

In most of the previous discussion, the Apollo space vehicle was used as an example to illustrate the importance of reliability concepts.   Problems in reliability are not confined to esoteric systems, however, but they are encountered in items in everyday usage.   Reliability is an important feature of household appliances, automobiles, telephones, power supplies, and so on, whether viewed from the vantage of the producer or the consumer.   Important decisions are based upon the reliability of the product.   For example, the five-year guarantee given by automobile manufacturers resulted from their determination of the reliability of the components falling under this guarantee.   Consumers often choose to purchase the brand of items whose failure rate is low.   High-reliability "consumer" products may be important from other than economic considerations.   An unreliable pacemaker inserted in a heart patient could result in his death.

To summarize then, obtaining systems that perform adequately for a specified period of time in a given environment is an important goal for both government and industry.   It has been frequently stated that the cost of maintenance and repair for such items as electronic equipment in their first year of operation often exceeds the purchase cost.   Hence study and use of the theory of reliability, which can be applied in the research, development, and production phases of a system to enable the user to evaluate and improve performance, is a worthwhile venture.   If reliability theory is to be useful, it must be quantitative in nature, because reliability must be demonstrable. Hence, probability and statistics play an important role in its development.

PROBLEMS

1. What are the two types of general problems which arise in reliability?

2. Explain the difference between a series and a parallel configuration of components.

3. (a) What is the status of the system of Figure 2 if A fails?
    (b) What is the status of the system of Figure 4 if A_1 fails? A_2 fails? A_1 and A_2 fail?

4. Consider Figure 3 and assume independent components.
    (a) If $R_{A_1} = .98$ and $R_{A_2} = .2$, what would the whole system's reliability be?
    (b) If the two independent components have the reliabilities given in (a) but are connected in series rather than in parallel, what would the whole system's reliability be?

5. Let the reliabilities of the independent components of Figure 4 be R_{A_1}, R_{A_2}, R_B, R_C, R_D, and R_E. What is the whole system's reliability in terms of the component reliabilities?

6. Consider the three independent components connected in parallel as shown:

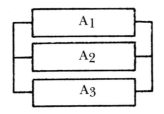

If the reliabilities of the three components are R_{A_1}, R_{A_2}, and R_{A_3}, what is the whole system's reliability in terms of the component reliabilities?

7. Why might it not be advisable to test reliability by performing a practice run with the final system?

8. How does *testing by attributes* differ from *life testing*?

9. Define the *hazard function*.

10. Refer to Figure 5. Why does the curve dip sharply in the beginning? Turn up at the end?

STATISTICS AND PROBABILITY APPLIED TO PROBLEMS OF ANTIAIRCRAFT FIRE IN WORLD WAR II*

E. S. Pearson *University College, London*

THERE ARE situations in which the statistician is called on to give assistance where exact observational data are so hard to come by that no refined techniques based on mathematical theory can be introduced to solve the problems under investigation. Nevertheless, the statistician's training with its understanding of the meaning of variation and correlation, of randomness and probability, and with its emphasis on the importance of adopting a critical outlook on all assumptions made should help him in handling what at first sight may seem a most intractable problem.

In 1939, at the beginning of World War II, my statistical group from

* This article is based on a talk given in October 1962 at the Eighth Conference on the Design of Experiments in Army Research Development and Testing. (Issued as ARO-D Report 63–2.) Permission for the talk was given by the British Ministry of Defence.

University College, London, was attached to the British Ordnance Board. This is an organization of some historical interest, as its origin can be traced back to an appointment made in 1414, the year of the battle of Agincourt. The Board had latterly become involved in certain aspects of the development and acceptance of weapons for the Army, Navy, and Air Force.   During the thirties, it had taken the initiative in putting in hand a variety of research projects in matters where information was sadly lacking.   One of these investigations, in which the initiative came from Col. A. H. D. Phillips, Superintendent of Applied Ballistics, concerned the problem of the lethality of antiaircraft weapons.   The first assignment of my group on joining the Board was to carry on and develop the work already in hand in this field.

THE ANTIAIRCRAFT PROBLEM

First let me try to put the problem into its setting of 30 years ago.   As far as the Ordnance Board group was concerned, we had not to consider problems relating to deployment of guns, acquisition of targets, handling of mass attacks, or other important tactical matters.   These were questions for the Antiaircraft Command and its Operational Research Section, formed in the summer of 1940.   Our work was closely related to the question of weapon design; for example, we had to try to understand more clearly the relationships of the "predictor" that controlled the gun-laying and the setting of the time fuse, the characteristics of the shell and its explosive filling, and the vulnerability of the enemy target to shell fragments.   Only then would it be possible to advise what improvements were feasible and likely to be worthwhile.

In this field of research where the terminal action in which we were interested might be taking place several thousand feet above ground, no overall experiment bringing in all the factors concerned was conceivable.   Consequently, it was essential to construct a mathematical model of the terminal engagement and then to consider how the parameters of this model might best be estimated.   As in so many problems of military or applied industrial science such a model even if simplifying the real situation, as it generally must, serves a necessary purpose.   It defines the relationships in the situation, shows how research investigation can be broken into separate pieces and emphasizes at what points lack of sure information is greatest and most hampering.   In 1939, the lack of information in the antiaircraft field was great indeed!

Let me first describe the model and the main headings under which gaps in knowledge had to be filled.   The problem was one in the field of *probability* because the type of answer we could hope to give was that under certain assumed conditions, there was a probability P that a single shell would destroy (or cripple) the aircraft at which it was fired.   As an alternative measure, the reciprocal of P $(1/P)$, the so-called *rounds per bird*, was often quoted. The problem was *statistical* because it was necessary to feed into the model certain numerical parameter values, derived from the analysis of various forms of observational or experimental data.

The form of the model was conditioned by the fact that, broadly speaking, the factors concerned could be classed under three headings:

(1) *The positioning errors,* that is to say, the distribution of likely positions around the target aircraft at which a fuse-initiated shell would burst.

(2) *The fragmentation characteristics of the shell.* In the problem of heavy antiaircraft guns firing at long range, the chance of a direct hit was small, and therefore damage must be done by fragments from the shell-casing, exploded some distance away. This was, of course, long before the day of guided missiles with homing devices.

(3) *The target vulnerability to shell fragments,* that is to say, the destructive power of fragments, according to their mass and velocity, on hitting the pilot or vital components of the aircraft.

CONSTRUCTION OF THE MATHEMATICAL MODEL

The Positioning Errors. In ground-to-ground firing with artillery shells, there was a good deal of information on the pattern of shell-strikes round the target. This pattern had been found to conform roughly to what is known as the *bivariate normal distribution,* with the density of strikes falling off as we get further from the center of concentration. For this distribution, which is found to represent well the familiar pattern of observed statistical data in studying problems where the error has two sources as here, the contours of equal density are roughly a series of concentric, coaxial ellipses as suggested in Figure 1, the major axis lying along the line joining gun to target, that

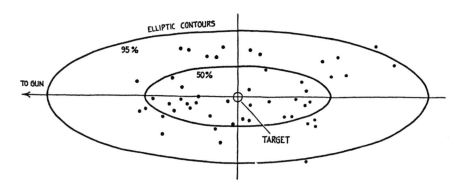

FIGURE 1

Typical chart showing fall of 50 shot

is, the largest errors being in range, not direction.[1]   Of course, there might be bias in that what I have termed the center of concentration is not on target, due perhaps to an error in range-finding or a miscalculation of corrections for wind.

It was a reasonable assumption that a similar distribution of errors in position would hold in three dimensions in the sky, although the errors would be much larger in magnitude.   However, there was no direct method of examining the error distribution as in the case of the scatter of shell impacts in two dimensions on the ground.   It is true that some records were available on the position of shell bursts round a towed target, obtained at antiaircraft practice camps, using a pair of kinetheodolites at each end of a baseline (i.e., theodolites taking continuous film records of angles and time).   But these records were not of much value because inevitably the towed target was moving more slowly than a free-flying aircraft and because in 1940 the only personnel whom the Army was prepared to spare for special trials were gunners in course of training.

It would be out of place here to describe the working of the mechanical "predictors" that controlled the firing of the gun, but a number of sources of error were involved in the early stages of the war, largely because it was necessary to have one moving pointer on certain dials followed manually by a second pointer.   Later, there was much improvement when a radar element was introduced into the predictor.   There was also the error involved in the running of the clockwork time-fuse, even when correctly set.   Some useful information on predictor accuracy was obtained in April 1940 from a trial in which a free-flying aircraft, rather than an aircraft-towed target, was followed by several predictor and gun crews simultaneously, and camera recordings of the output dials were synchronised with kinetheodolite tracking of the target.

However, when German aircraft began to come over England later in 1940, it was at once clear that the aiming errors under operational conditions were considerably greater than those estimated from trials.   We were up against the problem of increased operator inaccuracy under battle stress.   The real targets also did not necessarily fly on a straight-line course (as the predictor mechanism assumed) unless on a final bombing run.

All that could be done, therefore, was to assume that under given conditons, the burst of the shell about the target would occur, on repeated firing, in a distribution described mathematically by the three-dimensional analogue of that suggested in Figure 1.   The model allowed for the degree of scatter in the direction along the shell trajectory and at right angles to this to be adjusted at will.

---

[1] The pattern, with only 50 rounds fired, will not of course be regular, but in the diagram it is seen that 26 out of 50 rounds fall within the theoretical 50% ellipse and 48 of 50 within the 95% ellipse.

The Fragmentation Problem.   Before the war, the standard trials for determining the fragmentation characteristics of shell were:

> (1) Fragmentation in a sandbag "beehive," the shell fragments being recovered, passed successively through various sizes of sieve and (above a certain minimum size) counted and weighed.
>
> (2) Trials to measure the dispersion and penetrating power of fragments by detonating the shell some five feet above ground in a surround of two-inch-thick wooden targets, placed in a semicircle of, say, 30, 60, 90, or 120 feet radius. The detonation was either at rest or obtained from firing the shell (fitted with a percussion fuse) at appropriate velocities against a light bursting-screen sufficient to trigger the shell on impact.

Before refinements were introduced, the number of perforations of the two-inch targets, or of "throughs," as they were termed, per unit area was taken as a comparative index of the damaging power of different types of shell. The target records showed that the main fragment zone of a shell lay between two cones whose axes were that of the shell axis and trajectory at time of burst; in addition there was a small subsidiary nose cone of fragments. If Figure 2 were rotated in a third dimension about the shell axis, it would map out these zones. The greater the velocity of the shell, the further forward the main fragment belt would be thrown.

In addition, analysis of trial data showed that the number of throughs per unit area, that is, 0, 1, 2, . . . , could be well represented by the terms of a well-known probability distribution, the *Poisson distribution,* whose form depended only on a single parameter, the average number of throughs per unit area. For example, if on the *average* there were two throughs per unit area, the Poisson law gives the chances in any one firing of their being 0, 1, 2, 3, 4, 5, 6 throughs as 0.14, 0.27, 0.27, 0.18, 0.09, 0.04, 0.01, ...

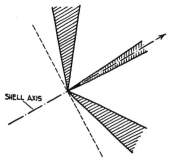

FIGURE 2
Fragment zones

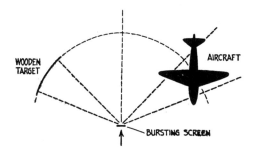

FIGURE 3
Plan of damage trial

Aircraft Vulnerability. In the earliest trials carried out shortly before the war, an aircraft and an arc of large two-inch thick vertical wooden screens were placed beyond and on opposite sides of a small burster-screen at which the shell (with percussion fuse) was fired at a prescribed velocity. Figure 3 illustrates the arrangement. In this way, it was possible to correlate the damage done to the aircraft with the density of throughs in a second, similarly constituted fragment stream. By noting and then painting round the fragment holes after each round was fired, the same targets could be used a large number of times, varying the aspect of attack and the distance of detonation as desired.

It was from the observed correlation of density of throughs and damage to the aircraft that it was possible to introduce into the model calculations a simplified *equivalent vulnerable target*. This was represented in the model by a sphere of a few feet in radius, such that its perforation by at least one lethal fragment (defined as a through) would result in a kill.

THE FUNCTION OF THE COMPLETED MODEL

This simple model, based on the three-dimensional normal distribution (for positioning errors), the Poisson distribution (for perforating fragments), and the equivalent vulnerable target, with bounding surfaces consisting of ellipsoids, cones, and a sphere, was amenable to computation, provided always that meaningful numerical values for the various parameters could be estimated. But the task of filling in these unsurely known elements was far from simple.

We could not hope to get an absolute value of the probability P defined above or of the rounds per bird that would correspond with observed results against enemy aircraft. In any case, this figure must vary from one type of engagement to another. But we believed that the model, particularly when improved by later refinements based on more sophisticated experiments, would be of value for comparative purposes. It should throw light on what might be achieved by improvement in predictor accuracy, in fuse mechanism, in modification of shell design (for example, by changes in wall thickness and use of more powerful fillings). This concept of constructing a mathematical model, the *changes* in whose end effects (rather than the *absolute* values) can be explored by altering the parameter values, is a basic one in operational research. And it was, of course, the need to introduce a scientific approach into problems of this kind in wartime which led to the postwar demand for more operational-research studies, particularly in industry.

FURTHER REFINEMENTS IN EXPERIMENTS AND MODEL

I have mentioned that the construction of a model and the critical testing of the assumptions on which it has been based, is of value in bringing to light serious gaps in knowledge. In the present case, one of our early puzzles was that when shells were burst in flight within the wooden target surround, the resulting pattern of perforations could not be accurately related to the pattern from a static burst, merely by adding the component forward velocity of the shell. Nor was it easy to link the distribution of fragment sizes from the sandbag collection with the number of perforations of the wood, using any simple assumptions about velocities and retardations. The essential need was for more basic physical experimentation; without this, generalization was impossible.

But such generalization was essential for it was not a practical proposition to ring all the possible changes in trials, of shell design, explosive filling, forward velocity of shell on detonation, and so on. Indeed the ultimate objective must be to predict the characteristics of the fragment distribution for any desired shell velocity from the drawing-board design and a knowledge of the particular explosive filling to be used.

Here we were lucky in getting help from a very skilled scientific team which had been working on explosives in our Safety in Mines Research Establishment at Buxton; these men initiated a program of research that gradually succeeded in disentangling the picture. Shells on which small letters were engraved in successive rings round the circumference were fired at rest, within a surround of strawboard against which small velocity measuring screens were placed. In this way, fragments subsequently collected and weighed could be identified with a particular zone of the shell, and velocities estimated either by direct measurement or, more crudely, from the depth of penetration into the strawboard.

It then became clear that the initial velocity of a fragment varied quite considerably with the part of the casing from which it came and that its size (or weight) also varied with position. To some extent this initial velocity could be related to the dimension of the cross-section of the part of the casing from which the fragment came. With information of this kind it began to be possible to relate the damaging power of a shell to its design and explosive filling.

In another direction the cruder trials, as illustrated in Figure 3 were supplemented by firing individual fragments from high-speed, small-bore guns against selected aircraft components tested in isolation.

Apart from the work in England, some very extensive and informative trials were carried out in the later stages of the war under the direction of a section of the U.S. Navy's Applied Physics Laboratory at Silver Spring, Md., and at its associated proving ground near Albuquerque.

Looking back after a number of years have passed, it seems to me that by 1944 we had really broken the back of the problem. It became possible to make recommendations with some confidence on a number of matters: on the optimum design characteristics of time-fused and proximity-fused shells, on the relative importance of case thickness and explosive filling, on what might be achieved by using methods to control the size of fragments, and on the relative gains to be won by improvement in fire control and in design of shells. Few such questions could have been answered in 1939.

It is a fact, of course, that much of the fundamental research bearing on military problems is only rounded off when it is too late to be of use in the war which provided the stimulus for the effort; and by the next war, the whole conditions of warfare are changed. But I think that the work I have described brought to light a number of principles capable of much wider application.

CONCLUSIONS

The reader of this essay may ask how far the theory and experiment that it has outlined was statistical, rather than contained within the fields of physics and mechanics. It is true that the statistical techniques involved were very elementary, but in order to pull the results together so as to provide comparative figures for the long-run chances of a lethal hit a model with its interpretation involving the theory of probability had to be introduced.

As an interesting corollary, when in the summer of 1944 the American radar proximity fuse was used in the British shell to fire against that ideal antiaircraft target, the V1 flying bomb with its straight-line course, subsequent calculations showed that the expected rounds per kill (rounds per bird) derived from the appropriate probability model, corresponded approximately to the actual operational results.

Finally it seems right to emphasize that in this peculiarly difficult field—the assessment of the operational performance of weapons—the statistician becomes the scientist who must merge his statistical identity into that of a group of men trained in several disciplines, claiming no undue weight for any one of them in the search for answers to the problems in hand.

There are, of course, many other military applications of statistical techniques, particularly in the field of reliability and quality control. In the modern version of the antiaircraft problem we should find much the same general treatment arising in estimating the effectiveness of ground to air guided weapons. With the high-speed computers now available, a much more sophisticated technique of handling the model is possible, but it still must take account of factors under the three headings listed earlier; the positioning and fuse errors, the damaging power of the warhead, and the vulnerability of the target. Although a few high-speed pilotless aircraft may be expended as targets, the number is likely to be far too small to get direct, accurate confirmation of more theoretical estimates of the chances of a kill. In the German V1 Flying Bomb of 1944, our gunners were provided with a standard target that, in one sense at least, cost us nothing!

PROBLEMS

1. Let $P = 1/3$. Calculate the *rounds per bird*.

2. List and describe the three factors concerned in setting up the mathematical model of antiaircraft fire.

3. The author states that the bivariate normal distribution is a good representation of the data where "the error has two sources as here." What are these two sources of error?

4. Give two reasons why the aiming error under actual operational conditions tended to be greater than during the trials.

5. Explain how the fragmentation characteristics of shells were determined before the war.

6. If the number of telephone calls per unit time is governed by a Poisson distribution with an average of two calls per unit of time, what is the chance of ≤ 4 calls in a given unit of time?

7. What is the physical significance of the ellipsoids, cones and sphere in the completed model?

8. How did the engraving of small letters on the shell casings help to advance the study?

ACKNOWLEDGMENTS

We wish to thank the following for permission to use previously published and copyrighted material:

Academic Press

Addison-Wesley Publishing Co., Inc., for permission to use the tables on pp. 210, 212, 216 and 217, from F. Mosteller and D. Wallace (1964), *Inference and Disputed Authorship.*

The American Association for the Advancement of Science for permission to publish the figure on p. 63, from Raymond Pearl (1938), "Tobacco Smoking and Longevity," *Science,* 87: March 4, pp. 216-217.

American Educational Research Association for permission to use Figure 3 on p. 162, from D. T. Campbell and J. C. Stanley (1963), "Experimental and Quasi-Experimental Design for Research on Teaching," in M. L. Gage, ed., *Handbook of Research on Teaching,* pp. 171-246.

American Journal of Public Health for permission to publish the table on p. 12, adapted from T. Francis et al. (1955), "An Evaluation of the 1954 Poliomyelitis Vaccine Trials—Summary Report," *American Journal of Public Health,* 45:5 (Part 2), pp. 1-63.

American Psychological Association and Seymour Rosenberg for permission to publish the figures on pp. 232 and 236, from S. Rosenberg, C. Nelson, and P. S. Vivekananthan (1968), "A Multidimensional Approach to the Structure of Personality Impressions," *Journal of Personality and Social Psychology,* 9, pp. 283-294. Copyright © American Psychological Association.

American Psychological Association for permission to publish the figures on pp. 164 and 166 and Figure 9 on p. 167, from D. T. Campbell (1969), "Reforms as Experiments," *American Psychologist,* 24:4 (April), pp. 409-429. Copyright © American Psychological Association.

American Psychological Association for permission to publish the tables on pp. 221 and 225, adapted from Norman Cliff (1959), "Adverbs as Multipliers," *Psychological Review,* 66:1, pp. 27-44. Copyright © American Psychological Association.

Atlantic Monthly and David D. Rutstein for permission to publish the figure on p. 7, from D. D. Rutstein (1957), "How Good is Polio Vaccine?" *Atlantic Monthly,* 199:48. *Copyright* © 1957 Atlantic Monthly Co., Boston, Mass.

Board of Education, City of Chicago, for permission to publish the map on p. 337, from the report of the Advisory Panel on Integration in the Public Schools, March 31, 1964.

The College Entrance Examination Board for permission to publish the essay beginning on p. 289, which draws heavily upon William H. Angoff (1968), "How We Calibrate College Board Scores," *The College Board Review,* No. 68 (Summer).

Columbia University Press for permission to publish the table on p. 146, from M. Rosenberg (1964), *The Pre-Trial Conference and Effective Justice.*

Interstate Commerce Commission, Bureau of Economics, for permission to reprint the table on p. 254 from its *Table of 105,000 Random Decimal Digits,* 1949.

The Law and Society Association for permission to publish the figures on p. 161 and Figure 10 on p. 167, from D. T. Campbell and H. L. Ross (1968), "The Connecticut Crackdown on Speeding: Time Series Data in Quasi-Experimental Analysis," *Law and Society Review,* 3:1, pp. 38, 42, 45.

Little, Brown and Company for permission to publish the tables on pp. 143, 144 and 146, from H. Kalven and H. Zeisel (1966), *The American Jury.*

The Macmillan Company for permission to use the figures on pp. 407 and 412 and the table on p. 411, from H. R. Alker, Jr. (1965), *Mathematics and Politics.*

New York City-RAND Institute for permission to publish the figures on p. 152, originally prepared for S. J. Press, report No. R-704.

Operations Research Society of America for permission to publish the table on p. 300, from G. R. Lindsey (1963), "An Investigation of Strategies in Baseball," *Operations Research,* 11, pp. 477-501.

Royal Statistical Society for permission to publish the tables on pp. 93 and 95, from M. S. Bartlett (1957), "Measles Periodicity and Community Size," *Journal of the Royal Statistical Society-A,* 120, pp. 48-70.

Sage Publications, Inc., for permission to publish the figures on pp. 161, 162 and 163, from H. L. Ross, D. T. Campbell, and G. V. Glass (1970), "Determining the Social Effects of a Legal Reform: The British 'Breathalyser' Crackdown of 1967," *American Behavioral Scientist,* 13:4 (March/April), pp. 493-509.

Simmons-Boardman Publishing Co. for permission to reprint the table on p. 256, from "Can Scientific Sampling Techniques Be Used in Railroad Accounting?" *Railway Age,* June 9, 1952, pp. 61-64.

University of Chicago Law Review for permission to publish the figure on p. 142, from H. Zeisel (1969), "Dr. Spock and the Case of the Vanishing Women Jurors," *University of Chicago Law Review,* 37, pp. 1-18.

University of Michigan Press for permission to publish the tables on p. 145, from Alfred Conrad et al. (1964), *Automobile Accident Costs and Payments.*

INDEX

485